Migration Processes in the Soil and Groundwater Zone

By
Prof.Dr.sc.techn.
Ludwig Luckner
and
Prof.Dr.sc.geol.
Wsewolod M. Schestakow

ENGLISH LANGUAGE EDITION

Translated from German, *Migrationsprozesse*
im Boden- und Grundwasserbereich

 LEWIS PUBLISHERS

Library of Congress Cataloging-in-Publication Data

Luckner, Ludwig
  [Migrationsprozesse im Boden- und Grundwasserbereich.
English]
  Migration processes in the soil and groundwater zone / by
Ludwig Luckner and Wsewolod M. Schestakow. — English
language ed.
      p. cm.
  Translated from German: Migrationprozesse im Boden- und
Grundwasserbereich.
  Includes bibliographical references and index.
  1. Groundwater flow—Mathematical models. I. Schestakov, V.
M. (Vsevolod Mikhaĭlovich) II. Title
TC176.L813 1991
551.49′01′5118—dc20   90-28977
ISBN 0-87371-302-8

Copyright
© VEB Deutscher Verlag für Grundstoffindustrie
Leipzig, 1991
Bundesrepublik Deutschland

LEWIS PUBLISHERS, INC.
121 South Main Street, Chelsea, Michigan 48118

PRINTED IN THE UNITED STATES OF AMERICA

The English edition of this book is dedicated to the Department of Land, Air, and Water Resources of the University of California at Davis, which gave me the opportunity to work as a visiting professor and scientist during the school year 1988/89 and teach selected chapters of this book. The assistance of Dr. D. R. Nielsen and the technical help of the department's staff provided invaluable support for creating this edition.

Dr. Ludwig Luckner is a full professor of Soil and Groundwater Science at the Dresden University of Technology, Water Science Division. He became first known as the head of the Joint Groundwater Research Group of the Berlin Institute for Water Management and the Dresden University of Technology. He is author of more than one hundred papers on soil- and groundwater problems and coauthor of several scientific textbooks: *Geohydraulics*, with K. F. Busch (Leipzig and Stuttgart); *Simulation of Geofiltration*, with W. M. Schestakow (Leipzig and Moscow); and *Migration Processes in the Soil and Groundwater Zone*, also with W. M. Schestakow (Leipzig and Moscow).

Professor Luckner leads both the postgraduate study in groundwater that has been under way at the Dresden University of Technology since 1976 and the international training course "Migration of Solute and Heat in the Soilwater and Groundwater Zone."

After educational and study stays in the USSR and the People's Republic of Bulgaria, he worked as a visiting professor and scientist at the Moscow State University and the Leningrad Mining Institute, at the City University of London in 1982, at the Komenius University Bratislava and at the Hydraulic Research Institute Nanking (China) in 1983, at the University of Puebla (Mexico) in 1985, and at the University of California at Davis, the Groundwater Modeling Center of Butler University (Indianapolis), and Cornell University (Ithaca) in 1986. He was engaged in the Regional Water Policies Project of the International Institute of Applied Systems Analysis (Austria) in 1983–1985, especially in the field of groundwater quality problems in regions with intensive open-cast lignite mining. During the school year 1988/89, he followed the invitation to work as a visiting professor and scientist at the University of California at Davis.

Professor Luckner acted for many years as deputy director for research at the Water Sciences division of the Dresden University of Technology and is now head of the Soil and Groundwater Science Department. In the field of mine drainage, he leads the Joint Open-Cast

Mine Dewatering Research Group of the Institute of Lignite Mining and the Dresden University of Technology, and the Joint Subsurface Water Treatment Research Group of the Research Center for Water, Technology and the Dresden University of Technology. Both groups are now integrated in the Joint Laboratory of Soil- and Groundwater Research.

Emil O. Frind is a professor at the Department of Earth Sciences and a member of the Centre for Groundwater Research at the University of Waterloo in Canada. He received his bachelor's and doctorate degrees in Civil Engineering and his Master's degree in Hydrology at the University of Toronto, completing his studies in 1971. Dr. Frind then joined the newly-founded groundwater group at the University of Waterloo, where he developed a groundwater modeling program as an integral part of the overall program in groundwater studies. Over the past 18 years, his research has covered a wide range of areas in mathematical model development and applications, focusing on the use of models for gaining insight into the multitude of physical and chemical processes that contribute to the migration problem in groundwater. Dr. Frind teaches courses at both the undergraduate and graduate levels on porous media flow theory, numerical methods, and mathematical modeling, and he has authored or co-authored over 60 research papers. His current interests include the migration of multiple interacting constituents in heterogeneous three-dimensional systems, the fate of organic contaminants in the vadose and saturated zones, the nature of dispersion processes, and the modeling of aquifer remediation.

Graham E. Fogg received a B.S. in Hydrology from the University of New Hampshire, an M.S. in Hydrology and Water Resources from the University of Arizona, and a Ph.D. in Geology from the University of Texas at Austin. From 1978 to 1988 he was a Research Associate at the Bureau of Economic Geology, University of Texas at Austin, performing research on characterization of the subsurface and modeling of subsurface fluid flow in projects dealing with high-level nuclear waste isolation, lignite mining hydrology, and petroleum recovery. Dr. Fogg is currently an Associate Professor of Hydrogeology in the Department of Land, Air, and Water Resources at the University of California, Davis. His research interests include geologic and stochastic characterization of aquifer/reservoir heterogeneity for mass transport and determination of effective porous media properties (scale-averaging) for numerical models. Dr. Fogg's teaching activities include courses in groundwater hydrology, geostatistics, groundwater modeling, and contaminant hydrology, as well as the development of a graduate hydrogeology degree program.

Thomas K.G. Mohr, who served as translator and facilitator for the English edition of this book, is a student of hydrogeology pursuing his Master's degree in the Hydrologic Sciences Program at the University of California at Davis. He is currently integrating aquifer testing, geochemistry, and modeling to optimize groundwater level control and groundwater quality protection at a municipal landfill site. Mr. Mohr has also performed field application of in situ anaerobic biodegradation of gasoline contaminated groundwater.

Table of Contents

Preface to the English Edition

The English edition is more than a simple translation of the German edition published in 1986 by VEB Deutscher Verlag für Grundstoffindustrie in Leipzig. More than a quarter of the material in the English edition is new and updated, and all other portions have been reworked. During my stay at the University of California (UC) at Davis, the scientific advice, support, and help of many well-known scientists contributed decisively to the improved content of the English edition, including

Donald R. Nielsen, UC Davis, Department of Land, Air, and Water Resources (LAWR) (whole edition and Part 1);
Kenneth K. Tanji, UC Davis, LAWR (Chapter 1.4);
Mark E. Grismer, UC Davis, LAWR (Sections 1.5.1 to 1.5.3);
Eric Conn, UC Davis, Department of Biochemistry and Biophysics (Section 1.5.4);
Miguel A. Mariño, UC Davis, LAWR (Chapter 2.1);
Emil O. Frind, University of Waterloo, Canada (Chapter 2.2);
Stephen E. Silliman, University of Notre Dame, Department of Civil Engineering (Part 3); and
Mohamed Alemi, UC Davis, LAWR (Section 3.2.2).

The scientific editorial work of Donald R. Nielsen (Part 1), Miguel A. Mariño (Chapter 2.1), Emil O. Frind (Chapter 2.2), Graham Fogg (Chapter 2.3), and Stephen E. Silliman (Part 3), and the important work of graduate students Tom Mohr (working with Donald R. Nielsen and myself on Part 1 and on the consistency of the English wording throughout the whole edition) and Jim Baumgartner (working with Miguel A. Mariño, Graham Fogg, and myself on Chapters 2.1 and 2.3) with the scientific editors and myself, finally resulted in this English edition of a book that is well known and highly ranked in the authors' home countries.

This edition is a comprehensive structured textbook with hundreds of equations and examples for the reader to actively use to aid in his or her understanding. The tutorial guidance of university teachers, seminar leaders, and other experts facilitates the course of study. The textbook provides the reader with a systematic, state-of-the-art theoretical framework of migration processes and enables the reader to solve comprehensive practical problems. Many applied examples with their feedback relations to theoretical fundamentals encourage the reader again and again to study migration problems in their whole complexity. The book is written based on the basic idea of "one receiver (reader or user), one transmitter (author or teacher)"; i.e., the book tries to integrate the highly specialized knowledge in the complex field of subsurface migration and offers a consistent text to the reader, student, or trainee in a well-balanced form. The textbook may serve as a study text in graduate courses at many universities in the United States and abroad, as a study text in postgraduate and professional training courses at different institutes such as the USGS, NWWA, EPA, IGWMC, and as a manual supporting specialists in their daily work in such fields as hydrology, geology, water management, waste management, agronomy, environmental protection, and mining.

Dr. Donald R. Nielsen
Professor
of Soil and Water Science at the
University of California, Davis

Dr. sc. techn. Ludwig Luckner
Professor
of Soil and Water Science at the
Dresden University of Technology

Preface to the German Edition (shortened)

Physically, chemically, and biologically caused transport, storage, exchange and transformation processes of heat and solutes in the soil- and groundwater zone are designated as migration processes. The term *migration* is used as a neutral term and does not imply any evaluation. By the term *subsurface pollution*, we understand quality changes implying potential effects on the present and future use of soil- and groundwater, and by *contamination* we mean pollution with real or potential health risks.

In many countries groundwater is the main source of drinking water supply. Soilwater recharges groundwater resources and supplies plants with water and nutrients. Good management of subsurface water resources and efficient protection against overuse and degradation require reliable knowledge of solute and heat migration in the soil- and groundwater zone. Future generations must be guaranteed sufficient water supplies free of health risks through regulated use and protection against contamination of groundwater resources.

The book presents an integral methodology of modeling migration processes, state-of-the-art analytical, numerical, and inverse solutions of migration models, and the technology of migration parameter estimation in laboratories and in the field, along with today's subsurface water monitoring methodologies.

The book strives to consider geological, hydrological, technical, mathematical, physical, chemical, and biological aspects equally, and to deal with data acquisition and data processing in a balanced way. It notably addresses students, practitioners, and scientists who are increasingly confronted with problems of using soil- and groundwater resources and protecting them against overuse and degradation. The book cites examples from the fields of water management, environmental protection, hydrogeology, hydrology, mining, reclamation, regional planning, etc.

The book provides a comprehensive overview of subsurface migration problems and addresses both theory and practice adequately. However, the study of the special aspects or the solution of special problems may require secondary reading of literature in the field of geology, engineering, physics, chemistry, microbiology, mathematics, and computing.

The authors would be especially grateful for any ideas or hints to further improve this book.

Introduction

SIGNIFICANCE OF MIGRATION

A worldwide increase in demand for water by communities, industries, and agriculture, resulting primarily from population growth and an improved standard of living, has placed great importance on water science and management practice. Unlike other raw materials, water has no substitute: it is absolutely essential. The future development of many countries will depend decisively upon how the increasing demand for water will be satisfied, and how the economic and environmental cost for that demand may be met. Hence, protection of soilwater and groundwater resources against depletion and degradation has developed into one of today's foremost problems worldwide.

For many countries, groundwater is the main source of drinking water supply. For example, on the European continent, Denmark and Austria rely on groundwater for 98% of their drinking water supply; Italy, 93%; Germany and Belgium, about 70% each; and the Netherlands and Czechoslovakia, 65% [8]. Soilwater, on the other hand, is the main source for plant nutrition, providing vegetation with water and nutrients. Soilwater not consumed by root uptake or evaporation ultimately percolates deeper into the subsurface, and comprises a large contribution to the recharge of groundwater resources.

Conjunctive use and management of surface and groundwater have proved increasingly necessary. Groundwater and surface water must be considered as inseparable components of an interactive system (see Figure 1). Any management of water quantity must also include consideration of water quality. It is therefore imperative that groundwater is used only to the extent and in a manner which also guarantees its sufficient

1

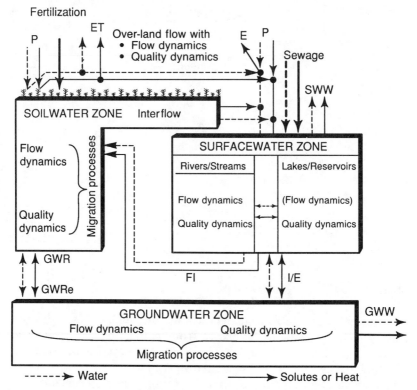

Figure 1. Interdependency of terrestrial water resources. *P*, precipitation; *ET*, evapotranspiration; *E*, evaporation; *SWW*, surface water withdrawal; *GWW*, groundwater withdrawal; *GWR*, groundwater rising; *GWRe*, groundwater recharge; *FI*, free infiltration; *I/E*, infiltration/exfiltration.

protection. The effective and risk-free use of water resources for present and future generations should be a common objective for all societies.

Two fundamental goals of subsurface water management must be met. The first fundamental goal, protection of soilwater and groundwater resources against overuse and ultimately against exhaustion, requires (see Figure 2)

- groundwater use only to the extent to which it is or can be recharged
- control or limitation of groundwater loss to the atmosphere or to surface waters
- wise and economical use of groundwater and soilwater, primarily for domestic water supply and vegetation needs

A second fundamental goal of subsurface water management should be the protection of soilwater and groundwater resources against degradation and contamination. The natural capacity of aquifers to purify

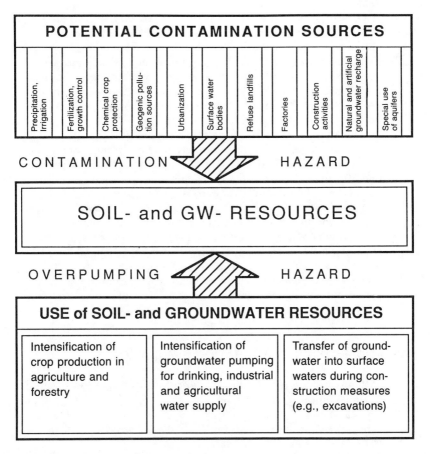

Figure 2. Pressures on soil- and groundwater resources caused by potential contamination sources and overuse.

subsurface water must not be destroyed by contamination from industrial, agricultural, and domestic sources. This is especially important because subsurface processes are extraordinarily slow, and harmful or toxic effects may only become evident several years or decades after a contaminant release. Rehabilitation of contaminated soils or aquifers, if possible at all, frequently requires many years, or even generations.

In that part of Germany that was formerly the German Democratic Republic, for instance, protecting groundwater resources against overuse and degradation is of great importance for several reasons [3]:

1. More than 75% of the domestic water is supplied by groundwater, and groundwater will remain the primary source for the foreseeable future.
2. The freshwater-saltwater interface is located at depths of about 100 to

200 m below the surface in many parts of the country, limiting ground-water use to the extent to which natural or artificial freshwater recharge occurs.

3. The quantities of groundwater resources used for drinking water, indus-trial water, and agricultural water supply are already at a very high level. The demand for drinking water increases by 2–3% annually, and exploitation of previously undeveloped groundwater resources has become increasingly complicated and expensive.

4. Due to drainage measures associated with brown coal mining, ground-water resources in mining areas have been heavily exploited. At present, groundwater withdrawal rates (1.8 billion m^3 mine water per year, or one-fifth of the area's stable water resources) for mining needs far exceed the withdrawal rates for domestic water supply.

5. Expanding industrialization and urbanization have led to an enormous potential for diffuse and point source contamination.

6. Superposition of intensive surface land use and use of subsurface water requires a comprehensive management strategy to minimize the impacts of one use upon another. Therefore, management of groundwater resources in Germany, as with most industrialized nations, must incor-porate a detailed understanding of these demands, which in turn requires a better understanding of migration processes.

MAIN OBJECTIVES OF MIGRATION SIMULATION

As with any process simulation, the objectives of migration simulation include four main topics:

1. verification of model concepts (hypothesis) using laboratory and in situ test results as a basis for comparison

2. scientific interpretation of observed migration phenomena (or scientifi-cally founded epignosis)

3. identification of parameters from measured data reflecting the observed phenomena

4. the scientific prognosis of future migration processes occurring under given or assumed conditions (scenario analysis)

Thus, scenario analysis is a prerequisite for designing effective methods and means for subsurface water management (i.e., for efficient computer-aided design) and for their effective use and, hence, process control (computer-aided management).

Process simulation presupposes the existence of usable process models. Such models always require the development of a flow model (quantity model) and a quality model. The two models are linked (see Figure 3). The kinematic viscosity ν and density ϱ of the water are functions of the quality model outputs: concentration c and temperature T. Outputs of the flow model, such as flow rate $\vec{v}$, alterations in storage

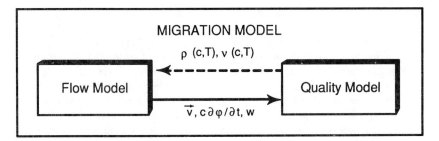

Figure 3. Block diagram of migration models.

$c\partial\phi/\partial t$, and internal flow sources and sinks w, impact the quality model. Therefore, the term *combined quantity-quality models* of subsurface water is sometimes used instead of the term *migration models.* If the impact of the quality model on the flow model can be neglected, the process is called *tracer migration.* This approximation brings significant simplifications and advantages for simulation and is therefore frequently applied.

The main fields of application of migration simulation are management and protection of soilwater and groundwater resources, and remediation of contaminated aquifers. Other engineering applications of migration simulation include onsite subsurface water treatment, leaching of ore deposits, and grouting dam foundations or creating other subsurface seals.

In many countries, important applications of migration simulation can be found in

- monitoring soilwater and groundwater quality. Simulation of migration processes supports the interpretation of observed changes in water quality; i.e., it supports the scientifically founded epignosis, and may help to determine the answers to the six "w" questions: **Why** is monitoring necessary? **Who** monitors? **What** is monitored? **Where? When? With which** techniques?
- assessing the formation of water quality in recharged groundwater resources, including prediction or control of water quality degradation due to acid precipitation, salt and pesticide-laden irrigation returns, land application of wastewater, and impact of urbanization on artificial recharge schemes
- characterizing the nature and effects of acid mine drainage from open cast mining of brown coal, and addressing the migration of geochemical weathering products of pyrite [4]
- predicting the migration of point-source and area-source industrial contaminants in groundwater, either as accidental or ill-advised releases, or as undetected or unreported leaks [6]
- reducing groundwater contamination due to agricultural practices by

analysis of irrigation and drainage with respect to transport of fertilizer salts and pesticides
- determining a scientific basis for the delineation of drinking water protection zones and establishing land-use restrictions for designated areas
- engineering design for total containment of hazardous waste and sanitary landfills, including specifications for liner requirements, leachate collection, and groundwater monitoring networks. Migration simulation is applied in the design of long-term remedial action plans for landfill leachate contaminant plumes created by short-sighted management practices of the past.
- treatment of groundwater in situ, for the removal of iron and manganese, treatment of acid mine water by reinfiltration, and storage of warm water in aquifers [5; 1]

REFERENCES FOR THE INTRODUCTION

1. Gringarten, A. C., P. A. Landel, A. Menjoz, and J.-P. Sauty. Stockage longue durée en nappe phréatique de calories a basse température pour l'habitat. Orléans: Bureau de recherches géologiques et miniéres, 1979.
2. Jirele, V. Die Bedeutung der Verunreinigung der Niederschläge für die Herausbildung der Grundwasserbeschaffenheit. Wiss. Konf. der Techn Univ. Dresden zur Simulation der Migrationsprozesse im Boden-und Grundwasser. Dresden: Techn. Univ., 1979, Bd. II, pp. 56–61.
3. Luckner, L. Aktuelle Aufgaben des Schutzes der Grundwasserressourcen der DDR bei ihrer intensivierten Nutzung. WWT 32 (1982) 1, pp. 20–23.
4. Luckner, L., and J. Hummel. Modelling and prediction of the quality of mine drainage water used for drinking water supply in the GDR. First Int. Mine Water Congress of the IMWA, Budapest, 1982, Proc. D, pp. 246–257.
5. Luckner, L., G. Muellerbuchhof, and N. Victor. Nutzung von Uferfiltrat und künstlichem Infiltrat als Wärmequellen für elektrisch angetriebene Kompressions-Wärmepumpen. WWT 33 (1983) 1, pp. 16–21.
6. Mironenko, V. A., et al. Ohrana podzemnyh vod v gornodobyvaésih rajonah. Leningrad: Nedra, 1980.
7. Reissig, H., D. Eichhorn, and R. Fischer. Beitrag zur unterirdischen Enteisenung. Wiss. Zeitschrift der Techn. Univ. Dresden 32 (1983) 1, pp. 163–166.
8. Zwirnmann, K.-H. Problems of ground water quality management. Wiss. Konf. der Techn. Univ. Dresden zur Simulation der Migrationsprozesse im Boden-und Grundwasser. Dresden: Techn. Univ., 1979, Bd. II, pp. 18–38.

Part 1

Methods of Mathematical Modeling

Model Concepts for Air, Water, and Solids

Process simulation and scenario analysis have their basis in mathematical modeling. This chapter focuses on the physical, chemical, and biological fundamentals of migration processes and their mathematical representation. This approach forms the basis of conceptual migration modeling and is a central theme within this text.

The subsurface is a multicomponent system (i.e., a mixture). The consolidated and unconsolidated mineral solids form a spatially fixed matrix, while fluids fill the voids. The following descriptions are developed with water as the wetting fluid and air as the nonwetting fluid. Petroleum, natural gas, and other subsurface fluids are also addressed by analogy.

1.1.1 CHARACTERIZATION OF MULTICOMPONENT SYSTEMS

Multicomponent systems may be divided into

- homogeneous systems (mixtures)
- heterogeneous systems (conglomerates)

Mixtures are more precisely called *mixphases*. A *phase* may be defined as a gaseous, liquid, or solid substance which is regarded as homogeneous in the system under consideration, whereby the "homogeneity" requires a scale of observation allowing component properties to be averaged. A phase which consists of different components (e.g., oxygen and nitrogen molecules) is called a mixphase. Different phases are separated by interfaces.

Mixphase and *component* are relative terms like the terms *system* and *element* in cybernetics. Usually, when one component of a mixphase dominates, the mixphase is called a *disperse system, dispersion,* or *solution,* with the disperse substances distributed in the dispersive medium or dispersant. The disperse substance is often also called a *solute;* the dispersive medium, as the dominant component, is then called a *solvent.* A dispersion may be

- coarse disperse (10^{-6} to 10^{-5} m, visible under an optical microscope)
- colloidal disperse (10^{-8} to 10^{-7} m, visible under a scanning electron microscope)
- molecular disperse (10^{-10} to 10^{-9} m, not visible, actually dissolved)

The disperse phase is called *coherent* if it is continuously distributed and *incoherent* if its distribution is discontinuous. Soil air or water in the unsaturated soil zone, for instance, may be dispersed coherently or incoherently. According to the state of the disperse phase and the dispersant, typical multicomponent systems can be distinguished as shown in Figure 1.1.

A primary consideration of contaminant or substance migration is the movement of specific solutes together with a mobile fluid solvent. Solutes that migrate through the spatially fixed rock matrix may be molecules, ions, complexes, particles, and microorganisms. These migrating species are hereafter called *migrants.* Each migrant travels with respect to the rock matrix and relative to the solvent, i.e., the flowing subsurface water or air.

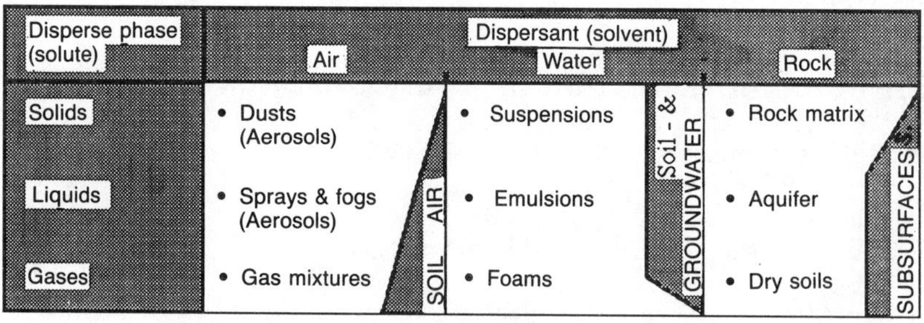

Disperse phase (solute)	Dispersant (solvent)		
	Air	Water	Rock
Solids	• Dusts (Aerosols)	• Suspensions	• Rock matrix
Liquids	• Sprays & fogs (Aerosols)	• Emulsions	• Aquifer
Gases	• Gas mixtures	• Foams	• Dry soils

SOIL - AIR   Soil - & GROUNDWATER   SUBSURFACES

Figure 1.1. Examples for typical multicomponent systems.

In order to describe the composition of a mixphase, the following extensive state variables may be used (see also Appendix 1):

mass

$$m = \sum_{i=1}^{N} m_i \qquad (m_i = \text{mass of the component i})$$

amount of substance

$$n = \sum_{i=1}^{N} n_i \qquad (n_i = \text{amount of the component i})$$

volume

$$V_o = \sum_{i=1}^{N} V_i \qquad (V_o = \text{total volume before mixing})$$

The total volume V of the mixphase after mixing may be different from V_o.

The terms *specific volume, specific amount of substance,* or *specific heat capacity* are used if the quantities are related to mass. Thus for example, specific volume is $v = V/m$. When quantities are related to the amount of substance present, they are called *molar mass, molar volume,* or *molar heat capacity* (e.g., molar mass $M = m/n$ or molar volume $V_m = V/n$) [1.13]. When related to volume these quantities are characterized by the term *content.*

The concentration of the component i in a mixphase may be expressed in terms of mass m_i, amount of substance n_i, or volume V_i of the component i in the mass m, the amount of substance n, or the volume V of the mixphase. The following terms of concentration are in use [1.13]:

- mass fraction $\quad w_i = m_i/m \qquad$ %, %o, ppm, . . .
- mole fraction $\quad x_i = n_i/n \qquad$ %, %o, ppm, . . .
- volume fraction $\quad \phi_i = V_i/V_o \qquad$ %, %o, ppm, . . .
- mass content $\quad \beta_i = \rho_i = m_i/V \qquad$ g/m³, mg/L, mg/dm³, . . .
  (partial density)
- mole content $\quad c_i = n_i/V \qquad$ mol/m³, mmol/L, . . .
  or molarity
- content of $\quad c_{i,eq} = c_{i,c} = n_{i,eq}/V \qquad$ mol/m³, mmol/L,
  equivalents
  or charges $\qquad$ . . .
- volumetric content $\quad \sigma_i = \theta_i = V_i/V \qquad$ %, %o, ppm, . . .
- mole ratio $\quad r_{i,j} = n_i/n_j \qquad$ %, %o, ppm, . . .
- mass ratio $\quad \xi_{i,j} = m_i/m_j \qquad$ %, %o, ppm, . . .
- volume ratio $\quad \psi_{i,j} = V_i/V_j \qquad$ %, %o, ppm, . . .

- molality $\qquad b_i = n_i/m_s$ $\qquad$ mol/kg solvent (for water, usually numerically equal to mole content)

Examples for equivalent units are

		w_i	x_i	ϕ_i
10^{-2}	1 %	1 cg/g	1 cmol/mol	1 cL/L
10^{-3}	1 %o	1 mg/g	1 mmol/mol	1 mL/L
10^{-6}	1 ppm	1 μg/g	1 μmol/mol	1 μL/L
10^{-9}	1 ppb	1 ng/g	1 nmol/mol	1 nL/L
10^{-12}	1 ppt	1 pg/g	1 pmol/mol	1 pL/L

According to the SI standard the concentration terms °dH (German degree of hardness) and val/L are no longer valid. These quantities are now expressed by mmol/L or mg/L Ca^{2+} and/or Mg^{2+}, respectively, or by the content of equivalents or charges:

$$c_{eq,i} = c_{c,i} = n_{eq,i}/V = n_i z_i/V$$

where $\quad z$ = effective valency

The concentration terms usually used in migration studies are limited to

- the mole content (molarity)
- the mass content (partial density)
- the mole fraction
- the molality

It is important to keep in mind that the common symbol c_i is used to represent each of the four types of concentrations:

$$c_i \doteq [i] = c_i \text{ or } p_i \text{ or } x_i \text{ or } b_i$$

It is further worth noting that the volume V in terms of the content refers to the subsurface as a whole expressed in $m^3 = m_v^3$ or dm_v^3 or only to the fluid phase in the subsurface preferentially expressed in L. The two ways of expressing volume result in quite different values of concentration. Therefore, one must pay attention to the type of concentration used in a report, paper, or book.

Each systems state description requires knowledge of the appropriate extensive and intensive state variable. The extensive state of a mixphase as a multicomponent system depends on the intensive state variables temperature T, pressure p, and composition n_i. The state function is often approximated by a nonhysteretic dependency and must be referring or related to an arbitrary reference or standard point $Z_o = f(T_o, p_o, n_{1,o}, n_{k,o})$:

$$\Delta Z = f(T, p, n_i, \ldots n_k) \qquad (1.1a)$$

where Z represents, for instance, the thermodynamic extensive state variables:

U = internal energy   S = entropy
H = enthalpy   F = free energy
G = Gibbs free energy   V = volume

Hence, it follows for the change of the systems state:

$$dZ = \left(\frac{\partial Z}{\partial T}\right)_{p,n_k} dT + \left(\frac{\partial Z}{\partial p}\right)_{T,n_k} dp + \Sigma\left(\frac{\partial Z}{\partial n_i}\right)_{T,p,n_{k \neq i}} dn_i \quad (1.1b)$$

Small letters (z, u, h, . . .) are often used for the extensive state variables when related to the amount of substance n (molar state quantities), volume V (state content quantities), or other extensive state variables. That means, extensive state variables do not become intensive ones by division by extensive state variables (see volumetric water content or concentration).

The chemical energy stored in each component of a mixphase at constant temperature and constant pressure is the enthalpy ΔH, where the delta indicates that this extensive state variable is referred to an arbitrary standard state (commonly 25°C, 1 bar). Enthalpy has two components, an internal component (entropy ΔS) and an external component (the Gibbs free energy ΔG). Entropy is a measure of organization or order; the entropy of gases is high and the entropy of solids is low. The change of free energy is the driving force in spontaneous chemical reactions. These three molar extensive state variables are linked by the following equation:

$$\Delta h = \Delta g + T\Delta s \qquad (1.1c)$$

1.1.2 MIXPHASE MODEL OF AIR

Air is a nondropping fluid, i.e., it is a gas. Gases have neither a defined shape nor a defined volume; they fill any space available, and their entropy is high. For migration models in the subsurface, air is regarded as a homogeneous fluid mixphase consisting of the components nitrogen, oxygen, argon, etc. Dry or "clean" atmospheric air consists of 78.09% N_2, 20.15% O_2, 0.93% Ar, 0.03% CO_2, and traces of other gases. These percentages are given as mole fractions x_i of the mixphase "air." The figure $x_{O_2} = 20.15\%$, therefore, states that 1.000 mol air contains 0.2015 mol O_2. If, for instance, gaseous H_2O or SO_2 is added to the air, the given mole fractions x_{N_2}, x_{O_2}, etc., decrease.

The pressure p of the mixphase results from the sum of the partial pressures of the components ($p = \Sigma p_i$), where the partial pressure p_i of the component i is obtained from the product of the mole fraction x_i and the total pressure p of the mixphase ($p = p_a$ = atmospheric pressure):

$$p_i = x_i p \quad (\text{e.g., } p_{O_2} = x_{O_2} p_a) \tag{1.2}$$

The ideal gas is of great importance as a model concept. Here, the three state variables V, p, and T are interrelated, as shown by Boyle, Gay-Lussac, and Avogadro (see also Equation 1.1a and Appendix 1):

$$V_m = RT/p \quad \text{or} \quad V = nRT/p = n_{O_2} RT/p_{O_2} \tag{1.3}$$

where

$\quad$ p $\quad$ = pressure of the gas in Pa = N/m^2 (1 atm = $1.013 \cdot 10^5$ Pa)
$\quad$ V_m = molar volume in m^3/mol
$\quad$ V $\quad$ = volume of gas-filled space in m^3, V = $V_m n$
$\quad$ n $\quad$ = amount of substance in mol (1 mol = $6.023 \cdot 10^{23}$ particles)
$\quad$ R $\quad$ = gas constant R = 8.315 Nm/(mol K) = 8.315 J/(mol K)
$\quad$ T $\quad$ = absolute temperature in K (0°C = 273.15 K)

The state of real gases deviates from that of the ideal gas as pressure increases and temperature decreases. In this case, the van der Waals equation of state or another equation of state for real gases must be used. For migration processes in the aerated soil zone, these deviations can be neglected, but they are of great importance for problems of subterranean gas storage.

The density or mass content of air ρ_a, assuming an ideal gas, is given (in kg/m^3) by (see also Appendix 1)

$$\rho_a = \frac{m}{V} = \frac{nM}{V} = \frac{pM}{RT} = \frac{p\Sigma(x_i M_i)}{RT} \tag{1.4a}$$

where

$\quad$ m $\quad$ = mass of air in kg
$\quad$ M $\quad$ = molar mass in kg/mol, M = m/n

The partial density of oxygen in air, ρ_{O_2} is given (in kg/m^3) by

$$\rho_{O_2} = [O_2] = \frac{m_{O_2}}{V} + \frac{n_{O_2} M_{O_2}}{V} + \frac{p_{O_2} M_{O_2}}{RT} + \frac{p x_{O_2} M_{O_2}}{RT} \tag{1.4b}$$

and the mole content of oxygen in air (in mol/m^3), c_{O_2}, by

$$c_{O_2} = [O_2] = n_{O_2}/V = \rho_{O_2}/M = p_{O_2}/(RT) \qquad (1.4c)$$

When air is regarded as an ideal gas, the volumetric content of oxygen $\sigma_{O_2} = V_{O_2}/V$ is equal to the mole fraction x_{O_2}.

Example calculations for dry air with $p_a = 1$ atm and $T = 273.15$ K result in

$$\rho_a = \frac{1.013 \cdot 10^5 \text{ Nm}^{-2} (0.78 \cdot 28 \text{ g/mol} + 0.21 \cdot 32 \text{ g/mol} + 0.01 \cdot 40 \text{ g/mol})}{8.315 \text{ Nm/(mol K)} \cdot 273.15 \text{ K}}$$

$$= 1293 \text{ g/m}^3 \text{ (g air per m}^3 \text{ of air-filled space)}$$
$$\rho_{O_2} = 300 \text{ g/m}^3 \text{ (g oxygen per m}^3 \text{ air-filled space)}$$
$$c_{O_2} = [O_2] = 300 \text{ g m}^{-3}/32 \text{ g mol}^{-1}$$
$$= 9.38 \text{ mol/m}^3 \text{ (mol oxygen per m}^3 \text{ air-filled space)}$$

Atmospheric moisture reduces the air density because gaseous water molecules are lighter than N_2 and O_2 molecules. This decrease is given by

$$\Delta\rho_a = (M_{a,dry} - M_{H_2O}) \, p_{H_2O}\phi/(RT) \qquad (1.5)$$

where

p_{H_2O} = water vapor pressure, equal to saturated partial pressure of water vapor in air (see p_v in Table 1.1)

M_{H_2O} = molar mass of water (18 g/mol)

$M_{a,dry}$ = molar mass of dry air (28.96 g/mol)

ϕ = relative atmospheric moisture (moisture divided by the saturation value of the moisture)

The mole fractions of soil air differ from those of atmospheric air. In comparison to air above the soil, soil-air typically has higher CO_2 (1–5%) and lower O_2 (0–20%) mole fractions and is fully H_2O saturated ($\phi \approx 1.0$).

The mixphase "air" contains mineral and organic solid particles and liquid particles as components in addition to the gaseous particles forming a pure mixture. Such mixtures of materials are usually called *aerosols*. Depending on the type and size of the particles in this "impure" mixture, they are designated as dust (dispersed solid particles ranging in size from 10^{-6} to 10^{-4} m) and fog (dispersed liquid particles).

For migration in the subsurface, the consideration of aerosols is not relevant except in connection with subterranean gas storage and the extraction of natural gas and petroleum. Investigations of O_2, H_2O, CO_2, CH_4, H_2S, NH_3, and vaporized hydrocarbons (volatile organic compounds) as migrants in the soil air are much more common.

1.1.3 MIXPHASE MODEL OF WATER

Water is a dropping fluid, i.e., it is a liquid. In the liquid state particles are difficult to compress but easy to deform. The entropy of water is substantially less in the liquid state than in the gaseous state. Therefore, a relatively large quantity of heat (the latent heat) is required to vaporize the liquid. The measure of a liquid's resistance to flow is its viscosity. Ideal liquids have a viscosity of zero. The cohesion in liquids, which manifests itself during the formation of droplets, is caused by the van der Waals attraction forces acting between molecules. Chemically "pure" water is a mixture of different H_2O molecules consisting of the different isotopes of oxygen and hydrogen. Of these molecules, 99.8% have the mass 18 g/mol ($^1H^{16}O^1H$). When the radioactive isotope tritium ^3H is incorporated into the structure, the $^3H^{16}O^1H$ molecule is formed. This isotopic water molecule is an effective tracer.

The structural model of the H_2O molecule is a tetrahedron, as shown in Figure 1.2 [1.42; 1.35]. The centers of positive and negative charges of this molecule do not coincide, i.e., they are not balanced as they would be for a nonpolar molecule such as O_2. The water molecule is therefore a dipole, with a dipole moment m of $6.13 \cdot 10^{-30}$ Coulomb $\cdot$ m. The dipole moments of other substances are lower than that of water: $\mu_{HCl} = 0.56\mu_{H_2O}$, $\mu_{NH_3} = 0.65\mu_{H_2O}$, and $\mu_{C_2H_3OH} = 0.95\mu_{H_2O}$.

Water should be regarded as a polymerized liquid of the formula $(H_2O)_n$. The degree of polymerization depends on temperature and pressure. Clusters of water molecules are aggregated polymolecules and occur in addition to single molecules. At 0°C water has, on average, the structural formula $H_{180}O_{90}$, and at 70°C, $H_{50}O_{25}$ [1.35].

Pure water also dissociates, as is evident by its electrical conductivity (see also §1.4.2):

$$2H_2O \rightleftarrows H_3O^+ + OH^- \tag{1.6}$$

For simplicity, H^+ is often written instead of H_3O^+. But in fact, the proton is always coupled to a H_2O molecule. The expression (in mol²/L²)

$$K_W = c_{H^+}c_{OH^-} = [H^+] \cdot [OH^-] \tag{1.7}$$

is called the *ion product* of water. The ion product depends upon temperature and is usually about $1.0 \cdot 10^{-14}$ mol²/L² at 25°C.

Table 1.1 presents important properties of water as a function of temperature.

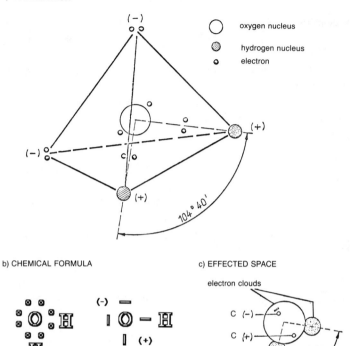

Figure 1.2. Different structural models of a water molecule where (−) and (+) symbolize the centers of the negative and positive charges.

In Table 1.1, the following notation is used:

η = dynamic viscosity (N m^{-2} s = Pa s = kg s^{-1} m^{-1})

ν = kinematic viscosity; $\nu = \eta/\rho$

ρ = density of gas-free water (kg m^{-3}); specific weight $\gamma = \rho g$; $g = 9.780(1 + 5.29 \cdot 10^{-3} \sin^2 \alpha - 5.9 \cdot 10^{-6} \sin^2 2\alpha)$, where α is the geographic latitude

σ = interface tension according to air moisture (Nm^{-2} = Pa)

β = isothermal compressibility; $\beta = -(1/V)\Delta V/\Delta p$ at atmospheric pressure

K_w = ionic product of water (mol^2/L^2)

$p_v = p_{H_2O}$ vapor pressure of water at saturation (Nm^{-2} = Pa)

q_v = specific latent heat of vaporization (kJ/g)

Compared to other liquids, water has a relatively high dielectric constant (78.25 at 25°C), high specific latent heats of vaporization (≈ 2.5

Table 1.1. Physical Properties of Water at Various Temperatures

Properties	±0°C	+4	+5	+10	+20	+30	+40
η $[10^{-3}Nm^{-2}s]$	1.7921	1.5674	1.5188	1.3077	1.0050	0.8007	0.6560
ν $[10^{-6}m^2s^{-1}]$	1.7944	1.5674	1.5189	1.3081	1.0068	0.8042	0.6612
ρ $[10^2kgm^{-3}]$	9.998	10.000	9.999	9.997	9.982	9.9566	9.922
σ $[10^{-2}Nm^{-1}]$	7.564	—	7.492	7.422	7.275	7.118	6.956
β $[10^{-10}m^2N^{-1}]$	5.098	—	4.928	4.789	4.591	4.475	4.422
K_w $[10^{-14}mol^2L^{-2}]$	0.1139	—	0.1846	0.2920	0.6809	1.469	2.919
p_v $[Nm^{-2}=Pa]$	610	813	872	1128	2338	4243	7376
q_v $[kJg^{-1}]$	2.500	—	2.489	2.477	2.453	2.430	2.406

Sources: Langguth and Voigt [1.49], Matthess [1.55], and Freeze and Cherry [1.27].

kJ/g) and melting (0.3337 kJ/g), and a high specific heat capacity (4.187 J/gK).

Soil- or groundwater is a disperse system, often called an *aqueous solution*. In such solutions pure water is the solvent (dispersant) and the dissolved constituents are the solutes (disperse phases). Due to its high dipole moment, water is a very effective solvent for many solid, liquid, and gaseous substances. The disperse phase is classified according to particle size as molecular dissolved, colloidal dissolved, or suspended substances. Solutions are distinguished according to their concentrations as

- dilute solutions (low concentration)
- concentrated solutions (high concentration)
- saturated solutions (substances will not dissolve further at constant temperatures)
- supersaturated solutions

Table 1.2 shows important solutes of soil- and groundwater, and Figure 1.3 shows their typical natural integrated frequency distributions.

Table 1.2. Important Solutes in Natural Waters

Mixture	Molecular-disperse				Colloidal-disperse	Coarse-disperse
Solution-system	True solution				Colloidal solution	Suspension, emulsion
Typical diameter	$10^{-10} - 10^{-9}$ m				$10^{-8} - 10^{-7}$ m	$10^{-6} - 10^{-5}$ m
	Electrolytes		Nonelectrolytes			
	Cations	Anions	Gases	Solids		
Principal solute > 1.0 mg/L	Na^+ K^+ Ca^{2+} Mg^{2+} (Fe^{2+}) (NH_4^+)	HCO_3^- Cl^- SO_4^{2-} NO_3^-	N_2 O_2 CO_2	$SiO_2 \cdot nH_2O$	$SiO_2 \cdot nH_2O$ clay particles	clay and silt particles, silicates and other minerals gas bubbles
Minor solutes > 0.1 mg/L < 10 mg/L	Fe^{2+} Mn^{2+} NH_4^+ Sr^{2+}	NO_2^- PO_4^{3-} HPO_4^{2-} $H_2PO_4^-$ I^- F^-, Br^-	H_2S NH_3 CH_4 He	organic compounds, metabolites	oxidohydrates of metals, such as $Fe_2O_3 \cdot nH_2O$, silicic acid silicates humic substances	oxidohydrates of Fe and Mn oils and fats organic substances such as organic humic substances
Trace solutes < 0.1 mg/L	Cu^{2+} Zn^{2+} Pb^{2+} Li^{2+} Tritium, etc.	HS^- S^{2-}	Rn	organic compounds	anthropogenic organic substances, such as tri- or tetra-chloroethylene or haloformic compounds	viruses germs microorganisms gases

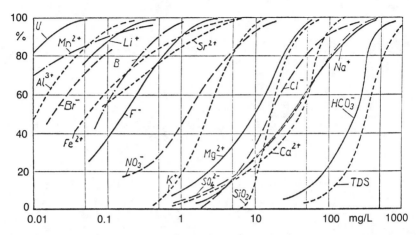

Figure 1.3. Typical natural integrated frequency distributions of water solutes.
Source: Everett [1.23].

Table 1.3 presents some supplementary information about the origin and impacts of these solutes. The type and amount of solutes present determine the quality of water; important processes affecting water quality within the hydrologic cycle are summarized in Figure 1.4.

Depending on the problem studied, migrants of interest in soil- and groundwater may be real or colloidally dissolved or emulsified solid, liquid, or gaseous particles; in tracer studies, they may also be tritium-carrying water molecules. Studies in coordination chemistry have shown that the elementary migrating particles dissolved in water are not free ions but complexes [1.90].

Complexes are relatively stable, electrically neutral or charged constituents that consist of metal ions (central atoms) and one or more ions or uncharged molecules (ligands). Free metal ions are hydrated in aqueous solutions and form aquatic complexes, which in turn lead to the formation of other types of complexes (see §1.4.2).

Knowledge of complexes present in soil- and groundwater is important for reliably interpreting and forecasting physicochemical processes. For example, the migration rate of metals in the subsurface depends significantly on their complexation state. While the total concentration of a substance can be determined by conventional analytical means, the concentrations of the different complexes containing this substance cannot. The concentrations of migrating complexes are therefore usually computed by means of thermodynamical equilibrium models using total substance concentrations measured conventionally (see §1.4.2). Special migrants in the aqueous phase are

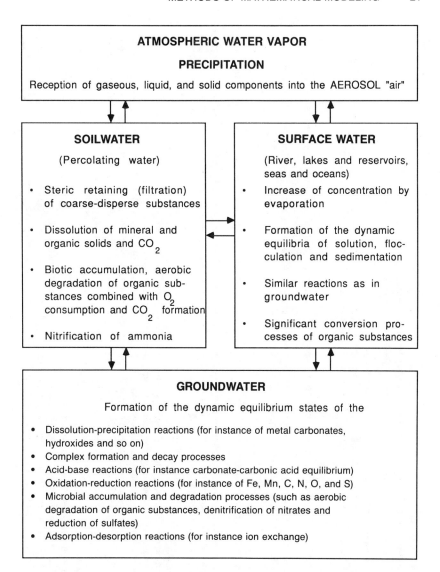

Figure 1.4. Selected processes of water quality genesis during the hydrologic water cycle.

- suspended migrants (such as viruses, bacteria, or substances used for the formation of subterranean seals)
- emulsified migrants (liquid particles insoluble in water, for instance, droplets of mineral oil, or "oil-in-water emulsion")
- small gas bubbles (occurring if the gas saturation concentration is exceeded)

Table 1.3. Important Natural Solutes in the Groundwater, Their Origin, and Their Impacts on Water Utilization

Silicon dioxide (SiO_2)

Origin: Feldspars, clay minerals, amorphous silicon dioxide, quartzes.

Impacts: In the presence of calcium and magnesium, silicon dioxide forms heat-insulating incrustations in boilers and steam turbines which are difficult to remove. Silicon dioxide may be added to soft water as a corrosion inhibitor in iron pipes.

Iron (Fe^{2+})

Origin: Volcanic rocks, hornblendes, biotites, pyrites, markasites, hematites, goethites, magnetites, sandstones, oxides, carbonates, clay minerals.

Impacts: Precipitation occurring in the presence of oxygen results in turbidity, staining of plumbing and instruments, washing machines and cookware, and affects the taste and esthetic quality of foods.

Manganese (Mn^{2+})

Origin: Metamorphic and sedimentary rocks, biotites, hornblendes; manganese in groundwater is mostly derived from the soil zone and from river sediments.

Impacts: Oxidation of manganese causes precipitation, with consequent impairment of water taste, deposition on foods during cooking, black staining of plumbing and washing machines, and increased maintenance expenditures in water supply systems (filter tanks). Industrial users usually require [Mn] < 0.2 mg/L.

Calcium (Ca^{2+}) and magnesium (Mg^{2+})

Origin:
Ca: Ca-hornblende, feldspars, fluorite, gypsum, augites, argonites, calcite, dolomite, clay minerals.
Mg: Mg-hornblende, olivines, augites, dolomites, volcanic rocks, clay minerals.

Impacts: Calcium and magnesium in the presence of bicarbonate, carbonate, sulfate, and silicon dioxide form heat-insulating incrustations in heat-exchangers and boilers, and buildups also reduce the cross-sectional area in pipelines. Both form soapy waters when in contact with fatty acids. A high magnesium content in drinking water can act as a purgative, especially for new users of the affected drinking water system.

Sodium (Na^+) and potassium (K^+)

Origin:
Na: Na-feldspars (albites), clay minerals, evaporites, halites.
K: K-feldspars (orthoclase, microcline), muscovites, biotites, and nephelines.

Impacts: In the presence of suspended material, a content of more than 50 mg/L of sodium and potassium results in the formation of foam and causes corrosion. Sodium and potassium carbonates promote the degradation of wood in cooling towers.

Carbonates (CO_3^{2-}) and bicarbonates (HCO_3^-)
Origin: Calcites, dolomites, biogenic sources.
Impacts: Upon heating, bicarbonate decomposes into carbon dioxide. Carbonate forms incrustations in pipes (cross-section reduction) and heat exchangers, particularly in the presence of magnesium and potassium. Therefore, many industrial users need water with a limited CO_3^{2-} and HCO_3^- content.

Sulfates (SO_4^{2-})
Origin: Ores, gypsum, anhydrite, barite.
Impacts: Sulfates with calcium form strongly adherent heat-insulating coatings. Many industries experience adverse effects when using water with sulfate contents > 250 mg/L. Water with $[SO_4^{2-}] > 500$ mg/L has a bitter taste, and >1 g/L acts as a purgative.

Chlorides (Cl^-)
Origin: Evaporites and sedimentary rocks, sea water intrusions in coastal aquifers, and estuaries.
Impacts: $[Cl] > 100$ mg/L results in a salty taste, >1 g/L in physiologically harmful effects. The food industry requires a Cl content < 250 mg/L. Several industrial users (textile industry, paper and rubber manufacturers) generally need water with a Cl content < 100 mg/L.

Fluorides (F^-)
Origin: Hornblendes, apatites, fluorites, mica.
Impacts: [F] between 0.6 and 1.5 mg/L promotes the formation of a strong tooth enamel, particularly in children, but [F] > 1.5 mg/L can result in brown mottled enamel on children's teeth, and [F] > 6 mg/L commonly causes tooth deformations.

Nitrates (NO_3^-)
Origin: Fertilizer and manure, atmosphere, legumes, plant remains, sewage sludge.
Impacts: $[NO_3^-] > 100$ mg/L results in a bitter taste in water and causes physiologically harmful effects. A concentration > 40 mg/L can cause methamoglobinemy in infants. Carcinogenic substances may be formed from NO_3^-. On the other hand, nitrate can strengthen steel used in high-pressure containers.

Dissolved Solids
Origin: Dissolution of minerals in water.
Impacts: Dissolved solids present at concentrations greater than 500 mg/L are not desirable for drinking water and water used for industrial processes. Specific industries requiring TDS < 300 mg/L include manufacture of textiles, paper, plastics, and rayon. Foam forms more easily when dissolved solids are present.

Source: Everett [1.23].

1.1.4 MIXPHASE MODEL OF THE ROCK MATRIX

In the subsurface, mineral solids form a spatially fixed solid mixphase, consisting of a skeletal framework with filmlike coatings. The structure of mineral solids has a definite spatial arrangement of solids and voids, often described by the terms *rock matrix* or *porous medium*. The voids in a rock matrix are commonly occupied by one or more fluid mixphases, which may be mobile or immobile. In the following discussion the term *rock* will often be meant to include the soil as well.

The structure of the rock matrix affects the fundamental properties of fluid flow such as porosity and permeability, upon which every fluid flow calculation rests, and also controls the inner surface or interface between fluid and solids in the subsurface. The size of this contact area per unit volume rock matrix is of great importance for the exchange of migrants between the fluid and the solid phases and is therefore important for the migration process as a whole.

Rocks are mineral aggregates, which within this text include both consolidated and unconsolidated materials (see Table 1.4). Therefore, minerals are the components of the mixphase rock matrix. Minerals determine the properties of this mixphase and strongly impact the migrants' fate in the subsurface.

Minerals are physically homogeneous solids with a definite chemical composition. Most minerals result from

Table 1.4. Classification of Rocks

	Petrographic classification		Characteristics	Examples
UNCONSOLIDATED ROCKS	WEATHERING PRODUCTS		extent and type of weathering, parent material	weathering braunite kaolin
	S E D I M E N T S	organic and chemical	parent material chemical behavior extent of genesis	chalk, sapropel, brown coal
		clastic — grain size	grain size	gravel, sand, silt, clay
CONSOLIDATED ROCKS			type of cementation	conglomerate, sandstone, mudstones, siltstones
		organic and chemical	parent material chemical behavior extent of genesis	limestone, gypsum, rocksalt, anhydrite, anthracite
	IGNEOUS		parent material microstructure	granite, diorite, syenite, basalt, quartzporphyry
	METAMORPHIC		parent material metamorphic type	clay schist, phyllite, gneiss, mica slate

- magmatic succession
- sedimentary succession (weathering, precipitation, and biogenic processes)
- metamorphic succession (formation of new minerals by the metamorphic transformation of existing minerals)

Minerals may be divided into nine classes [1.81]:

    I. Native Elements
   II. Sulfides
  III. Halides
   IV. Oxides, Hydroxides
    V. Nitrates, Carbonates, (Borates)
   VI. Sulfates, (Chromates, Molybdates, Tungstates)
  VII. Phosphates, (Arsenate, Vanadate)
 VIII. Silicates
   IX. Organic Substances (Salts of Organic Acids, Solid Hydrocarbons, Resins)

The principal components of rocks are formed from 40 to 50 minerals. Silicates (feldspars, pyroxenes, hornblendes, micas, olivines, and clay minerals) and oxides (quartz) are the predominant rock-forming minerals. Table 1.5 presents a partial list of silicates and oxides. More comprehensive compilations of rock-forming minerals are given in Strunz [1.81], Rösler [1.66], and Rösler and Lange [1.67].

The chemical composition of minerals may be expressed as a stoichiometric formula. The occurrence of elements in rocks of the continental crust given as mass fraction w_i in% or volumetric content σ_i in% is shown in the following compilation (the first number represents w_i, the second σ_i in %):

O (46.71/94.24)	Si (27.69/0.51)	Al (8.07/0.44)
Fe (5.04/0.37)	Ca (3.65/1.04)	Na (2.75/1.21)
K (2.58/1.88)	Mg (2.08/0.28)	

Minerals are crystalline solids that have an orderly internal structure. The constituents of crystals (atoms, ions, molecules) are spatially arranged in a periodic crystal lattice. A lattice may be regarded as the blueprint of a crystal, or a regular pattern of points in which each point has an environment identical to any other point in the pattern. This pattern is referred to as the *coordination* of constituents arranged in a crystal lattice. The smallest repeating unit of a crystal lattice is designated as a unit cell. The corners of a unit cell are formed by the gravity centers of the constituents (see Figure 1.9a).

Table 1.5. Examples of Important Minerals, Their Density ρ, and Typical Occurrence

			Density ρ (g/cm^3)	Occurrence
		NONSILICATES		
ELEMENTS	Graphite	C	2.25	in metamorphics
SULFIDES	Pyrite	FeS$_2$	5.2	in carbon formations
HALIDES	Rock salt	NaCl	2.2	in evaporites
	Fluorite	CaF$_2$	3.2	in volcanic rocks
OXIDES	Quartz	SiO$_2$	2.65	in all rocks
HYDROXIDES	Hematite	Fe$_2$O$_3$	5.2	in many rocks
	Magnetite	Fe$_3$O$_4$	5.2	in many rocks
	Goethite	FeOOH	3.4	formed during weathering
CARBONATES	Calcite	CaCO$_3$	2.7	in metamorphic and
	Dolomite	CaMg(CO$_3$)$_2$	2.9	sedimentary rocks
SULFATES	Barite	BaSO$_4$	4.5	in veins of metal ore
	Anhydrite	CaSO$_4$	3.0	in sedimentary rocks
	Gypsum	CaSO$_4 \cdot$ 2H$_2$O	2.4	in sedimentary rocks
PHOSPHATES	Apatite	Ca$_5$(F,OH,Cl)(PO$_4$)$_3$	3.2	in volcanic and other rocks
		SILICATES		
NESOSILICATES	Olivine	(Mg,Fe)$_2$SiO$_4$	3.3	in basic volcanic rocks
	Kyanite	Al$_2$[OSiO$_4$]	3.7	in metamorphics
INO- & CYCLOSILICATES	Hornblende	Ca$_2$(Mg,Fe)$_5$ [(OH)$_2$/Si$_8$O$_{22}$)]	3.1	in volcanic rocks and metamorphics
PHYLLOSILICATES	Muscovite	KAl$_2$[(OH)$_2$ AlSi$_3$O$_{10}$]	2.9	in all rocks
	Biotite	K(Mg,Fe)$_3$[(OH)$_2$ AlSi$_3$[O$_{10}$]	2.8–3.2	in volcanic rocks and metamorphics
	Chlorite	(Mg,Fe)$_5$Al[(OH)$_8$ /AlSi$_3$O$_{10}$]	2.5–2.9	in metamorphics and volcanic rocks
	Kaolinite	Al$_4$[(OH)$_8$/Si$_4$O$_{10}$]	2.1–2.6	weathering product, in sediments
	Montmorillonite (Smectite)	(Al,Mg)$_2$[(OH)$_2$ Si$_4$O$_{10}$] Na$_{0.33}\cdot$H$_2$O	about 2.5	weathering product, in sediments
	Illite	(K,H$_3$O)Al$_2$[(OH)$_2$/ AlSi$_3$O$_{10}$]	2.7–2.9	weathering product, in sediments
TECTOSILICATES	Orthoclase	KAlSi$_3$O$_8$	2.53–2.56	in many rocks
	Quartz	SiO$_2$	2.65	in all rocks
	Plagioclase		2.61	in many rocks
	Albite	NaAlSi$_3$O$_8$	to	(most frequently
	Anorthite	CaAl$_2$Si$_2$O$_8$	2.76	encountered mineral)

Source: Grunert and Schneider [1.34].

In addition to the coordination of constituents in a crystal lattice, the way in which interstices between large structural elements are filled by smaller ones is of particular interest. The size of the constituents in the crystal lattice, represented as mutually contiguous spheres, is the most important feature controlling the structure of a mineral.

The atomic or ionic radius controls the size of a chemical element

[1.66, p. 742; 1.67, p. 56; 1.47]. The ionic radius r of an ion decreases as the positive charge increases. The following are characteristic examples (r in 10^{-10} m):

Pb^{4-} r = 2.15	Se^{2-} r = 1.93	O^{2-} r = 1.32
Pb^{0} r = 1.74	Se^{0} r = 1.40	O^{0} r = 0.6
Pb^{2+} r = 1.32	Se^{3+} r = 0.78	O^{6+} r = 0.1
Pb^{4+} r = 0.84	Se^{4+} r = 0.50	

A variety of attractive forces are responsible for the chemical bonding of the atomic, ionic, and molecular constituents of the crystal lattice. These include (see also Figure 1.5 and Sommer [1.80])

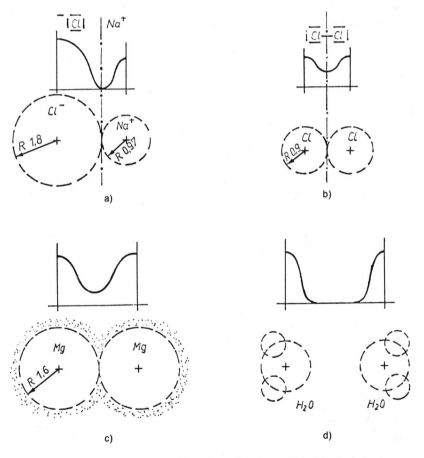

Figure 1.5. Examples of the four bond types of elements and characteristic electron density distributions (electron clouds): *(a)* ionic bond, *(b)* atomic bond, *(c)* metallic bond, *(d)* van der Waals bond (molecular bond).

- the van der Waals bond (molecular bond)
- the ionic bond (heteropolar bond, present in more than 80% of the minerals, especially the oxides and halides)
- the atomic bond (covalent or homopolar bond)
- the metallic bond

Figure 1.5 shows that the electron density distributions or electron clouds are partially overlapping. The model of mutually penetrating electron clouds is therefore considered to be a better representation of reality than the model in which electron clouds only contact each other at the edges (see Figure 1.2c and 1.9b).

The thermal conductivity of minerals is directly related to the type of chemical bonding between atomic, ionic, or molecular constituents of the crystal lattice. Thermal conductivity is greatest for metallic bonds and lowest for atomic bonds. The lattice energy is a measure of the sum of the binding forces in crystal lattices.

Geometric size relations of the lattice elements control the formation of coordination groups. These groups are designated according to their coordination number (see Figures 1.6 and 1.7). For example, in a NaCl crystal, each Na cation is surrounded by six Cl anions, and each anion is surrounded by six Na cations. The sodium and chloride ions are said to be in a six coordination. Lines connecting the centers of the six coordi-

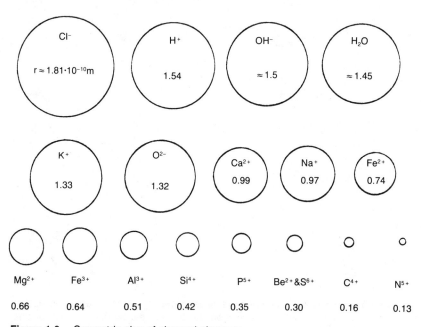

Figure 1.6. Geometric size of charged elements.

Coordination number	Coordination type			Radius ratio	Example
3			Triangle	0.155 . . . 0.255	$(NO_3)^{1-}$ $(CO_3)^{2-}$
4			Tetrahedron	0.255 . . . 0.414	$(SO_4)^{2-}$ $(PO_4)^{3-}$
5			Octahedron	0.414 . . . 0.732	$(SiO_4)^{4-}$ $Mg(OH)_6$ $Al(OH)_6$

Figure 1.7. Important coordination types, their graphic representation, and examples of their occurrence. *Source:* Rösler [1.66].

nated anions form an octahedron, giving rise to the additional term, octahedral coordination.

Most of the coordination groups possess a small cation (central atom) positioned between several large anions (ligands). The three, four, and six coordinations are the most important structural elements of the rock-forming minerals. The coordination tetrahedra and octahedra which comprise the unit cells of silicate minerals are in four-one or six-one coordination (see Figures 1.7 and 1.8). If relatively small cations and large anions form such cells, the polarity and the dipole moment of this arrangement are always high (see also Figure 1.2c). Summarizing these different models of mineral structure, Figure 1.9 shows several ideal structural models of the mineral pyrite.

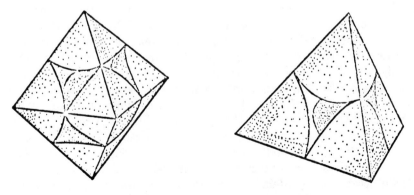

Figure 1.8. Octahedral and tetrahedral coordination polyhedron.

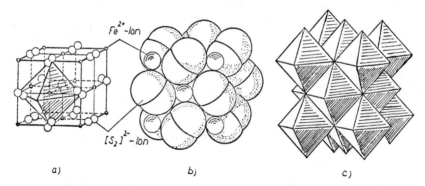

Figure 1.9.   Ideal structural models of pyrite FeS_2 as an example: *(a)* lattice model, *(b)* space-filling model, *(c)* model of coordinated polyhedrons (unit cells).

Naturally occurring minerals do not possess an ideal structure but rather exhibit distinct lattice defects. These defects are called *disorder* when atomic elements are lacking or fixed at wrong positions, or *disstructure* when larger regions are disordered. In addition, natural crystals do not generally have an exact stoichiometric composition. Therefore, their gross and structural formulas should be regarded as models only.

An important consideration for migration processes in the soil- and groundwater zone is the ability of lattice constituents to be replaced with or substituted by other ions. Ionic substitution in the crystal lattice is also referred to as *diadochy*. This may occur when the ratio of the atomic or ionic radii, $\Delta r/r$, is less than 0.15. The ion with the smaller radius is preferentially substituted into the crystal lattice, or in the case of equal radii, the ion with the higher positive charge is substituted. When the charges of substituted ions are balanced, *coupled substitution* is said to occur. A detailed discussion is given in Rösler [1.66] and Rösler and Lange [1.67].

The property of ionic substitution is most important for the ion exchange ability and the ion exchange capacity of the rock matrix. Ion exchange and sorption capacity, which are primary factors affecting the migration of solutes in the subsurface, are of particular interest when studying the clay minerals. Clay minerals are layer silicates, which are structurally classified as dioctahedral and trioctahedral two-layer, three-layer, and four-layer silicates. Layer silicates are composed of alternating SiO_4 tetrahedral layers and $Al(OH)_6$ octahedral layers [1.68].

In the SiO_4 tetrahedral layers, the oxygen atoms occupy the corners of the tetrahedron, surrounding the silicon atom (see Figure 1.7). If the Si atoms are partially substituted by Al atoms, aluminosilicates are formed. In the octahedral layer (also shown in Figure 1.7) the OH⁻ ions in the six corners are partially replaced by O atoms. If there are bivalent cations

(Mg^{2+}) in the interstitial voids between the oxygen atoms and hydroxide groups, the octahedral layer will be completely developed, i.e., it will form a trioctahedral layer structure. With trivalent cations (Al^{3+}) only 2/3 of the octahedral sites are occupied, and a dioctahedral layer structure will form (see Figure 1.10).

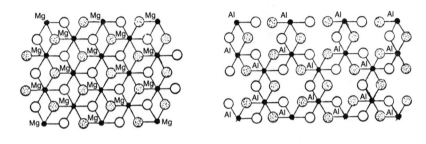

●   Mg and Al cations, respectively

○   OH⁻ and O, respectively, located above the cations

◉   OH⁻ and O, respectively, located below the cations

Figure 1.10. Structural model of tri- and dioctahedral layers of clay minerals in the plane of projection.

Unoccupied sites can then be filled by K^+ ions, which fit perfectly in size.

Figure 1.11 shows the structure of layer silicates. The four most important clay minerals are (see also Table 1.5)

- kaoliniteanhydrated dioctahedral two-layer silicate
- illiteanhydrated dioctahedral three-layer aluminosilicate
- montmorillonitehydrated dioctahedral three-layer silicate
- chloriteanhydrated trioctahedral four-layer silicate

Since adjacent tetrahedral layers of three-layer silicates are equally charged, they preferentially incorporate cations, water molecules, and hydrated cations. This insertion between layers causes the important swelling properties of three-layer silicates.

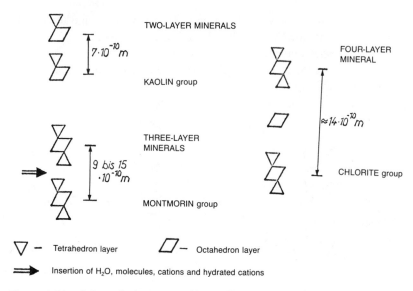

Figure 1.11. Schematical structure of layer silicates.

The ion exchange capacity of clay minerals can be consequently attributed (1) to the negative layer charges originating from the substitution of cations with higher valency by cations with lower valency (e.g., the substitution of Si^{4+} by Al^{3+} in tetrahedral layers and Al^{3+} by Mg^{2+} in octahedral layers) and (2) to the amphoteric properties of side and fracture surfaces of clay minerals which, depending on pH, are charged either positively or negatively [1.27; 1.64].

Let us consider a montmorillonite crystal with a structural formula $EC(Al_{3.33}, Mg_{0.67})\,[(OH)_2/SiO_{10}]_2$ where EC( ) represents the exchangeable cations. The molecular mass of this clay is 734 g. The substitution of the 0.67 mol aluminum with 0.67 mol magnesium yields a cation exchange capacity of

$$CEC_c = 0.67/734 = 91.5 \cdot 10^{-5} \text{ mol/g} = 91.5 \text{ mmol}_c/100 \text{ g}$$

Clay minerals are characterized by a high cation exchange capacity (CEC) and a low anion exchange capacity (AEC), with the exception of kaolinites, which also have a high AEC. Preferentially exchanged anions in clays are PO_4^{3-}, SO_4^{2-}, and to a lesser extent Cl^- and NO_3^-. Anion exchange takes place primarily at fracture surfaces. In clay minerals, anions are usually bound to Al hydroxides, but may also be bound to the colloids of amphoteric iron and manganese below the isoelectric or zero point of charge (see also Figure 1.46). Finally, Table 1.6 shows the ion exchange capacity of clay minerals, humic acids, and other organic substances.

Table 1.6. Ion Exchange Capacity of Clay Minerals, Humic Acid, and Organic Substances at pH = 7

	Cation Exchange Capacity mmol/100 g	Anion Exchange Capacity mmol/100 g
Kaolinite	3 . . . 15	5 . . . 10
Illite	10 . . . 50	—
Montmorillonite	80 . . . 150	20 . . . 30
Chlorite	100 . . . 50	—
Humic acid	100 . . . 500	—
Organic substances	up to 300	—

Note: Exchange capacity = equivalents of the exchangeable ions or number of charges (expressed as mol_{eq} or mol_c) per mass unit of the dry rock; take note of the exchange capacity being pH dependent.

1.2

Subsurface Model Concept

In investigations of migration processes, the model chosen depends largely on the observation scale of the multicomponent subsurface system. The problem of selection and justification of an appropriate scale of consideration is introduced in §1.2.1. At the local and regional level, two principal concepts are of particular interest. The first assumes that only one fluid mixphase is mobile in the rock matrix (discussed in §1.2.2), and the second that several mutually immiscible fluid mixphases are mobile in the rock matrix (discussed in §1.2.3). Finally, two limiting cases of fluid distribution and fluid flow are discussed in §1.2.4.

1.2.1 SCALE OF OBSERVATION

Investigations of migration problems are conducted at the molecular, microscopic, local, and regional investigation levels. At the molecular scale of observation, the molecules are the system and the atomic particles (protons, electrons, etc.) are the system elements. This concept was used in Chapter 1.1 to discuss structural properties of gaseous, liquid, and solid substances.

At the microscopic scale of observation, the mixphases air, water, or rock form systems, and the atoms, molecules, and ions are the indivisible elements of the system. Due to the exceedingly large number of these elements (one mole contains $6.0232 \cdot 10^{23}$ particles) it is usually more convenient to choose a volume as the systems element containing a sufficiently large number of molecules, atoms, or ions such that only their mean statistical behavior is relevant. By applying this approach, the mixtures air and water become mixphases, or continua. A volume enclosing such a continuum molecular mixture is called a *representative elementary volume (REV)* of the fluid phase. The REV must be large compared to the mean free path of molecules caused by Brownian motion. Similarly, the smallest time interval of observation Δt, the representative elementary time interval (RET), must be long compared to the time necessary for the molecules to travel this mean distance.

Typical migration parameters and state variables at the microscopic or *pore scale*, where the fluid continuum is limited by pore walls, are fluid density, viscosity, and concentration of the migrants in the fluid. In the continuum approach, material characteristics and dependent variables of the REV are lumped in a "physical point." Thus all the systems parameters and state variables are considered to be smooth functions of the spatial and temporal coordinates.

The degree of integration is again increased when migration processes are investigated at the local level. Here, the real spatial distributions of phases and interfaces are ignored and replaced by their statistical averages. The system "air-water-solids" is then replaced by a structureless, fictitious continuum called *subsurface* or soil.

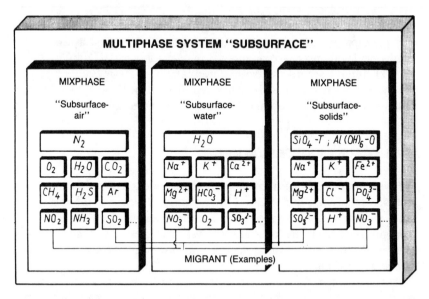

Figure 1.12.   The three-level model approach of the multiphase system "subsurface."

In the multiphase system subsurface, the different mixphases—solids, air, and water—are intermixed, and each has an apparently continuous distribution across the entire volume considered. The same model approach was used for each of the mixphases individually (see §1.1.2 to 1.1.4). Depending on the objective of an investigation, one or more of the components forming the mixphases of the multiphase system are the migrants (see Figure 1.12). This three-level model approach is a fundamental concept in mathematical modeling of migration.

At the local scale of investigation, it is again necessary to select and define a REV and RET as elements of the subsurface (see Figure 1.13).

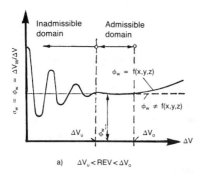 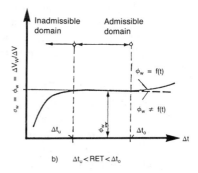

Figure 1.13. Defining the representative elementary volume REV and the representative elementary time interval RET at the local scale of observation with an example of partially saturated soil (σ_w is the volumetric content of the wetting phase at the point P = P(x,y,z,t).) *See also* Bear [1.4].

"Representative" in this sense refers to the necessity of finding an element which is uniform for all migration state and process variables. Examples include flow rate $\vec{v}_i$, reaction rate, ionic activity, and volumetric phase content $\theta_i = \sigma_i$.

In selecting the REV at the local scale of investigation, two conditions must be met. On the one hand, the REV must be large compared to the characteristic dimensions of the systems heterogeneity, such as pore size and distance between flow channels. This is required to allow the use of local statistical averages of geometry, kinematics and kinetics, and physical, chemical, and biological processes. On the other hand, the REV must be small compared to the characteristic site dimensions, such that the local statistical averages may be regarded as continuous subsurface parameters and variables. Migration processes may then be represented by means of differential equations. This second condition is similar to that of the finitization of continuous field problems, akin to space discretization in the finite element method. Accordingly a REV must be smaller than a finite element.

The estimation of statistical averages is a problem not only with respect to space but also with respect to time. The RET must be long enough in duration such that time-dependent variables may be averaged within the REV. The RET is determined primarily by the time required to attain the equilibrium states in the REV. Justification of REV and RET selection is therefore comparable to estimations of expectation values in stochastic process analyses [1.14].

The magnitudes of REVs and RETs are also important in selection and design of measuring instruments for the subsurface. Each such instrument permits an observer to look into the system under investigation as through a window. Measured quantities may be without meaning if the window of the instrument is smaller than the REV, or if the measuring time is shorter than the RET. Measuring instruments improperly scaled to the REV and RET will lead to problems in monitoring of migration processes.

Matters pertaining to selection and justification of the REV and RET are of even greater consequence at the regional scale of investigation. Here, the solid particles of the local scale are replaced by low-permeability zones such as silt lenses in an aquifer, and pore channels are replaced by laterally continuous strata of coarse sand or gravel. This regional scale of consideration must be used with particular care because of the long RET and the geometric window problems.

Figurative models have proven to be useful tools when building mathematical migration models at local and regional scales. A three-stage sequence of figurative models is recommended, with an increasing degree of abstraction as shown in Figure 1.14 [1.51; 1.53].

In the first stage, the figurative model must reflect the migration process at the level immediately below the level of investigation (see Figure 1.12). Therefore, if the level of investigation is the local level, the first figurative model should be a schematic representation of the processes taking place at the microscopic or pore scale level. Figure 1.14a shows such a schematic representation characterizing the distribution of solids, pore channels, and pore spaces filled with mobile and immobile fluids. At the regional scale of consideration, detailed geologic cross sections would serve as the first figurative model.

The primary aim of first stage figurative models is to aid in model concept development and understanding of model approximations. Figurative models must be characterized by a high degree of clarity to allow effective communication between model builders and model users familiar especially with the subsurface characteristics.

At the second stage, the figurative model should represent the averaged multiphase system. Here, the real distribution of mixphases described in the first stage is projected onto a unit volume corresponding to the REV. In this manner all real phases and interfaces are reflected as representative averages (see Figure 1.14b). In this figure each phase of the REV should also be represented by a lumped element to reflect the

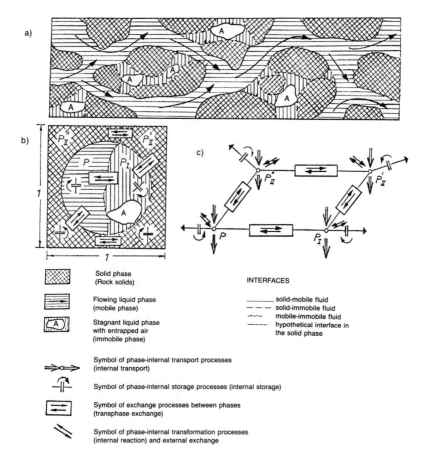

Figure 1.14. Example of a three-stage succession of figurative models in mathematical modeling of migration processes in the soil- and groundwater zone. *Source:* Luckner and Nitsche [1.51].

extensive state variable storage (S) because each phase is able to store matter or heat. The intensive state variables (P) representing temperature, pressure, or activities in the REV must also be lumped for each phase of the REV. Both S and P are linked by the fundamental Equation 1.1a: $S = f(P_i)$. Phase-internal reactions, transport processes, and exchange of migrants across interfaces should also be represented symbolically in the second figurative model, and each element in this figure (see Figure 1.14b) should be derivable from phenomena illustrated in the first model stage (Figure 1.14a).

Finally, in stage three, the figurative model of stage two is abstracted to the extent that migration processes can be directly described with mathematical models. The block diagram shown in Figure 1.14c is particularly useful for this purpose.

1.2.2 THE "MOBILE/IMMOBILE" TWO-PHASE MODEL APPROACH

The "mobile/immobile" two-phase model assumes that all of the constituents in the subsurface can be assigned to either a mobile fluid mixphase or the immobile rock matrix. Accordingly, there are two categories of models: single fluid flow models and multifluid miscible flow models.

Using the two-phase model approach, Figure 1.14c becomes a two-nodal scheme; one node represents the mobile phase and the other the immobile phase. To proceed with this analysis, it is necessary to clarify which solid, liquid, and gaseous components are considered dispersed in the mobile fluid phase, and which are bound to the rock matrix. For example, if the soil air is the only mobile fluid in the subsurface, one must decide which gases form the mixphase model.

Processes that immobilize matter include stearic restraint, electromolecular adsorption, chemical adsorption, and biological fixation. Hence, sand grains are frequently coated by clay shells and biofilms, and pore spaces are frequently filled with humic particles. Most solids are hydrophilous, i.e., water molecules are preferentially adsorbed and are coated by shells of immobile water molecules. These molecules have an entropy level comparable to that of ice crystals (see Figure 1.14a and 1.15a).

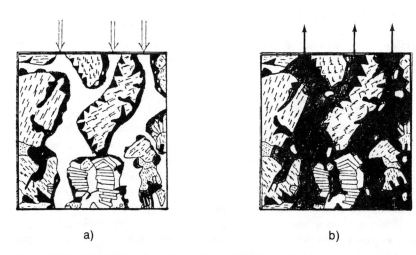

a)   b)

Figure 1.15. Distribution of mobile and immobile phases in the two-phase model: *(a)* "gas-rock" and *(b)* "water-rock." *Source:* adapted from Czolbe, Kretzschmar, and Kühnel [1.15].

In the presence of soil air, liquid pore water is partially immobilized in the rock matrix by capillary forces. Immobilized water may be enriched with several different components.

Wetting properties are the adsorption properties of solid surfaces with respect to liquid water and gas molecules, and may vary even at the pore scale (see Figure 1.16).

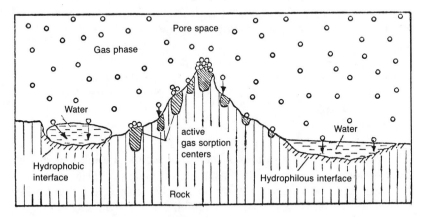

Figure 1.16.   Immobilization of molecules onto walls of larger pores. *Source:* adapted from Kretzschmar and Czolbe [1.48].

In the groundwater zone, liquid water is usually the only mobile mixphase, (see the figurative model in Figure 1.15b). This concept is often used to investigate the formation and transformation of the aqueous mixphase groundwater with its solutes. As before, one must assign components to the mobile and immobile phase models. If the migrants investigated only occur in the mobile phase, i.e., if they are not exchanged between the mobile and immobile phase, then Figure 1.14c is reduced to a one-nodal scheme. Migrants which are not exchanged between the mobile and immobile phases such as tritium or chloride may be used as tracers.

1.2.3 MULTIFLUID IMMISCIBLE FLOW MODEL APPROACH

Air and water are typical examples of mutually immiscible fluids. The microscopic domains of the three mixphases air, water, and rock solids are separated by interfaces (see §1.2.1). Tensions acting upon these interfaces depend on material properties of the mixphases. The *interfacial tension* is defined as the internal pressure difference of the two contiguous phases, and the *surface tension* as the interfacial tension of a phase in equilibrium with its own vapor phase.

The internal pressure of a phase is generated by van der Waals attraction forces. In the interior of the phase these electromolecular forces are balanced, but molecules at the interface are unbalanced. The unbalanced attraction forces towards the phase center result in an internal pressure and a stress at the interface (see Figure 1.17).

Internal pressure and interfacial tension characterize the ability of a phase to do work per unit area. Interfacial tension is therefore expressed in units of energy per unit area, i.e., in $Nm/m^2 = N/m$. Examples of interfacial tensions between common materials at 20°C follow:

water-air	$\sigma_{W,A} = 72.8 \cdot 10^{-3}$ N/m
soap solution–air	$\sigma_{S,A} \approx 30 \cdot 10^{-3}$ N/m
mineral oil–air	$\sigma_{O,A} \approx 20$ to $50 \cdot 10^{-3}$ N/m
glass-air	$\sigma_{G,A} \approx 200$ to $300 \cdot 10^{-3}$ N/m
mercury-air	$\sigma_{Hg,A} \approx 500 \cdot 10^{-3}$ N/m

Each system strives toward its energy minimum. Liquids do this by deformation. In a vacuum, for instance, the water droplet takes on a spherical shape. A water droplet on a glass plate displaces air, and water molecules wet the glass in order to minimize the system energy ($\sigma_{G,W} <$

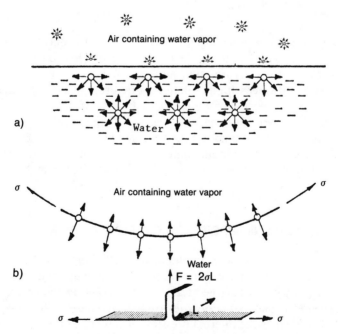

Figure 1.17. Molecular attraction forces *(a)* in the phase interior and *(b)* at the interface or surface.

$\sigma_{G,A}$). For the same reason mineral oil is displaced by water on glass ($\sigma_{G,W} < \sigma_{G,O}$), although not as spontaneously.

Silicate rocks have a wettability similar to that of glass. In the water-air-silicate rock system, water molecules are the wetting substance and air molecules the nonwetting fluid. For the case of liquid water and mineral oil in silicate rocks, oil is the nonwetting fluid. For air and mineral oil in all types of rocks, oil is the wetting fluid and air the nonwetting fluid. Mercury will not wet glass, and paraffin is not wetted by water. The energy minimum of both two-phase systems is reached when mercury or water minimizes its own surface area. Therefore, in order for water to wet organic soils, such as coal mine spoils, detergents or other surfactants are used to lower the internal pressure and surface tension.

Three phases such as quartz, water, and air can only be mutually contiguous along contact lines. Equilibrium of forces along contact lines is described by Young's Equation (see Figure 1.18a).

$$\sigma_{S,F} = \sigma_{S,L} + \sigma_{L,F} \cos \delta \qquad (1.8)$$

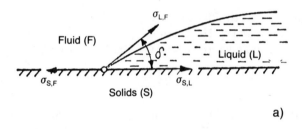

a)

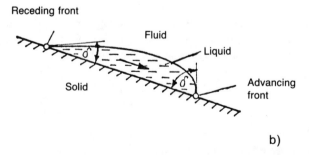

b)

Figure 1.18.   Liquid droplet *(a)* on a horizontal solid surface (static case) and *(b)* on an inclined solid surface (dynamic case).

The contact angle δ between the fluid-liquid interface and the surface of the solid phase is a measure of wettability. The contact angle δ is hysteretic, i.e., it is larger at the advancing front of the wetting liquid phase than at the receding front (see Figure 1.18b).

In the absence of specific data, the following values may be used as first estimates for sands, gravels, and sandstones:

water-air	$\sigma_{W,A} = 70 \cdot 10^{-3}$ N/m	$\delta \approx 0°$
water-oil	$\sigma_{W,O} \approx 25 \cdot 10^{-3}$ N/m	$\delta \approx 0°$
oil-air	$\sigma_{O,A} \approx 40 \cdot 10^{-3}$ N/m	$\delta \approx 30$

Capillary pressure may be determined from these interfacial tensions, and is defined as the pressure difference between the nonwetting and wetting fluid phase at a point in the subsurface:

$$p_c = (p_{nw} - p_w)|_{rock\ matrix} \tag{1.9}$$

According to Laplace's Law, the capillary pressure p_c is given by:

$$p_c = \sigma_{w,nw} (1/r_1 + 1/r_2)$$

where the curvature of the interface is expressed as two orthogonal circular arcs with radii r_1 and r_2. Hence, in the case of a capillary tube where $r = r_1 = r_2$ and the contact angle $\delta = 0°$, Laplace's Law becomes

$$p_c = 2\sigma_{w,nw}/r \tag{1.10a}$$

The same result can be obtained from evaluating the equilibrium of the capillary force and the weight of water in a vertical tube (see also Figure 1.19a):

$$2\pi r\sigma_{w,nw} = \pi r^2 g\rho_w h_c \tag{1.10b}$$

taking into account that capillary pressure head h_c may be expressed as $h_c \approx p_c/(g\rho_w)$. For the general case, $p_c = g(\rho_w - \rho_{nw})h_c$ (see Figure 1.19b). Finally, when the contact angle $\delta \neq 0°$, the capillary pressure p_c is given by

$$p_c = (2\sigma_{w,nw} \cos \delta)/r \tag{1.10c}$$

An oil film occurring at the air-water interface will reduce capillary pressure and, hence, capillary height:

$$(\sigma_{w,o} \cos \delta_{w,o} + \sigma_{a,o} \cos \delta_{a,o}) < \sigma_{w,a} \cos \delta_{w,a} \tag{1.10d}$$

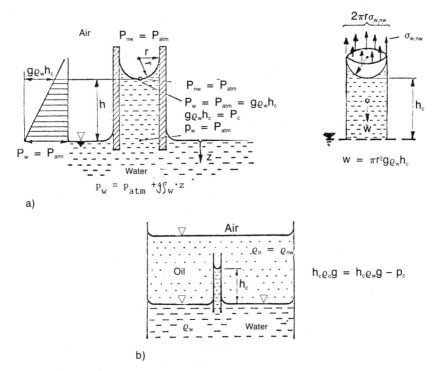

Figure 1.19. Capillary pressure in a vertical tube: *(a)* for the air-water interface for $\delta = 0°$, *(b)* for the water-oil interface.

Table 1.7 presents order of magnitude estimates of capillary pressures in unconsolidated rocks.

Table 1.7. Order of Magnitude of Capillary Pressure in Unconsolidated Rocks

Rock	Pore Radius r (in m)	Air/Water p_c (in kPa)	Oil/Water p_c (in kPa)	Air/Oil p_c (in kPa)
Clay	10^{-7}	1400	600	700
Silt	10^{-6}	140	60	70
Sand	10^{-5}	14	6	7

If the voids in the rock matrix are considered as channels of different cross-sectional radii as shown in Figure 1.20, it is evident that the degree to which voids are filled with wetting and nonwetting fluid phases is a function of capillary pressure. In this text, ϕ symbolizes porosity of the rock matrix ($\phi = \sigma_v$—volumetric content of voids), s stands for the degree of saturation of pores, and θ for the volumetric fluid content.

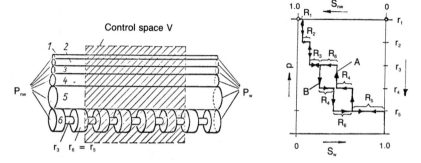

Figure 1.20. Tube bundle as a model of interconnected pores of a soil: *A*, drainage (drainage or drying curve); *B*, wetting (imbibition or wetting curve).

Thus:

$$\phi = (V_{nw} + V_w)/V \quad \theta_w = V_w/V \quad \theta_{nw} = V_{nw}/V \quad (1.11a)$$

$$\phi = \theta_{nw} + \theta_w \quad\quad V = V_{nw} + V_w + V_{solid} \quad (1.11b)$$

$$s_{nw} = \theta_{nw}/\phi \quad\quad s_w = \theta_w/\phi \quad s_{nw} + s_w = 1 \quad (1.11c)$$

When capillary pressure p_c is initially zero, we assume that the control space V shown in Figure 1.20 is completely filled by the wetting fluid ($s_w = 1$ and $\theta_w = \phi$). Now, if p_c is increased by either increasing p_{nw} (pressure method) or decreasing p_w (suction method), the nonwetting fluid advances in tube 5 first, and displaces the wetting fluid. Consequently we find that in the control space $s_w < 1$ and $s_{nw} > 0$. In tube 6, the interface is stopped at the first bottleneck. There, the wetting liquid is displaced only when $p_{nw} - p_w$ becomes larger than the potential capillary pressure p_c. Finally, when $p_c > \sigma_{w,nw} \cos \delta/r_1$, the wetting phase is completely displaced from the control space ($s_w = 0$ and $s_{nw} = 1$).

If the capillary pressure p_c is now reduced, the control space will be refilled with the wetting fluid phase. This process is called *wetting* or *imbibition*. The wetting or imbibition curve $s_w = f(p_c)$ does not coincide with the drainage or drying curve. The degree of saturation $s = f(p_c)$ is a hysteretic state function according to the fundamental Equation 1.1a. The domain of Darcian flow for both immiscible fluids exists between the main drainage curve (MDC) and the main wetting curve (MWC), i.e., the MDC and MWC limit the mobility domain of both fluids at any capillary pressure [1.54].

The hysteretic retention function has the shape shown in Figure 1.21 for both consolidated and unconsolidated rocks. Even at a high capillary pressure there is a residual water content, i.e., a residue of the wetting phase $\theta_{w,r}$ remains in the rock matrix when $p_c \to \infty$. The cause of this phenomenon is the discontinuous or incoherent distribution of the wet-

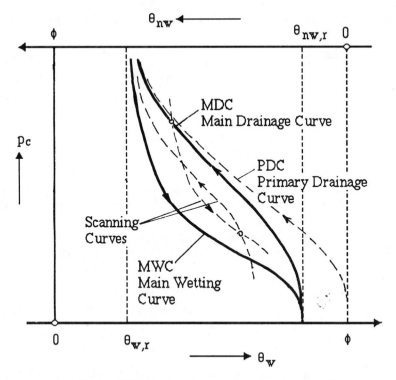

Figure 1.21. Schematic fluid retention diagram of a subsurface sample.

ting fluid in the rock matrix as p_c approaches high values. Incoherently distributed fluids are immobile as a phase. When the process is reversed, imbibition cannot completely saturate the pores with the wetting fluid. As p_c approaches zero, the volumetric content of the nonwetting fluid θ_{nw} approaches $\theta_{nw,r}$ (residual saturation of the nonwetting fluid), because under this condition the nonwetting fluid in the rock matrix becomes incoherently distributed [1.54].

For most applied problems the scaled state function Equation 1.12 reflects the hysteretic hydraulic properties of the subsurface with sufficient accuracy. The storage coefficient or storage capacity is obtained from the derivative $\partial\bar{\theta}/\partial p_c$ multiplied by the scaling term $\phi - A - B$ (see Equation 1.1b).

$$\bar{\theta} = \frac{\theta_w - A}{\phi - A - B} = \left(\frac{1}{1.0 + (\alpha p_c)^n}\right)^m \tag{1.12}$$

Table 1.8 provides definitions of the four constants A, B, α, and n. As exponent $m = 1 - 1/n$ may be chosen. All of the parameters in Equation 1.12 can be derived from drainage curve data (usually from main or

Table 1.8. The Parameter Vector of Equation 1.12 for Various Wetting and Drying Curves

	MDC	PDC	MWC	SDC	SWC
A	$\theta_{w,r}$	$\theta_{w,r}$	$\theta_{w,r}$	$\theta_{w,r}$	Eq. 1.13b
B	$\theta_{nw,r}$	0	$\theta_{nw,r}$	Eq. 1.13a	$\theta_{nw,r}$
α	α_d	α_d	α_w	α_d	α_w
n	n_d	n_d	n_w	n_d	n_w

Note: MDC = main drying curve; PDC = primary drying curve;
  MWC = main wetting curve; SDC = scanning drying curve;
  SWC = scanning wetting curve.

primary drainage curve data) and from any scanning wetting curve data. These data can be obtained from laboratory soil sample tests [1.52; 1.54].

For the scanning curves, the scaling parameters A and B (see Table 1.8) take on variable values according to the following equations:

Drainage: $(\delta p_c/\delta t > 0)$

$$B = \phi - \theta_{w,r} - (\theta_{wo} - \theta_{w,r})[1.0 + (\alpha_d p_{co})^{n_d}]^{m_d} \qquad (1.13a)$$

Imbibition: $(\delta p_c/\delta t < 0)$

$$A = \frac{\phi - \theta_{nw,r} - \theta_{wo}[1.0 + (\alpha_w p_{co})^{n_d}]^{m_w}}{1 - [1.0 + (\alpha_w p_{co})^{n_w}]^{m_w}} \qquad (1.13b)$$

where $P(\theta_{wo}, p_{c\,o})$ is the starting or reference point.

When the degree of saturation lies between the main drainage curve and the main wetting curve, both fluids are coherently distributed, and both are therefore mobile as a phase. Consequently, if two flow gradients exist, both phases will flow simultaneously. Each fluid will form and flow through its own channel network. The walls of the channels are formed by fluid-solid and fluid-fluid interfaces. The geometric shape of the networks, the channel cross sections, and therefore the permeability coefficients for each fluid phase are usually considered to be nonhysteretic functions of the volumetric fluid content.

Darcy's Law is the equation governing flow in both wetting and nonwetting fluid:

$$\vec{V}_w = - K_w [(g\rho_w)^{-1} \text{ grad } p_w + \text{ grad } z] \qquad (1.14a)$$

$$\vec{V}_{nw} = - K_{nw} [(g\rho_{nw})^{-1} \text{ grad } p_{nw} + \text{ grad } z] \qquad (1.14b)$$

Permeability is a function of fluid properties, rock-matrix properties, and the volumetric fluid content. Frequently, relative permeability coefficients are used according to Equations 1.15a and 1.15b:

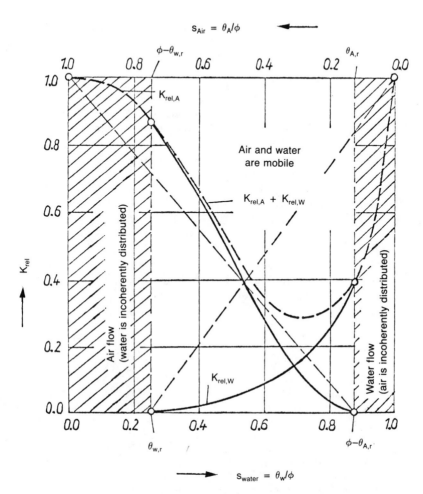

Figure 1.22. Characteristic functional relations of the relative permeabilities of the rock matrix with air as the nonwetting fluid phase and water as the wetting fluid phase.

$$K_{rel,w} = K_w/K_{w,o} \qquad \text{where } K_{w,o} = K_w(\theta_{wo}) \qquad (1.15a)$$
$$K_{rel,nw} = K_{nw}/K_{nw,o} \qquad \text{where } K_{nw,o} = K_{nw}(\theta_{wo}) \qquad (1.15b)$$

The shapes of relative permeability functions for water and air in the rock matrix are shown in Figure 1.22. Values of relative permeabilities can be experimentally measured in soil and rock samples, or easily approximated using Mualem's and van Genuchten's approach (see [1.54]):

$$K(\theta_w) = K_{wo}\,(\bar{S}_w/\bar{S}_{wo})^{1/2}\left[\frac{1-(1-\bar{S}_w^{1/m})^m}{1-(1-\bar{S}_{wo}^{1/m})^m}\right]^2 \qquad (1.15c)$$

where $\bar{S}_w = (\theta_w - \theta_{w,r})/(\phi - \theta_{w,r})$ and $K_{wo} = K(\theta_{wo})$

Permeability values measured at $\theta_{wo} < \phi$ have proven especially valuable for the calculation of capillary conductivity [1.54].

The residual saturation of the wetting phase for the case "air displaces water (A → W)" and "air displaces mineral oil (A → 0)" may be estimated by the following quantities (1 darcy $\approx 10^{-12}$ m^2 and refers to $K_w \approx 10^{-3}$ m/s for a fully water-saturated rock matrix at 10°C):

k	(k = Kν/g)	10^{+2}	10^{+1}	10^{+0}	10^{-1}	10^{-2} darcy
$\theta_{w,r}$	(A → W)	0.05	0.15	0.35	0.70	0.90
$\theta_{w,r}$	(A → 0)	0.05	0.05	0.25	0.45	0.75

A rough estimate for $\theta_{nw,r}$ is $\theta_{nw,r} \approx 0.5\,\theta_{w,r}$.

Similar relations are applicable to three immiscible fluid phases in rocks. This is shown in Figure 1.23 with the example of the typical multiphase system "gas-water-oil-solids."

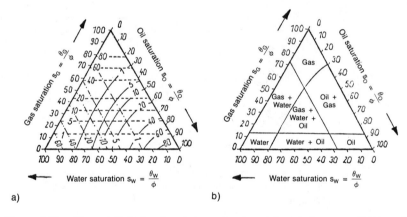

Figure 1.23. Characteristic functional dependencies of the relative permeabilities of gas, water, and mineral oil as fluid phases in a medium sand: (a) relative permeabilities; (b) mobility domains. *Source:* Busch and Luckner [1.11].

1.2.4 CONCEPTS FOR LIMITING CASES OF FLUID DISTRIBUTION AND FLOW

In some cases, neither the single fluid phase model approach nor the multiple immiscible fluid phase model approach corresponds closely to observed fluid flow behavior. A less abrupt change from immiscible fluid flow to mixphase fluid flow sometimes dominates in natural systems. A characteristic transition from multiphase immiscible to mixphase fluid flow is illustrated in Figure 1.24.

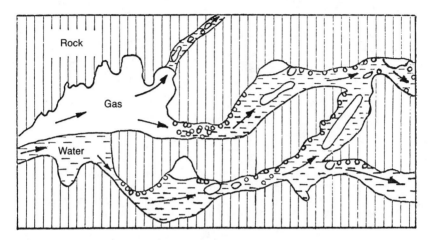

Figure 1.24. Pore-scale model of phase distribution if water and gas are simultaneously flowing in a pore channel. *Source:* Kretzschmar and Czolbe [1.48].

Multiphase immiscible fluid flow is more likely to occur (see left-hand part of Figure 1.24) when interfacial tension between fluid phases is high. Many model experiments have shown that immiscible fluid flow may be assumed to be present if the interfacial tension between fluids is

$$\sigma > (10 \text{ to } 20) \cdot 10^{-3} \text{ N/m}$$

and mixphase fluid flow may be expected (see the right-hand side of Figure 1.24) if

$$\sigma < (10 \text{ to } 20) \cdot 10^{-3} \text{ N/m}$$

There is a strong likelihood for transition forms of both immiscible and single miscible fluid flow to occur between these two cases, in the range of $(10 \text{ to } 20) \cdot 10^{-3}$ N/m.

A further characteristic problem of the transition between multiphase and mixphase flow is illustrated in Figure 1.25. In this figure, oil globules are shown which are larger than the average pore size. They are distributed as isolated islands in the groundwater zone and may migrate under the influence of capillary pressure gradients where variability in pore size and aquifer materials exists.

If one assumes the existence of a sudden macroscopic discontinuity of

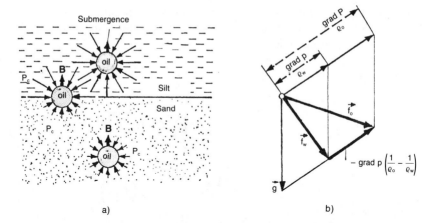

Figure 1.25. Forces acting upon suspended oil droplets in water saturated porous media: *(a)* oil particles in the groundwater zone (oil is enriched in the upper part of an aquifer); *(b)* specific forces f (f = force/mass) acting on water, oil, or gas particles if grad p_c = 0. See also Hubbert [1.36; 1.37] and De Wiest [1.89].

fluid distribution between two fluids in the subsurface, independent of their miscibility, a completely different limiting case results. This approach is of great practical importance to many migration investigations.

In groundwater hydraulics, the actual continuous phase transition from the fluid phase air to the fluid phase water, and vice versa, is commonly approximated by an abrupt discontinuity at the fictitious interface called the *groundwater surface.* Another example of the macroscopic discontinuity approach is the assumption of an abrupt phase transition between the miscible phases freshwater and saltwater.

The concept of sharp fronts between fluid phases is an important investigation tool and a good approximation for oil and water, contaminated and uncontaminated water, freshwater and saltwater, and many other two-fluid systems in the subsurface, provided the size of the phase transition zone is small in comparison with the region under consideration.

The effective mean capillary pressure $\bar{p}_c$ at the fictitious water-air interface and the mean height of the capillary fringe $\bar{h}_c$, respectively, are typical examples of an abrupt phase transition approach and are given by

$$\bar{p}_c = (1/K_{w,o}) \int K_w dp_c \quad \text{and} \quad \bar{p}_c = \bar{h}_c g \rho_c \quad (1.16)$$

where K_w is the capillary conductivity of water and $K_{w,o}$ is the conductivity when the soil is fully water saturated.

In Figure 1.26a, an equation for the approximate interface at points A and B is given by

$$H_{A1}g\rho_1 + z_{A}g\rho_1 \pm \bar{p}_c = H_{A2}g\rho_2 + z_{A}g\rho_2$$
$$\underline{- [H_{B1}g\rho_1 + z_{B}g\rho_1 \pm \bar{p}_c = H_{B2}g\rho_2 + z_{B}g\rho_2]}$$
$$\Delta H_1 g\rho_1 + \Delta z g\rho_1 + 0 = \Delta H_2 g\rho_2 + \Delta z g\rho_2$$

$$\Delta z = \frac{\Delta H_2 \rho_2 - \Delta H_1 \rho_1}{\rho_1 - \rho_2} \qquad (1.17)$$

In the special case of a nonflowing fluid, such as fluid phase 2 in Figure 1.26b, where ($\Delta H_2 = \Delta h_2 = 0$), the position of the interface in the subsurface ($\Delta H_1 = \Delta h_1$) may be obtained by the following equation:

$$\Delta z = \frac{\rho_1}{\rho_2 - \rho_1}\Delta H_1 \qquad (1.18)$$

When $\rho_2 > \rho_1$ (see Figure 1.26a), the interface is elevated toward the direction of flow, and when $\rho_2 < \rho_1$ (see Figure 1.26b), the interface is depressed toward the direction of flow. In the case of air as fluid 2 in Figure 1.26b, $\rho_1/(\rho_2 - \rho_1) = -1.0$ holds, and consequently, $\Delta z = -\Delta H_1$ (free surface with respect to the flow direction of fluid 1). A typical estimate for $\rho_1/(\rho_2 - \rho_1)$ of a saltwater–freshwater interface is $+40$, and -6 for a water–mineral oil interface.

Figure 1.27 depicts further characteristic types of contaminated liquid distributions in the soil- and groundwater zone. However, it should be remembered that the stability of a macroscopically abrupt interface is not guaranteed when the front mobility M is much greater than one:

$$M = (K_1/\rho_1)/(K_2/\rho_2) > 1 \qquad (1.19a)$$

or, using the specific permeabilities $k = K\eta/(\rho g)$ in m²:

$$M = (k_1/\eta_1)/(k_2/\eta_2) > 1 \qquad (1.19b)$$

where in the case of fluid 1 displacing fluid 2, K_1 is the K value of fluid 1 at residual saturation of fluid 2, and K_2 is the K value of fluid 2 at residual saturation of fluid 1 [1.4]. The formation of fingers, or interfingering of fluid 1 and fluid 2, results in destruction of interfaces and is a consequence of the instability of fronts (see Figure 1.28).

If the specific permeabilities k of two fluids are assumed equal, i.e., $k_1 \approx k_2$ (see Busch and Luckner [1.11]), the mobility M is given by the ratio of the dynamic viscosities of both:

$$M \approx \eta_2/\eta_1 \qquad (1.19c)$$

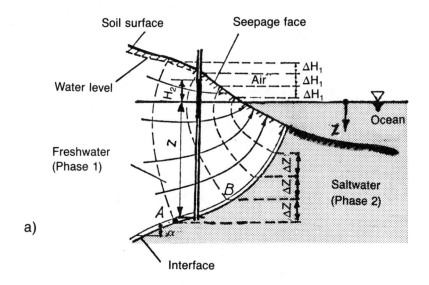

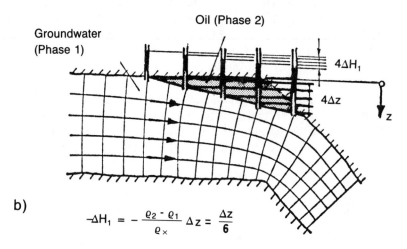

Figure 1.26. Approximations of abrupt interfaces: *(a)* freshwater-saltwater interface; *(b)* oil-groundwater interface. *Source:* Busch and Luckner [1.11].

Accordingly, an interface of water (fluid 1) displacing air (fluid 2) has the mobility of $M = \eta_a/\eta_w \approx 1/70$ and is consequently considered stable (i.e., advancing infiltration fronts in dry soils are stable). The interface of air (fluid 1) displacing water (fluid 2) with $M \approx 70$, or the interface of gas displacing mineral oil, should be considered as unstable. Gas-water displacement in the subsurface storage of gas is also considered unstable. Unstable fronts result in mixing and simultaneous flow of immiscible

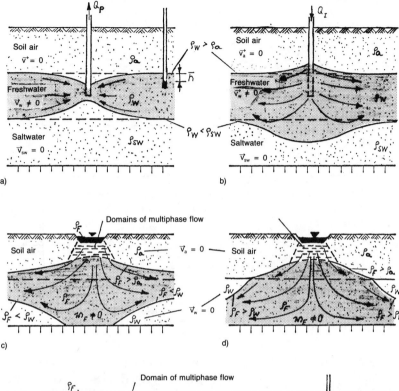

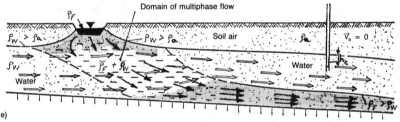

Figure 1.27.   Typical steady-state positions of abrupt interfaces: *(a)* saltwater upconing caused by freshwater flowing over resting saltwater; *(b)* spreading of freshwater over resting saltwater (similar to the formation of freshwater lenses under ocean islands); *(c)* and *(d)* infiltration of water with different density than the density of resting groundwater in the aquifer; *(e)* same as *(d)* but for flowing groundwater.

fluids. The dependency of front-stability criteria on the rate at which an interface shifts is more precisely discussed in Luckner and Tiemer [1.53] and Stugren [1.82].

The stabilities of different fronts are also apparent from the fact that the residual content of the wetting fluid $\theta_{w,r}$ is larger than residual content

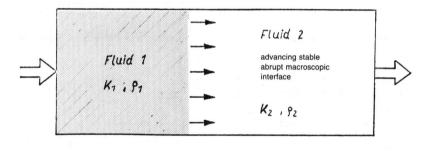

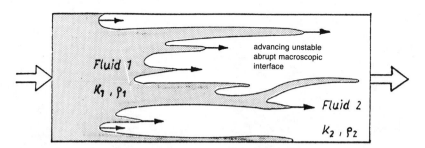

Figure 1.28. Finger formation caused by advancing interfaces in porous media.

of the nonwetting fluid $\theta_{nw,r}$. Both values, $\theta_{w,r}$ and $\theta_{nw,r}$, are changed by the number of wetting front passages (displacement cycles). The permeability of rocks with respect to both fluid phases may also decrease with the number of displacements (mutual throttling) [1.15]. Another result of displacement cycles is the decrease of capillary pressure at the interface, as shown in Equation 1.16.

1.3

Mass Transport in the Subsurface

The transport of migrants in the subsurface proceeds by

- molecular diffusion (§1.3.1)
- convection or advection (§1.3.2.)
- hydrodynamic dispersion (§1.3.2)

Transport processes due to molecular diffusion occur independently of convection. Hydrodynamic dispersion, however, is largely attributed to convection and is extremely scale dependent. This problem is discussed in detail in §1.3.3. Finally, limitations to the applicability of the hydrodynamic dispersion model approach are explained in §1.3.4.

1.3.1 MOLECULAR DIFFUSION

Molecular diffusion is primarily due to Brownian motion in solids, liquids, and gases. It is also caused by osmotic forces, thermal diffusion, and electroosmosis. If a concentration gradient of the migrant i or a gradient of another extensive state variable exists in a mixphase, the migrant moves in the direction of decreasing concentration because, due to Brownian motion, more molecularly dispersed migrants move from a volume element with concentration c_i or specific heat W/V toward the neighboring volume element with a smaller concentration or specific heat content than in the other direction. The mass flux density $\vec{f}c_M$ of the migrant i in a mixphase j is consequently proportional to dc_i/dl (where dl is the distance between neighboring volume elements). This may be

Table 1.9. Molecular Diffusion Coefficients of Selected Substances at p = 1 bar

In Air			In Water		
Diffusing substance (migrant)	D_M (m²/s)	Temperature (°C)	Diffusing substance (migrant)	D_M (m²/s)	Temperature (°C)
O_2	$1.8 \cdot 10^{-5}$	0	O_2	$1.3 \cdot 10^{-9}$	10
	$2.0 \cdot 10^{-5}$	12		$2.0 \cdot 10^{-9}$	20
H_2O	$2.8 \cdot 10^{-5}$	16	Na^+	$1.3 \cdot 10^{-9}$	20
NH_3	$2.0 \cdot 10^{-5}$	0	Cl^-	$2.0 \cdot 10^{-9}$	20
I_2	$0.8 \cdot 10^{-5}$	20	$NaCl$	$1.2 \cdot 10^{-9}$	18
H_2	$7.4 \cdot 10^{-5}$	12	NH_4^+	$1.8 \cdot 10^{-9}$	15
$HCOOH$	$1.3 \cdot 10^{-5}$	0	$HCOOH$	$1.1 \cdot 10^{-9}$	12
HC_3COOH	$1.1 \cdot 10^{-5}$	0	CH_3COOH	$0.9 \cdot 10^{-9}$	12
CH_3COOH	$1.3 \cdot 10^{-5}$	0	CH_3OH	$1.3 \cdot 10^{-9}$	20

expressed for isothermal and isobaric conditions by Fick's first law as follows:

$$\vec{fc}_{M,i,j} = -D_{M,i,j} \text{ grad } c_{i,j} \qquad (1.20)$$

where $\vec{fc}_{M,i,j}$ = mass flux density of migrant i in the mixphase j due to molecular diffusion, e.g., in kg/(m² s)

$D_{M,i,j}$ = molecular diffusion coefficient of migrant i in the mixphase j in m²/s

The order of magnitude of $D_{M,i,j}$ is approximately 10^{-5} m²/s in gases, 10^{-9} m²/s in liquids, and $< 10^{-14}$ m²/s in solids. Table 1.9 shows typical values for $D_{M,i}$ in the mixphases air and water. $D_{M,i,j}$ increases for increasing temperatures and decreases as the molecular mass of a migrant i decreases.

Osmotic transport expressed as mass flux density is given in simplified form by the following equation:

$$\vec{fc}_{osm} = -D_{osm} \text{ grad } c \qquad (1.20a)$$

where $D_{osm} = D_1 - (1 - \partial\rho/\partial c)D_2$

In the case of NaCl in water, for example, $D_1 = 1.61 \cdot 10^{-9}$ m²/s, $D_2 = 1.93 \cdot 10^{-9}$ m²/s and $\partial\rho/\partial c = 0.7$. The quantity D_{osm} is then approximately $1.0 \cdot 10^{-9}$ m²/s [1.69].

The transport due to thermal diffusion expressed as mass solute flux follows [1.30]:

$$\vec{fc}_{TD} = -D_{TD} \text{ grad } T \qquad (1.20b)$$

where $D_{TD} \approx 2 \cdot 10^{-9}$ m²/(s K)

Similarly, electroosmotic transport expressed as mass flux density is described by

$$\vec{fc}_{EO} = -D_{EO} \text{ grad } U \qquad (1.20c)$$

where U is the electrical potential in volts

The electroosmotic diffusion coefficient D_{EO} ranges from about (1 to 8) · 10^{-9} m²/(V s) [1.57]. This transport phenomenon is of practical significance in the electroosmotic drainage of low-permeability unconsolidated rocks. Pressure diffusion may also play a role in solute transport at greater depths [1.76].

In multiphase subsurface systems, the diffusion coefficients for a particular mixphase diminish according to the volumetric content of the mixphase and to the tortuosity of its transport paths. In addition, an electromolecular retardation factor must be taken into account caused by charged solid surfaces. Thus the molecular diffusion coefficient for the groundwater zone [1.70] is given by

$$D_o = \chi\eta\phi D_M \qquad (1.21a)$$

where χ = tortuosity (ranging from 0.5 to 0.7 for unconsolidated rock and from 0.25 to 0.50 for solid rock)
η = electromolecular retardation factor (ranging from 0.9 to 1.0 for gravels and sands, from 0.4 to 0.5 for silts, and about 0.2 for clay)
ϕ = porosity

Using this expression, the mass flux density of the migrant i due to molecular diffusion in the multiphase system subsurface is given by

$$\vec{fc}_{M,i} = -D_{o,i} \text{ grad } c_i \qquad (1.21b)$$

The value of the molecular diffusion coefficient for a migrant diffusible in both soil air and soilwater, such as molecular oxygen, may be estimated by an integral value $D_{o,i}$, which is not equal to the sum of the individual molecular diffusion coefficients (see also §1.6.1). Soil laboratory tests are usually necessary to determine such integral values. Resulting test data mostly fit the graph shown in Figure 1.29.

Molecular diffusion of migrants distributed in the fluid mixphase in suspended or emulsified form should not be confused with migrants which are molecular-disperse distributed (see Table 1.12 in a later chapter). Molecular transport of suspended or emulsified matter is usually neglected in molecular diffusion modeling because of the shorter free pathways of Brownian motion caused by the greater masses of these particles.

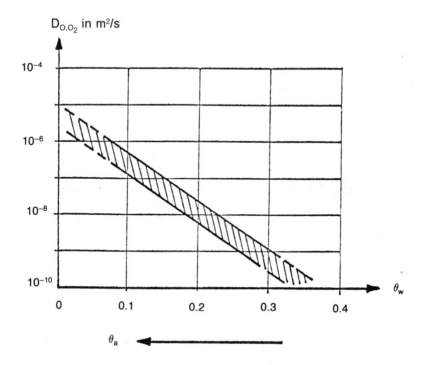

Figure 1.29. The molecular diffusion coefficient of oxygen in sandy soils as a schematic function of water and air content.

1.3.2 CONVECTION AND HYDRODYNAMIC DISPERSION

Transport of migrants in the fluid phase of the multiphase system subsurface is primarily caused by two phenomena:

1. *convection* (used synonymously with advection): the transport of migrants due to bulk flow of a fluid mixphase in the subsurface, i.e., the mean statistical motion of all its components (sometimes called the bulk flow approach)
2. *hydrodynamic dispersion:* the motion of subject migrants relative to the convective motion (the bulk flow) of the mixphase

In the multiphase system subsurface, the convective transport of a migrant i in all fluid mixphases j is given by (see also Figure 1.12)

$$\vec{fc}_{c,i} = \sum_j \vec{v}_j \, c_{i,j} = \sum_j \sigma_j \, \vec{v}_{a,j} \, c_{i,j}$$

$$(1.22)$$

where $\vec{v}_j$ = volume flux density of the mixphase j (often called *flow rate, specific discharge, Darcy velocity,* or *Darcy flux*)

$\vec{v}_{a,j}$ = average pore fluid velocity of the mixphase j (also referred to as *average phase velocity* $v_{a,j} = v_j/\sigma_j$)

σ_j = volumetric content of the flowing fluid phase j

The volume flux density of the mixphase j is given by Darcy's Law (see also Equation 1.14):

$$\vec{v}_j = -K_j \left(\frac{1}{\rho g} \text{ grad } p + \text{grad } z\right) \tag{1.23}$$

For $\vec{v}_j = \vec{v}$ in Cartesian coordinates, Equation 1.23 can be written as

$$v_x = -\frac{K_x}{\bar{\rho}} \frac{\partial h_o}{\partial x} \; ; \; v_y = -\frac{K_y}{\bar{\rho}} \frac{\partial h_o}{\partial y} \; ; \; v_z = -K_z \left(\frac{1}{\bar{\rho}} \frac{\partial h_o}{\partial z} - \frac{1}{\bar{\rho}} + 1\right)$$

$$h_o = \frac{p}{\rho_o g} + z \; ; \; p = (h_o - z)\rho_o g \text{ and } \bar{\rho} = \rho/\rho_o$$

where ρ_o represents a reference quantity.

In this relationship, the dependency of permeability on density, $K = K(\rho)$, must be accounted for. Laboratory experiments are commonly used to estimate the required data $K(\rho)$. Actual flow rates $\vec{v}_j$ are not often measured, but are frequently determined by geohydraulic calculations or simulations.

The average migration or transport velocity $\vec{u}_i$ of the migrant i in the multiphase system subsurface can be derived from the flow rates $\vec{v}_j$ by means of a mass balance for the REV. If the exchange of migrants between mixphases is in thermodynamic equilibrium, then

$$\vec{u}_i = \overline{dl}/dt = \sum_j \vec{v}_j / \sum_j ca_{i,j} \tag{1.24}$$

where $\overline{dl}$ = the average distance traveled by the migrant i and $\sum_j ca_{i,j}$ = the effective volume-related storage capacity of the multiphase system subsurface with respect to migrant i. Hence, transport of a tracer in flowing groundwater where water is the only fluid phase may be described by

$$\vec{u}_i = \vec{v}_w/\phi = \vec{v}/\phi$$

with $\theta_w = \phi$.

Velocities of individual migrants vary about the average migration velocity $\vec{u}$. The transport paths traveled by these migrants in time interval dt scatter longitudinally and transversely about the average path $d\vec{l}$. Therefore, some migrants can always be expected to travel ahead or behind the average migration velocity $\vec{u}_i$. This relative motion is usually described by means of Fick's first law, which like Brownian motion is considered to be purely random, and therefore normally distributed:

$$\vec{fc}_{D,i,j} = -\bar{\bar{D}}_{D,i,j} \text{ grad } c_{i,j} \qquad (1.25a)$$

where $\bar{\bar{D}}_{D,i,j}$ is the hydrodynamic dispersion coefficient of migrant i in the mixphase j. The bulk flow of all constituents may be described by the sum:

$$\vec{fc}_{D,i} = -\sum_j \sigma_j \vec{fc}_{D,i,j} \qquad (1.25b)$$

where $\vec{f}c_{D,i}$ is the mass flux density of migrant i in the multiphase system subsurface due to hydrodynamic dispersion and σ_j is the volumetric content of the mixphase j (for instance $\sigma_j = \theta_w$, where θ_w is the water-filled porosity).

The hydrodynamic dispersion coefficient $\bar{\bar{D}}_D$ is usually represented by a second-order tensor. Its two orthogonal principal axes are the longitudinal axis in the direction of $\vec{v}$ and $\vec{u}$, and the transverse axis perpendicular to $\vec{v}$. Accordingly, the terms *longitudinal* and *transverse dispersion coefficients* $D_{D,l}$ and $D_{D,tr}$, respectively, are used. As with transient flow problems, the direction and magnitude of $\vec{v}$ and $\vec{u}$ are constantly changing; therefore the orientation of the principal axes of the dispersion tensor and values $D_{D,l}$ and $D_{D,tr}$ are also changing.

It has been shown by numerous experiments (see, for instance, Bear [1.4], Freeze and Cherry [1.27], Luckner and Tiemer [1.53], and Klotz and Moser [1.44]) that the dispersion coefficient may be approximated with sufficient accuracy by

$$\sigma D_{D,l} \approx \delta_l v \text{ and } \sigma D_{D,tr} \approx \delta_{tr} v \approx (0.1 \text{ to } 0.2) \, \delta_l v \qquad (1.26)$$

where $v = |\vec{v}|$ – flow rate.

The dispersion coefficient depends quadratically on $\vec{u}$ or $\vec{v}$ only when the influence of convection is small compared with the influence of molecular diffusion expressed by small values of $\delta v / D_M$ [1.21]. Commonly the parameter δ is called *dispersivity* and the ratio $\delta v / D_M$ the *Peclet number*.

In the x,y-coordinate system, the mass flux density due to dispersion $\vec{f}c_D$ becomes

$$\vec{fc}_D = -\sigma\overline{\overline{D}}_D \text{grad } c = -\begin{pmatrix} fc_{D,x} \\ fc_{D,y} \end{pmatrix} = -\sigma\begin{pmatrix} D_{xx} & D_{xy} \\ D_{yx} & D_{yy} \end{pmatrix}\begin{pmatrix} \partial c/\partial x \\ \partial c/\partial y \end{pmatrix} \quad (1.27)$$

where $D_{xx} = (D_{D,l} v_x^2 + D_{D,tr}v_y^2)/(v_x^2 + v_y^2)$

$D_{xy} = D_{yx} = [(D_{D,l} - D_{D,tr})v_y v_x]/(v_x^2 + v_y^2)$

$D_{yy} = (D_{D,tr}v_x^2 + D_{D,l}v_y^2)/(v_x^2 + v_y^2)$

If the direction of $\vec{v}$ and $\vec{u}$ coincides with the x-axis, then $v_y = 0$, and consequently, $D_{xy} = D_{yx} = 0$, $D_{xx} = D_{D,l}$, and $D_{yy} = D_{D,tr}$.

1.3.3 SCALE PROBLEMS OF HYDRODYNAMIC DISPERSION

At the local scale of investigation (see §1.2.1), statistical kinematic deviations of single migrants from expected values $\bar{u}$ and $\bar{x} = \bar{u}t$ may be caused by (see Figure 1.30)

- variation of flow velocities across pore channels
- variation of cross-sectional areas of different flow channels resulting in different flow velocities
- variation of flow velocities due to different lengths of flow paths in the labyrinth of the channel network
- transverse propagation of migrants due to sharing and joining of flow channels ($\alpha^* \approx 3°$)

According to this interpretation, hydrodynamic dispersion is generally caused by the heterogeneity of permeability and storage capacity in the subsurface [1.52].

At the regional scale of investigation, the shaded domains in Figure 1.30 may also represent subsurface domains of diminished permeability such as silt lenses, which are separated by interconnected domains of preferential flow such as coarse sand layers or gravel stringers. Hydrodynamic dispersion and dispersivity δ are thus scale dependent in real aquifers. Therefore, δ-values obtained in laboratory soil columns are not immediately transferable to larger-scale field conditions. Similarly, it is

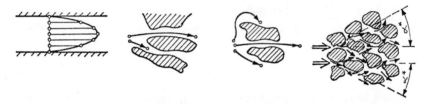

Figure 1.30. Causes of hydrodynamic dispersion at the local scale of observation.

impossible to transfer δ-values determined from migration pumping tests to regional migration problems.

The dispersivity δ is a characteristic length of the subsurface, comparable to the characteristic length used in the Reynolds number (δ is used in the Peclet number $Pe^* = \delta v/D_M$ or in the cell Peclet number of spatially discrete models $Pe \approx \Delta x/\delta$). This characteristic length δ represents a geometric measure of the permeability and storage heterogeneities in the subsurface and has a significant impact on the migration process under consideration.

Let us consider the following hydrodynamic dispersion model, derived from laboratory tests:

$$\sigma D_{D,L} = 1.8\nu \, (vd_{50}/\nu)^{1.2} \quad \text{and} \quad \sigma D_{D,tr} = 0.11\nu \, (vd_{50}/\nu)^{0.7} \quad (1.28a)$$

where d_{50} is the particle diameter in m taken from the particle size distribution curve at 50%, and ν the kinematic viscosity in m²/s. If these equations are roughly approximated by

$$\sigma D_{D,l} \approx 2d_{50}v \text{ and } \sigma D_{D,tr} \approx 0.1d_{50}v \quad (1.28b)$$

the following characteristic lengths result:

$$\delta_l \approx 2d_{50} \quad \text{and} \quad \delta_{tr} \approx 0.1d_{50} \quad (1.28c)$$

This interpretation may be generalized and transferred to subsurface heterogeneities formed by different permeability domains. In migration-pumping tests, migrants travel a distance of several meters, and δ-values can be expected in the decimeter range. In regional migration processes, migrants travel several kilometers, and δ-values in the decameter range are likely. Thus, the problem of δ determination may be reduced to a geometric analysis of the subsurface heterogeneities through which migrants actually travel.

As Figure 1.31 shows, the sampling method used to determine the parameter δ may have a significant influence on the value obtained. For example, a point sample taken from one layer, or a vertically averaged (mixed) sample taken from the whole aquifer, will yield rather different values for dispersivity. In migration-pumping tests (see §3.3.3.2), mixed samples result in δ-values that are usually one order of magnitude larger than those obtained from point samples. If K_i and m_i are normally distributed and δ_{tr} equals zero, then, for mixed samples, δ may be estimated by

$$\delta_l = 0.5 \, (\sigma/\overline{Km})^2 \quad (1.29a)$$

where $\overline{KM}$ is the average of the larger transmissivities, and σ their standard deviation.

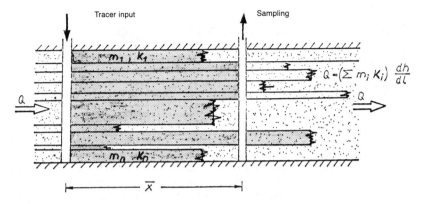

Figure 1.31. Impacts of sampling on the estimation of hydrodynamic dispersion.

A more complex parameter model of total aquifer dispersivity (mixed sample) has been proposed by Gelhar [1.28]:

$$\delta_l = l^2\sigma^2/(3\delta_{tr}) \tag{1.29b}$$

where σ = relative standard deviation of the logarithmically distributed local permeability coefficient K ($\sigma = \sigma_{lnK}/\overline{lnK}$)

δ_{tr} = local transverse dispersivity

l = parameter of the covariance function (correlation distance)

Gelhar also estimated the impact of $\overline{x}$ on δ_l by

$$\delta_l(\overline{x}) = \delta_l(1 + l^\nu((2\nu + 1)^2 - 2)\text{erfc } \nu^{1/2} - 2(\nu/\pi)^{1/2}(2\nu + 1)) \tag{1.29c}$$

where $\nu = \delta_{tr}\overline{x}/l^2$.

Different K-values on different stream lines of equal length also contribute to hydrodynamic dispersion in an aquifer with vertically averaged properties (see Figure 1.31). Stream lines that are locally deformed with respect to idealized one-dimensional parallel flow also cause hydrodynamic dispersion even when the K-values are constant in the considered region. Some typical examples are shown in Figure 1.32.

Our current knowledge of the causes of hydrodynamic dispersion scale effects, as introduced above, enables us to empirically generalize previous test results using the following parameter model [1.50]:

$$\delta_l = f(\overline{x}) = \alpha\overline{x}^\beta \tag{1.30}$$

where $\alpha = 0.03\ \overline{x}^{+0.3}$

$\beta = \overline{x}^{-0.075}$

Figure 1.33 illustrates this model. Detailed tests described in Pickens and Grisatz [1.60] result in $\delta_l = 0.1\ \overline{x}$ for $10^0 < \overline{x} < 10^3$, and tests

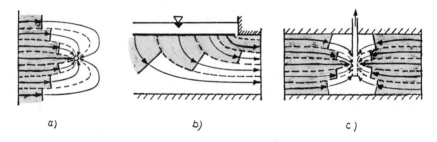

a)   b)   c)

Figure 1.32. Causes of hydrodynamic dispersion of larger scales due to locally deformed stream lines: *(a)* stream line deformation around a well or drain; *(b)* stream line deformation during infiltration from partially penetrating surface water bodies; *(c)* stream line deformation of flow toward a partially penetrating well.

carried out in relatively coarse and heterogeneous aquifers in the vicinity of the Alps yield a dispersivity value of $\delta_l = 0.3\,\bar{x}^{\,0.72}$ for $10^1 < \bar{x} < 10^3$ [1.44]. These test results support the model described by Equation 1.30.

Further investigations of the dispersivity function described by Equation 1.30 should (1) illuminate the relationship between α and the standard deviation σ_K^* of $K^* = K^n$ with $0 < n \le 1$ in an extended migration region characterized by $\bar{x}$ and (2) explain the decrease of β with increas-

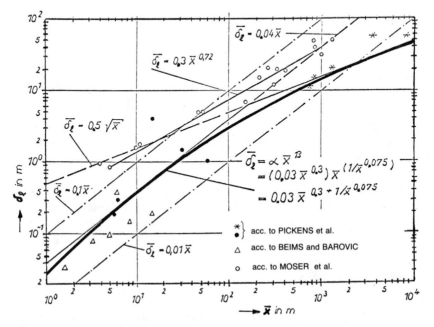

Figure 1.33. Functional dependence $\delta = \delta(x)$.

ing x̄ based on the approach of t to the RET (representative elementary time) in some migration regions.

For all practical purposes, it may be assumed that Equation 1.30 yields the expectation values of longitudinal dispersivity. Variations from the expectation values are primarily caused by deviations from "mean values" of the permeability K, the Darcy flux v, the thickness of the aquifer m, and the standard deviation σ_K. When one or more of these parameters are larger than reasonable mean values, longitudinal dispersivity δ_l will take on a value larger than that predicted by Equation 1.30.

1.3.4 APPLICABILITY OF THE HYDRODYNAMIC DISPERSION MODEL APPROACH

For a larger scale REV (representative elementary volume), the practical applicability of the dispersion model approach is restricted by the assumptions made in §1.3.2 and §1.3.3:

1. Heterogeneities, velocities, and distances traveled by migrants are normally distributed.
2. An averaged state is given in each of the mixphases of the REV.
3. The superposition principle applies to hydrodynamic transport and other migration processes.

The assumed normal distribution of heterogeneities and kinematic characteristics (assumption 1) gives rise to uncertainties which are shown in Figure 1.31 and 1.32. These figures demonstrate that the assumption of normally distributed migrant-transport characteristics is clearly a very rough approximation. In more precise investigations, modeling of each layer becomes necessary, because it is often inappropriate to extend a single uniform REV over the entire thickness of an aquifer. Similar conclusions can be made for the stream tube approach in Figure 1.32.

The derivation of hydrodynamic dispersion from permeability heterogeneities is treated in many current scientific journals. For instance, discrete stationary flow models with stochastic autocorrelated permeability parameter models are discussed in Smith and Schwartz [1.77; 1.78; 1.79]. However, these models were unable to confirm normal distribution of the kinematic characteristics of migration and the stability of the dispersivity value δ. This discrepancy was shown to be mainly due to inadequate spatial averaging over the stationary flow field, and consequently the assumed conditions were not sufficiently fulfilled throughout the REV. Sensitivity analyses on such models have also shown that the model-aided forecast of migrant arrival times at a particular location is very uncertain. Similar results were obtained using models that have only two discrete K-values (K = 0 and K = K$_o$) which are randomly distributed over the finite elements of the investigated migration region [1.74].

In several investigations of migration, the following parameters turned out to be especially significant:

- the standard deviation of the normal distribution of ln K [1.7; 1.52]
- the anisotropy of permeability K$_h$/K$_v$
- the distance actually traveled by the migrants

Other investigations in this field adhere to the "classical" model approach of hydrodynamic dispersion, and appear to show that this approach is practicable and appropriate even at the regional scale of consideration [1.29]. Further research is needed.

Assumption 2 may cause uncertainties if larger domains of high and low permeability and storage capacity are considered as the quasi-homogeneous medium subsurface in migration models (see Figure 1.34). Such a model approach requires not only a very large REV, but also a very large RET to establish an average systems state such as average concentrations in different fluid mixphases of the REV. This problem is discussed in more detail in Schestakow [1.72].

If we consider the stratified aquifer shown in Figure 1.34a as a quasi-homogeneous aquifer, we would obtain, according to Schestakow [1.70]:

$$D_{D,l} = \delta_t v^2 = \frac{\phi_2^2 \, m_2^3 \, v^2}{3\phi^2 \, (m_1 + m_2)D_o} \text{ for } t > \text{RET} \approx \frac{\phi_2 m_2^2}{5D_o} \quad (1.31)$$

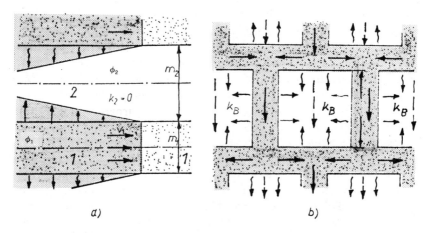

Figure 1.34. Macroscopic heterogeneities aquifers with deterministic structure: *(a)* regular layer structure; *(b)* regular block structure. *Source:* Schestakow [1.71].

where D_o = molecular diffusion coefficient of the less-permeable layer 2
m_1, m_2 = thicknesses of layers 1 and 2
$v = v_1 m_1/(m_1 + m_2)$ averaged Darcy flux, $v_2 = 0$
$\phi = (\phi_1 m_1 + \phi_2 m_2)/(m_1 + m_2)$ averaged porosity
δ_t = characteristic time duration which is time dependent for t < RET [1.71]

The practical consequences of Equation 1.31 should be illustrated by the following example:

$$m_2 = 0.1 \text{ m}; \quad \phi_2 = 0.5; \quad D_o = 10^{-10} \text{ m}^2/\text{s}$$

results in RET $\approx$ 100 days. With $v_1 = 10^{-5}$ m/s, $v_2 = 0$, $\phi_1 = 0.3$, and t = 100 days, the tracer migrants must travel a distance of $\bar{x} = ut = v_1 t/\phi_1$ $\approx$ 300 m before the conditions necessary for the quasi-homogeneous medium subsurface assumption are fulfilled. If, on the other hand, the permeability of layer 2 is two orders of magnitude less than the permeability of layer 1, then $v_2 = v_1/100$, and instead of D_o, $D_{D,tr} = \delta_{tr} v_1/100$ is used in Equation 1.31. With $\delta_{tr} = 5 \cdot 10^{-2}$ m, we obtain an RET $\approx$ 2 days. The distance which the migrants now travel before the quasi-homogeneous condition is valid would only amount to $\bar{x} \approx$ 6 m.

For the case shown in Figure 1.34b, the calculated value of the hydro-dynamic dispersion coefficient for t > RET is given according to Schestakow [1.71] by

$$D_{D,l} = \delta_l v = (1 - \chi) v^2/\gamma \qquad (1.32)$$

with

$$\delta_l = l(1 - \chi)/(K_B/K_\Sigma)$$

where γ = transport coefficient in the blocks; $\gamma = f(l, D_o, K_B)$ in s^{-1}
l = characteristic block length
χ = relative saturation of pore channels
K_B = K-value of less-permeable blocks
K_Σ = K-value of entire fissured porous medium

The uncertainties resulting from assumption 3 arise from superposition of transport due to hydrodynamic dispersion and transport due to convection, including all other subprocesses:

$$\begin{bmatrix} \text{Hydrodyn.} \\ \text{Dispersion} \end{bmatrix} = 0$$

+ [Convection]   = [Storage] + [Exchange] + [Internal Reaction] + . . .

_____

$$[\text{Convection}] + \begin{bmatrix} \text{Hydrodyn.} \\ \text{Dispersion} \end{bmatrix} = [\text{Storage}] + [\text{Exchange}] + [\text{Internal Reaction}] + . . .$$

All impacts of storage, exchange, and internal reaction processes in the REV upon hydrodynamic dispersion are ignored in this model approach. Under these conditions the integral transport process in the multiphase system subsurface may be described using Equations 1.21b, 1.22, and 1.25b as

$$\vec{fc}_i = -\bar{D}_{o,i} \overline{\text{grad } c_i} + \Sigma_j (\vec{v}_j c_{i,j} - \sigma_j \bar{\bar{D}}_{D,i,j} \text{ grad } c_{i,j})$$ (1.33)

In most considerations, molecular diffusion is simply superimposed on hydrodynamic dispersion of the mobile phase. The matrix elements D_{xx} and D_{yy} must then be incremented accordingly (see Equation 1.27). The symbol $\bar{\bar{D}}$ is used for the resulting tensor. Thus, we obtain the integral transport equation of migration processes in its most common form:

$$\vec{fc}_i = \Sigma_j (\vec{v}_j c_{i,j} - \sigma_j \bar{\bar{D}}_{D,i,j} \text{ grad } c_{i,j})$$ (1.34)

Instead of Fick's first law, diffusion/dispersion can also be represented by a statistical model which may be superimposed on the convective migrant transport model as shown previously. For the example shown in Figure 1.35, the following equation is obtained directly from the two-dimensional density function of the normal distribution (see also Equation 2.45b) [1.22]:

$$\phi(x,y,t) = \phi(x,t) \cdot \phi(y,t)$$

$$c(x,y,t) = \frac{c_o V_o}{2\pi \sqrt{\sigma_x \sigma_y}} \exp \left[-\frac{(x - ut)^2}{2\sigma_x^2} - \frac{(-y_o)^2}{2\sigma_y^2} \right]$$ (1.35)

where $c_{max} = c_o V_o / (2\pi \sqrt{\sigma_x \sigma_y})$
$V_o = \dot{V} \Delta t$ tracer volume at the moment $t_o = 0$
σ_x, σ_y = standard deviation in x- and y-direction

This solution corresponds to results obtained using Fick's first law (i.e., Equation 1.34) and the following functional connection between σ and δ:

$$\sigma = \sqrt{2t\delta u} = \sqrt{2\bar{x}\delta}$$ (1.36)

where $\bar{x} = ut$. Substituting the terms of Equation 1.36 into Equation 1.35 gives [1.53]

$$c(x,y,t) = \frac{c_o V_o}{4\pi\bar{x} \sqrt{\delta_l \delta_{tr}}} \exp \left[-\frac{(x - ut)^2}{4\delta_l ut} - \frac{(y - y_o)^2}{4\delta_{tr} ut} \right]$$

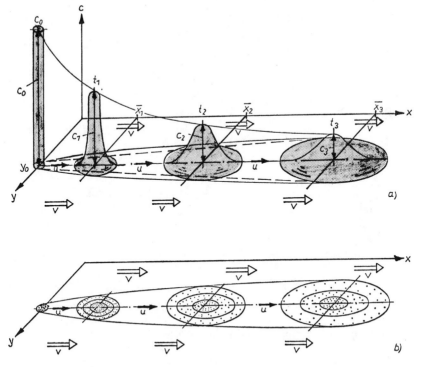

Figure 1.35. Tracer propagation in a uniform aquifer with v_x = const., v_y = 0, and pulse-type injection (e.g., a time-limited tracer injection in a monitoring well).

A generalization of this approach is the random walk method. For the case shown in Figure 1.35, let us imagine a thousand "walking elements" assembled at point (x_o, y_o), each of which contains the "solute load" $c_o V_o / 1000$. If all the elements begin "walking" at the same time and are allowed to travel over the time interval $\Delta t = t_1$, each of these elements will cover the distance $x_1 = \bar{x}_1 = ut_1$, due to convection. This position (x_1, y_o) is the expected value, around which the actual positions of the thousand traveling elements at the time t_1 will scatter. In order to account for the statistical component (Gaussian distribution) of the motions, the position of the walking elements is obtained by (see also Rao et al. [1.63] and Simmons [1.75])

$$x(t_1) = x_o + \bar{x}_1 + \epsilon \sqrt{2\bar{x}_1 \delta_1}; \qquad y(t_1) = y_o + 0 + \epsilon \sqrt{2\bar{x}_1 \delta_{tr}} \qquad (1.37)$$

A generalization of this approach is shown in Figure 1.36.

FORMER POSITION

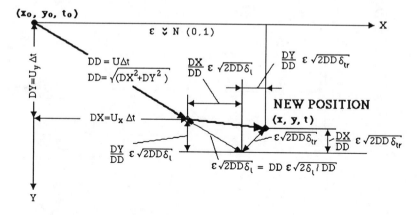

New position $(t_0 + \Delta t)$	Former position	Convection	Longitudinal dispersion	Transverse dispersion
x =	x_o	+ DX	+ DX $\varepsilon\sqrt{2\delta_l / DD}$	+ DY $\varepsilon\sqrt{2\delta_{tr}/DD}$
y =	y_o	+ DY	+ DY $\varepsilon\sqrt{2\delta_l / DD}$	+ DX $\varepsilon\sqrt{2\delta_{tr}/DD}$

Figure 1.36.   General random walk. *Source:* Prickett [1.62].

1.4

Phase-Internal Conversion Processes

In the mixphases air, water, and rock solids of the multiphase system subsurface (see Figure 1.12), the migrants are subjected to numerous conversion processes. The mathematical formulation of these processes is based upon the principles of thermodynamics (see §1.4.1).

1.4.1 FUNDAMENTALS OF CHEMICAL THERMODYNAMICS

1.4.1.1 Chemical Potential

The driving force of chemical reactions is given by Σ $(\partial G/\partial n_i|_{T,p,n_{k \neq i}})dn_i$ according to the fundamental Equation 1.1b. The term in parentheses is called chemical potential μ_i and is not an intensive state variable but rather a capacity (a chemical capacity — see Equation 1.1b). Chemical potentials μ_i are equivalent to slope coefficients of the systems state function G. Therefore, the magnitude of μ_i is equal to the partial molar Gibbs free energy of the component i in a mixphase at constant temperature T, constant pressure p, and unchanged composition. All n_k are constant with the exception of the considered component n_i:

$$\mu_i = (\partial G/\partial n_i)_{T,p,n_{k \neq i}} \text{ in J/mol} \tag{1.38a}$$

It is useful to split the chemical potential μ_i into two terms:

$$\mu_i = \mu_i^o + RT \ln a_i = f(p,T,[i]) \tag{1.38b}$$

where the pressure- and temperature-dependent chemical standard potential μ° is equal to the molar standard Gibbs free energy of formation Δg_f°, and RT ln a_i reflects the concentration-dependent term.

The other terms introduced in this equation are R = gas constant = 8.314 J/(Kmol); T = absolute temperature; $(RT)_0$ = 2.479 kJ/mol at T = 298.15 K (25°C); and a_i = activity of the migrant i, a dimensionless quantity.

The activity a_i = [i] of a component or migrant i in the mixphase water or air is given by definition as

$$a_{i,w} = \{i\}_w = \gamma_i[i]_w \tag{1.39a}$$

and

$$a_{i,a} = \{i\}_a = x_i = f_i \tag{1.39b}$$

for most natural systems, where γ_i is the activity coefficient of the migrant i (as $\{i\}$ is dimensionless, γ_i has the inverse dimension of [i]), and f_i is the fugacity of the migrant i, a dimensionless quantity.

By definition, the activity of all pure condensed phases at their standard states is equal to one (e.g., $\{H_2O\}$ = 1, $\{minerals\}$ = 1). The value of the standard potential μ_i° (standard partial molar free energy) is given by the chemical potential at the activity $\{i\}$ = 1, as represented in Equation 1.38b. Table 1.10 presents μ_i°-values of several selected substances at p = 1 bar and T = 298.15 K (25°C). Pure elementary substances in their most stable form have a standard potential equal to zero. For the proton H^+, the electron e^-, and gases at p = 1 bar, the standard potential μ_i° = 0. The standard potential μ_i° is relatively independent of pressure but is sensitive to changes in temperature.

For example, for an ideal system where γ_i = 1 L/mol or 1 kg/mol, f_i = p_i/p = x_i, the carbon dioxide concentration in water $[CO_2]_{aq}$ = 10^{-5} mol/L ≈ 10^{-5} mol/kg, and the mole fraction of CO_2 in adjacent air x_{CO_2} = 10^{-3} we would obtain

$$\mu_{CO_2} (aq) = -386.5 + 5.71 \log (10^{-5}) = -415.0 \text{ kJ/mol}$$

$$\mu_{CO_2} (g) = -394.6 + 5.71 \log (10^{-3}) = -411.0 \text{ kJ/mol}$$

Therefore, CO_2 dissolves in water until the two μ-values become equal. This would be attained at $[CO_2]_{aq}$ = 5.12 · 10^{-5} mol/L if x_{CO_2} = 10^{-3} remains constant.

The activity coefficient γ_i is dependent upon the ionic strength I of the solution (see Figure 1.37), which is calculated from

$$I = 0.5 \Sigma [i] z_i^2 \tag{1.40}$$

Table 1.10. Chemical Standard Potential $\mu° = \Delta g_f^o$ and Standard Molar Enthalpy of Formation Δh_f^o (both at p = 1 bar, T = 298.15 K) of Selected Materials

Formula	State	$\mu° = \Delta g_f^o$ [kJ/mol]	Δh_f^o [kJ/mol]
	Calcium		
Ca	c	0	0
Ca^{2+}	aq	−555.4	−543.3
CaF_2	c	−1162.7	−1215.4
$CaSO_4$	c	−1321.2	−1433.6
$CaSO_4$	aq	−1295.9	
$CaCO_3$	c	−1129.5	−1207.9
$Ca(HCO_3)_2$	aq	−1728.3	
	Chlorine		
Cl_2	g	0	0
Cl^-	aq	−131.3	−167.6
	Iron		
Fe	c	0	0
Fe^{2+}	aq	−85.0	−87.9
Fe^{3+}	aq	−10.6	−47.7
Fe_3O_4	c	−1014.9	−1117.9
$Fe(OH)_2$	c	−483.9	−568.6
FeS_2, pyrite	c	−150.7	−178.0
$FeCO_3$	c	−674.3	−748.2
$Fe(OH)_3$	c	−695.0	
	Fluorine		
F^-	g	−276.7	−329.3
	Potassium		
K	c	0	0
K^+	aq	−282.4	−251.4
KOH	aq	−439.9	
$KAlSi_3O_8$	c	−3583.9	
	Carbon		
CO_2	g	−394.6	−393.4
CO_2	aq	−386.5	
HCO_3^-	aq	−587.4	−691.6
CO_3^2	aq	−528.5	
$H_2CO_3 = CO_2 \cdot H_2O$	aq	−623.8	
	Magnesium		
Mg	c	0	0
Mg^{2+}	aq	−456.3	−462.3
$MgSO_4$	aq	−1198.8	
	Manganese		
Mn	c	0	0
Mn^{2+}	aq	−227.8	−219.0
	Sodium		
Na	c	0	0
Na^+	aq	−262.0	−239.8
NaCl	aq	−393.3	

Table 1.10 continued.

Formula	State	$\mu° = \Delta g_f^°$ [kJ/mol]	$\Delta h_f^°$ [kJ/mol]
	Oxygen		
O_2	g	0	0
OH^-	aq	−157.4	−230.1
H_2O	g	−228.7	−242.0
H_2O	l	−237.3	−286.0
H_2O_2	aq	−131.8	
	Sulfur		
S	c	0	0
SO_2	g	−300.4	−297.1
SO_3	g	−370.5	−395.4
SO_3^{2-}	aq	−486.1	−624.7
SO_4^{2-}	aq	−742.5	−908.1
H_2S	g	−33.1	−20.2
H_2S	aq	−27.4	
HSO_4^-	aq	−753.4	−886.3
H_2SO_4	aq	−742.5	
	Nitrogen		
N_2	g	0	0
NO	g	86.7	90.4
NO_2^-	aq	−34.5	
NO_3^-	aq	−110.7	−206.7
NH_3	g	−16.6	−46.2
NH_3	aq	−26.6	
NH_4^+	aq	−79.5	−132.9
	Hydrogen		
H_2	g	0	0
H^+	aq	0	0

Source: Matthess [1.55].
Note: c = crystalline, aq = aqueous solution, l = liquid, g = gaseous.

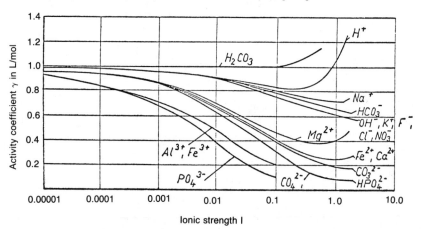

Figure 1.37. Activity coefficient γ as a function of the ionic strength I for selected inorganic compounds in soil-and groundwater. *Source:* Freeze and Cherry [1.27] and Matthess [1.55].

where z is the valence of the ion i, and [i] is its concentration in mol/L. Hence, the ionic strength I of groundwater containing the six major ions (principal solutes) follows from

$$I = 0.5 \; ([Na^+] + 4[Mg^{2+}] + 4[Ca^{2+}] + [HCO_3^-]$$
$$+ [Cl^-] + 4[SO_4^{2-}])$$

For calculations involving the ionic strength of dissolved neutral particles (e.g., O_2, N_2, neutral complexes), γ may be assumed as $\gamma \approx 1$ for all practical purposes. For $I < 0.1$ the use of the Kielland Table (table of ionic activities arranged by size of ions) is valid (see D'Ans and Lax [1.18] and Möbius and Dürschen [1.58]). An approximate value of γ for low ionic strengths in the soil- and groundwater may also be obtained from the following equation [1.35]:

$$-\log \gamma_i = 0.5 \; z_i^2 \sqrt{I} \; (1 + \sqrt{I})^{-1} \qquad (1.41)$$

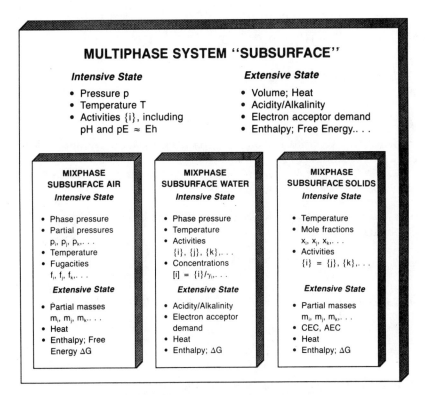

MULTIPHASE SYSTEM "SUBSURFACE"

Intensive State
- Pressure p
- Temperature T
- Activities {i}, including pH and pE ≈ Eh

Extensive State
- Volume; Heat
- Acidity/Alkalinity
- Electron acceptor demand
- Enthalpy; Free Energy.. . .

MIXPHASE SUBSURFACE AIR
Intensive State
- Phase pressure
- Partial pressures p_i, p_j, p_k, · · ·
- Temperature
- Fugacities f_i, f_j, f_k, · · ·

Extensive State
- Partial masses m_i, m_j, m_k, · · ·
- Heat
- Enthalpy; Free Energy ΔG

MIXPHASE SUBSURFACE WATER
Intensive State
- Phase pressure
- Temperature
- Activities {i}, {j}, {k}, · · ·
- Concentrations [i] = {i}/γ_i, · · ·

Extensive State
- Acidity/Alkalinity
- Electron acceptor demand
- Heat
- Enthalpy; ΔG

MIXPHASE SUBSURFACE SOLIDS
Intensive State
- Temperature
- Mole fractions x_i, x_j, x_k, · · ·
- Activities {i} = {j}, {k}, · · ·

Extensive State
- Partial masses m_i, m_j, m_k, · · ·
- CEC, AEC
- Heat
- Enthalpy; ΔG

Figure 1.38. Thermodynamic state variables of the multiphase system "subsurface."

Summarizing, Figure 1.38 characterizes the thermodynamic state of the multiphase system subsurface by the fundamental relationships discussed above and in §1.1.1.

1.4.1.2 Chemical Reactions

The chemical driving force of each reaction

$$\text{Reactants (R)} \underset{r''}{\overset{r'}{\rightleftharpoons}} \text{Products (P)} \tag{1.42a}$$

(e.g., $aA + bB \rightleftharpoons cC + dD$ is given by

$$F_{Ch} = \Sigma(\partial G/\partial n_i|_{T,p,n_{k \neq i}})dn_i = \underset{P}{\Sigma} \nu_i\mu_i - \underset{R}{\Sigma} \nu_j\mu_j \tag{1.42b}$$

This force is equal to the free-energy change $F_{Ch} = \Delta_r G$. Incorporating this statement and the statement of Equation 1.38b, we obtain

$$\Delta_r G = \underset{P}{\Sigma} \nu_i\mu_i - \underset{R}{\Sigma} \nu_j\mu_j = \underset{P}{\Sigma}(\nu_i\mu_i^\circ + RT \ln(\Pi(\{i\}^{\nu_i})))$$

$$- \underset{R}{\Sigma}(\nu_j\mu_j^\circ + RT \ln(\Pi(\{j\}^{\nu_j}))) = \Delta_r G^\circ + RT \ln(\Pi\{i\}^{\nu_i}/\Pi\{j\}^{\nu_j}) \tag{1.43a}$$

where $\Delta_r G^\circ = \underset{P}{\Sigma} \nu_i\mu_i^\circ - \underset{R}{\Sigma} \nu_j\mu_j^\circ$ \hfill (1.43b)

Therefore,

1. if $\Delta_r G < 0$, the reaction proceeds from left to right, i.e., reaction products are formed
2. if $\Delta_r G > 0$, the reaction proceeds from right to left, i.e., reactants are formed from the reaction products

For $\Delta_r G = 0$, chemical reaction equilibrium will be established and $\Delta_r G^\circ$ is obtained from Equation 1.43a as

$$\Delta_r G^\circ = -RT \ln(\underset{P}{\Pi}\{i\}^{\nu_i}/\underset{R}{\Pi}\{j\}^{\nu_j}) \tag{1.44}$$

The argument of the logarithmic function is the thermodynamic equilibrium constant K:

$$K = \underset{P}{\Pi} \{i\}^{\nu_i}/\underset{R}{\Pi} \{j\}^{\nu_j} = K(p,T) \tag{1.45a}$$

It follows that $\Delta_r G^\circ = -RT \ln K$ and, therefore,

$$K = \exp(-\Delta_r G^\circ/(RT)) \tag{1.45b}$$

with $RT = 2.479$ kJ/mol.

Since K is obtained from $\Delta_r G^\circ$, and hence from μ°, K is subject to the same pressure and temperature conditions p° and T° as μ°. As with the standard potential, K does not vary significantly with changes in pressure. Given the thermodynamic equilibrium constant K at standard conditions for a particular reaction, K of the same reaction at another temperature may be determined by van't Hoff's differential equation:

$$d \ln(K_T/dT) = \Delta_r H^\circ/(RT^2)$$

where $\Delta_r H^\circ$ is the change of the molar enthalpy of formation for the chemical reaction under standard conditions (see also Table 1.10):

$$\Delta_r H^\circ = \sum_P \nu_i \Delta h^\circ_{f,i} - \sum_R \nu_j \Delta h^\circ_{f,j}$$

Assuming $\Delta_r H^\circ$ is constant over the temperature interval under consideration, the solution of this equation is

$$\log K_T = \log K_{T^\circ} - \frac{\Delta_r H^\circ}{2.3R} \left(\frac{1}{T} - \frac{1}{T^\circ} \right) \qquad (1.45c)$$

The quantity of product formed per unit time is the reaction rate r, which may be assumed to be proportional to $\Delta_r G$ when the equilibrium state of the system is approached:

$$r = r' - r'' \approx k' \sum_R \nu_j \mu_j - k'' \sum_P \nu_i \mu_i = -k\Delta_r G \qquad (1.46a)$$

where r = reaction rate in $mol/(s m^3_v)$
            k = rate constant in $mol/(J s m^3_v)$
           m^3_v = cubic meter space of the subsurface

This model of reaction kinetics, however, is not commonly used. Usually it is simply supposed that the number of reaction-effecting particle collisions increases with increasing concentrations of the reactants $c_j = [j]$. The reaction rate is then determined for the most part by the number of collisions:

$$r = k \prod_R [j] \qquad (1.46b)$$

e.g., $r = k[A][B]$.

The reaction rate constant k is a function of temperature. If the temperature is increased by 10 K, the reaction usually proceeds at a two- or threefold rate, but the ratio r'/r'', and hence equilibrium, remain practically unaffected.

The reaction rate model described by Equation 1.46b assumes that the forward reaction rate r' is dominant. If the influence of the backward reaction rate r'' increases, r should be used as follows:

$$r = k' \prod_R [j] - k'' \prod_P [i] \tag{1.46c}$$

e.g., $r = k' [A][B] - k'' [C][D]$.

The disadvantage of these reaction-kinetic models 1.46b and 1.46c, however, is that they do not strive toward an equilibrium state, i.e., if $\Delta_r G = f([j],[i]) \to 0$ and hence $K_T \to \prod\{i\}^{\nu i}/\prod\{j\}^{\nu j}$, the reaction rate does not approach zero. For Equation 1.46c this can be only achieved by regarding the equilibrium constant K_T as

$$K_T = \prod_P [i]/\prod_R [j] = k'/k'' \tag{1.46d}$$

Thus Equation 1.46c can be written as

$$r = k' \left(\prod_R [j] - \prod_P [i]/K_T \right) \tag{1.46e}$$

e.g., $r = k'([A][B] - [C][D]/K_T)$, where K_T is obtained using Equations 1.45b and 1.45c.

If we would start from employing the thermodynamic expression for K according to Equation 1.45a, we obtain

$$r = k' \left(\prod_R \{j\}^{\nu j} - \prod_P \{i\}^{\nu i}/K_T \right) \tag{1.46f}$$

e.g., $r = k'(\{A\}^a\{B\}^b - \{C\}^c\{D\}^d/K_T)$.

However, it is frequently necessary to give an exact description of the reaction kinetics not only for $r \to 0$ (close to equilibrium) but also at larger reaction rates r occurring further from the equilibrium state. In most real cases r does not tend to ∞ but to a limiting value $r_m = r_{max}$. This condition may be met best by the following reaction-kinetic model:

$$r = r^* r_m/(r^* + r_m) \tag{1.46g}$$

where r^* is given by Equations 1.46a, 1.46e, or 1.46f. For $r^* \to 0$, Equation 1.46g approaches Equations 1.46a, 1.46e, or 1.46f; for $r^* \to \infty$, Equation 1.46g yields $r \to r_m$. When $r^* = k^* c = (r_m/k_m)c$, the system described by 1.46g is known as the Michaelis-Menten reaction kinetics.

A shortcoming of these reaction-kinetic models is that they do not consider that a definite activation energy and enthalpy are necessary to initiate the reaction (see Figure 1.39). Activation energies in the subsurface environment can be considerably lowered by the presence of biocatalysts. These catalysts, produced by microorganisms (see §1.5.4), frequently increase the rate constant k by several orders of magnitude, while the ratio of forward and backward rate constants k′/k″ remains practically unaffected.

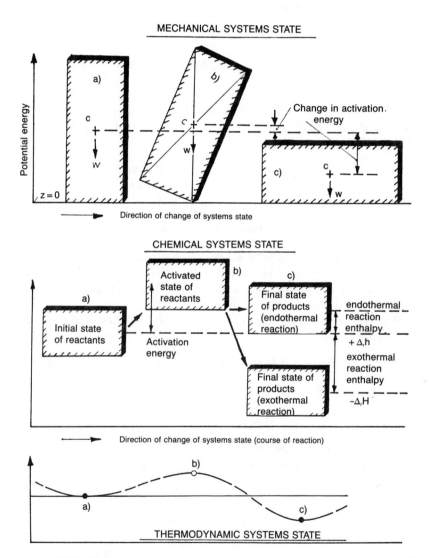

Figure 1.39. Energy change in mechanical and chemical systems: *(a)* metastable state; *(b)* unstable state; *(c)* stable state. *Source:* Stumm and Morgan [1.83] and Sommer [1.80].

Finally, the mathematical description of thermodynamic processes also requires calculation of the electric charge balance. For soil- or groundwater analyses the electric charge balance error ϵ should not exceed 5%:

$$\epsilon = 100\% \ (\Sigma(z_c[C]) - \Sigma(z_A[A]))/(\Sigma(z_c[C]) + \Sigma(z_a[A])) \quad (1.47)$$

where [C] is the concentration of cations and [A] the concentration of anions.

1.4.2 ASSOCIATION AND DISSOCIATION OF PARTICLES IN SOLUTION

Association (coupling) and dissociation (separation) of migrants in soil- and groundwater is an interactive process involving three possible forms:

1. molecule $\rightleftarrows$ cation + anion
2. complex $\rightleftarrows$ subcomplexes
3. floccule $\rightleftarrows$ colloids

1.4.2.1 Electrolytic Dissociation and Activity of Water

The interactive process between neutral molecules and ions is called *electrolytic dissociation*. True electrolytes are those which are already composed of ions in the solid state (e.g., NaCl), while potential electrolytes only dissociate in water.

Water dissociates into the ions H_3O^+ and OH^-, and these ions reassociate to form water:

$$2\ H_2O \rightleftarrows H_3O^+ + OH^- \quad (1.48a)$$

which, in simplified form, can be written as

$$H_2O \rightleftarrows H^+ + OH^- \quad (1.48b)$$

From Equation 1.43b and Table 1.10, the change in Gibbs free energy at $T = 298.15$ K may be calculated:

$$\Delta_r G^\circ = 0.0 - 157.4 - (-237.3) = 79.9 \ kJ/mol \ water$$

and from Equation 1.45b the equilibrium constant follows:

$$K = 1.005 \cdot 10^{-14}$$

For $\{H_2O\} = 1$, consequently, the dissociation constant K_w at 25°C and 1 bar is given by (see also §1.1.3 and Table 1.1):

$$K_w^o = \{H^+\}\{OH^-\} = 1.005 \cdot 10^{-14} \qquad (1.49a)$$

In general, K_w may be determined at any given temperature from van't Hoff's relation (Equation 1.45c) and Table 1.10:

$$\log K_w \approx -13.998 - \frac{55.9 \cdot 10^{+3}}{2.3 \cdot 8.314} (T^{-1} - 3.354 \cdot 10^{-3}) \qquad (1.49b)$$

Thus at 10°C (283 K), $\log K_w$ amounts to –14.53 and K_w to $0.295 \cdot 10^{-14}$ (compare values in Table 1.1).

The *pH value* is defined as the negative decimal logarithm of the proton activity:

$$pH = -\log\{H^+\} = -\log[H^+] - \log \gamma_{H^+}$$

and as

$$pH = -\log K_w + \log[OH^-] \qquad (1.50)$$

For pure water in its standard state, it follows from the mass and electric charge balance that $[H^+] = [OH^-]$, and hence at 25°C for $\gamma_{H^+} = \gamma_{OH^-} = 1$ that $\{H^+\} = [H^+] = \sqrt{10^{-14}} = 1.00 \cdot 10^{-7}$, and that $pH = 7$ (and at 10°C $pH_{10} = 7.27$). Upon addition of bases or acids to water (see §1.4.3), the activity of protons $\{H^+\}$ and hydroxyl ions $\{OH^-\}$ is changed, and therefore the value of pH is changed. The proton activity $\{H^+\}$ can be measured using glass electrodes.

1.4.2.2 Complex Formation and Decay Processes

Free metal ions Me^{m+} can only occur in the gaseous state and at high temperatures. In aqueous solutions, metal ions occur in hydrolyzed form as aquocomplexes (see §1.1.3):

$$Me^{m+} + n\ H_2O \rightleftarrows Me(H_2O)_n^{m+} \qquad (1.51)$$

A corresponding model for aquocomplexes is presented in Figure 1.40. The number of water molecules (dipoles) bound to a metal ion is related to the size of the participating ions (compare Figures 1.6 and 1.7). For example, the beryllium ion is smaller than the aluminum ion and therefore binds fewer water molecules than the aluminum ion:

$$Be^{2+} + 4\ H_2O \rightleftarrows Be(H_2O)_4^{2+}$$

$$Al^{3+} + 6\ H_2O \rightleftarrows Al(H_2O)_6^{3+} \qquad (1.52)$$

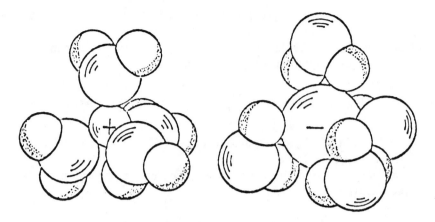

Figure 1.40. Models of a hydrated cation and a hydrated anion.

The aquocomplex $Al(H_2O)_6^{3+}$ is stable at about pH = 4, while the neutral aquocomplex $Al(OH)_3^0$ dominates at pH between 6 and 8, and the negative charged complex $Al(OH)_4^-$ at pH > 8. The proton H^+ also forms aquocomplexes. These complexes exhibit a chain structure H_3O^+, $H_5O_2^+$, $H_7O_3^+$, $H_9O_4^+$, etc.

In the sequence of complex formation, water dipoles are substituted by other ligands contained in the solution. Frequently, aquocomplexes are only the initial configuration in forming higher types of complexes:

$$Me(H_2O)_n^{m+} + L^{k-} \rightleftarrows Me(H_2O)_{n-1}L^{(m-k)+} + H_2O \qquad (1.53)$$

e.g.,

$$Fe(H_2O)_6^{3+} + Cl^- \rightleftarrows Fe(H_2O)_5Cl^{2+} + H_2O$$

Usually, however, the water molecules bound to the ion are not reflected in equations of complex formation and decay, and aquocomplexes are written as free ions [1.84; 1.64].

If the complex formation and decay processes are regarded as chain reactions (third step of complex formation):

$$Me + L \rightleftarrows MeL, \quad MeL + L \rightleftarrows MeL_2, \quad MeL_2 + L \rightleftarrows MeL_3, \ldots,$$

individual complex stability constants are given by

$$K_1 = \frac{\{MeL\}}{\{Me\}\{L\}}, \quad K_2 = \frac{\{MeL_2\}}{\{MeL\}\{L\}}, \ldots, \qquad (1.54a)$$

and the gross stability constants β_i for the whole reaction by [1.84; 1.64; 1.26]

$$\beta_1 = K_1, \qquad \beta_2 = K_1 \cdot K_2, \ldots, \qquad \beta_n = \prod_{i=1}^{n} K_i \qquad (1.54b)$$

In the classification of ligands, uni- and multidental ligands are distinguished. Unidental ligands (e.g., H_2O, OH^-, Cl^-, S^{2-}) have only one ligator atom and occupy one coordination site in the complex. In comparison, multidental ligands (e.g., humic acids) have several ligand atoms and occupy several sites in the complex. They surround the central atom (metal ion) similar to the claws of a crab surrounding its victim. Such ligands are called *chelants* and the complexes *chelates* [1.64].

Complexes containing several central atoms are termed *multinuclear complexes,* and if several different ligands are bound in coordination at one complex, this group is designated as a *mix-ligand complex.*

However, the chemical analysis of water samples usually yields only the integral concentration c_i of the analyzed substance i. Therefore, this value $c_i = [i]$ reflects the sum of all the different complexes and ions containing the substance i:

$$c_{i,TOTAL} = \Sigma c_{i,\text{free ions}} + \Sigma c_{i,\text{inorg. complexes}} + \Sigma c_{i,\text{org. complexes}} \qquad (1.55)$$

This total concentration must be analyzed with respect to the contributions of the different migrating species, i.e., the species which are actually subjected to transport, storage, exchange, and transformation processes in the subsurface. For instance, the migrants SO_4^{2-}, $CaSO_4^0$, $MgSO_4^0$, and $NaSO_4^-$, commonly called sulfate ion pairs or sulphate complex pairs, contribute chiefly to the total concentration of SO_4^{2-} measured in solution. Various computer programs based on thermodynamics of complex formation and decay are used to determine the influences of each complex upon the total concentration. They calculate the composition by means of different numerical routines [1.61; 1.84; 1.25; 1.87; 1.86; 1.26].

1.4.2.3 Coagulation and Flocculation

In addition to molecular-dissolved substances, soil- and groundwater also contain inorganic and organic colloidal and suspended constituents (see Table 1.2). In the following text, these constituents are simply designated as *colloids.* They can be classified as hydrophilic (wettable) or hydrophobic (nonwettable).

Whereas hydrophobic colloids form true interfaces with water (see §1.2.3), hydrophilic colloids are surrounded by several shells of water molecules bound by attraction forces (see Figures 1.41a and 1.43). The cations are fixed in the inner shell. Electrostatic attraction of the negatively charged colloids links these cations with their hosts (see Figure

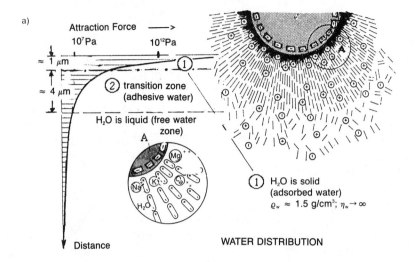

a)

Attraction Force ——>

10⁷Pa   10¹²Pa

≈ 1 μm

≈ 4 μm

transition zone
(adhesive water)

H₂O is liquid (free water
zone)

A

① H₂O is solid
(adsorbed water)
$\varrho_w \approx 1.5$ g/cm³; $\eta_w \rightarrow \infty$

Distance

WATER DISTRIBUTION

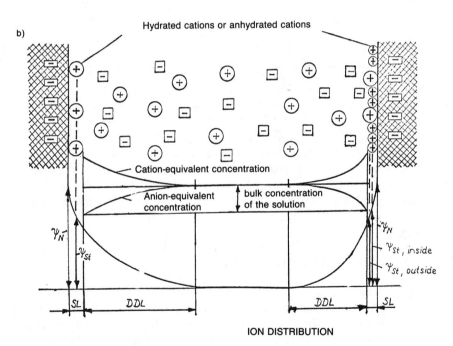

b)

Hydrated cations or anhydrated cations

Cation-equivalent concentration

Anion-equivalent
concentration

bulk concentration
of the solution

ψ_N

ψ_{St}

ψ_N

$\psi_{St, inside}$

$\psi_{St, outside}$

SL   DDL   DDL   SL

ION DISTRIBUTION

Figure 1.41. Formation of the electric double layer and attractive forces on the water-colloid interface: ψ_N = Nernst attraction potential; ψ_{St} = Stern attraction potential.

1.41b). The shell adjacent to the surface of the colloids is called the *Stern* or *Helmholtz layer*. It is followed by the diffuse double layer, also called the *Gouy-Chapman layer*. This layer extends into the bulk solution to a point where charge equilibrium between anions and cations is attained. The thickness of the diffuse double layer decreases with increasing concentration, i.e., with increasing availability of cations. Therefore, the permeability of soils to water is a function of soil salinity.

Colloids coagulate if the repulsive forces are weaker than the attractive forces. The more closely colloids approach each other due to Brownian motion, the more the attractive forces become effective. If trivalent cations such as Fe^{3+} and Al^{3+} penetrate into the inner shell and replace there the monovalent cations Na^+, K^+, and others, then shielding of the negative surface charges is more effective. Under these conditions, remarkable shrinking of the diffuse double layer occurs and the effective repulsive forces decrease.

The linking of coagulated colloids by bridging and other polymerization mechanisms is usually designated as *flocculation,* and the particles formed are called *agglomerates, flocculus,* or *flocs.* Coagulation and flocculation frequently result in precipitation, stearic retaining (filtering, sieving), and in adsorption.

Coagulation and flocculation should be understood as a reversible association-dissociation process which proceeds toward a thermodynamic equilibrium state. The inversion of coagulation and flocculation, i.e., the disintegration of coagulated colloids, is called *peptization* [1.1].

1.4.3 ACID-BASE REACTION PROCESSES

Chemical reactions characterized by proton transfer are designated as acid-base reactions and are of particular importance in the subsurface. Their definition has been continuously refined and improved. The best known is Brönsted's definition, a further development of the Arrhenius concept [1.1; 1.83]. Starting from the fundamental definition

$$\text{Acid (HA)} \rightleftarrows \text{Base (B)} + H^+ \tag{1.56}$$

acids are molecules or ions donating protons ($HA \rightleftarrows A^- + H^+$), and bases are molecules or ions accepting protons ($B + H^+ \rightleftarrows BH^+$).

Consequently, acids act as proton donors and bases as proton acceptors, forming corresponding pairs. Upon proton donation, each acid is transformed into its corresponding base ($A^- \mathrel{\hat=} B$) and may be retransformed by proton acceptance ($BH^+ \mathrel{\hat=} HA$). Typical examples of corresponding acid-base pairs are

$$H_3O^+ \quad \rightleftarrows H_2O \quad + H^+ \qquad\qquad HCO_3^- \rightleftarrows CO_3^{2-} \quad + H^+$$
$$H_2O \quad\; \rightleftarrows OH^- \; + H^+ \qquad\qquad H_3PO_4 \rightleftarrows H_2PO_4^- \quad + H^+$$
$$NH_4^+ \quad\; \rightleftarrows NH_3 \;\; + H^+ \qquad\qquad H_2PO_4^- \rightleftarrows HPO_4^{2-} \quad + H^+$$
$$CO_2 \cdot H_2O \rightleftarrows HCO_3^- + H^+ \qquad\qquad HPO_4^{2-} \rightleftarrows PO_4^{3-} \quad + H^+$$

These examples show that there are substances which may act as both proton donor and proton acceptor (e.g., H_2O, HCO_3^-, $H_2PO_4^-$, HPO_4^{2-}). These substances are called *ampholytes*.

The complete proton transfer, called *prototropy*, requires linking of two corresponding acid-base pairs:

$$H^+ \text{ donation by the base } HA_I : HA_I \qquad\qquad \rightleftarrows B_I + H^+ \quad (1.57a)$$

$$H^+ \text{ acceptance by the base } B_{II} : B_{II} + H^+ \quad \rightleftarrows HA_{II} \qquad (1.57b)$$

$$\Sigma: \text{ Protolytic reaction} \qquad : HA_I + B_{II} \rightleftarrows B_I + HA_{II} \; (1.57c)$$

The equilibrium constants of both partial reactions and the total reaction are given by

$$K_{AI} = \{H\}^+\{B_I\}/\{HA_I\} \qquad K_{AII} = \{H^+\}\{B_{II}\}/\{HA_{II}\} \quad (1.58a)$$

$$K = \{HA_{II}\}\{B_I\}/(\{B_{II}\}\{HA_I\}) = K_{AI}/K_{AII} \qquad\qquad (1.58b)$$

$$\log K = \log K_{AI} - \log K_{AII} \text{ where } pK = -\log K \qquad\qquad (1.58c)$$

Example:

$$I: \quad H_2O \qquad\qquad \rightleftarrows OH^- + H^+ \qquad \Delta_r g^\circ = + 79.9 \text{ kJ/mol}$$

$$II: \quad NH_3 + H^+ \quad \rightleftarrows NH_4^+ \qquad\qquad \Delta_r g^\circ = - 52.9 \text{ kJ/mol}$$

$$\Sigma: \quad H_2O + NH_3 \quad \rightleftarrows OH^- + NH_4^+ \qquad \Delta_r g^\circ = + 27.0 \text{ kJ/mol}$$

$$K_I = K_{AI} = 1.0 \cdot 10^{-14} \quad K_{II} = 1/K_{AII} = 1.85 \cdot 10^9$$

$$K = 1.86 \cdot 10^{-5} \text{ and } pK = 4.73$$

For a corresponding acid-base pair, expressed as ($B \hateq A^-$, $BH^+ \hateq$ HA), the ratio $\{A^-\}\{BH^+\}/(\{HA\}\{B\})$ equals one and, consequently,

$$K_A K_B = \{H_3O^+\}\{OH^-\} = \{H^+\}\{OH^-\}$$
$$= K_{H_2O} = K_w^\circ \qquad\qquad (1.59a)$$

At 25°C and 1 bar, the following important equation results:

$$pK = -\log K_B - \log K_A = 14 \qquad\qquad (1.59b)$$

Example:

$HCO_3^- + H_2O \rightleftarrows CO_3^{2-} + H_3O^+$	$CO_3^{2-} \rightleftarrows HCO_3^- + OH^-$
$\Delta_r g° = 58.9$ kJ/mol	$\Delta_r g° = 21.0$ kJ/mol
$K_A = 4.80 \cdot 10^{-11}$	$K_B = 2.09 \cdot 10^{-4}$
$K = K_A K_B = 1.005 \cdot 10^{-14}$	$-\log K_A - \log K_B =$ $10.32 + 3.68 = 14.0$

For the solvent water, which has an activity equal to one, the acid constant K_A yields a measure of acid strength and the base constant K_B a measure of base strength:

$$(1.60)$$

$$HA + H_2O \rightleftarrows A^- + H_3O^+ \qquad B + H_2O \rightleftarrows BH^+ + OH^-$$
$$K_A = \{H_3O^+\}\{A^-\}/(\{HA\}\{H_2O\}) \qquad K_B = \{BH^+\}\{OH^-\}/(\{B\}\{H_2O\})$$

The following classification of acids and bases is commonly used (see Ackermann [1.1] and Table 1.11):

very strong	acid or base	$\log K_A$ or $\log K_B < -1.74$
strong	acid or base	$-1.74 \leq \log K_A$ or $\log K_B < 4.5$
medium	acid or base	$4.5 \leq \log K_A$ or $\log K_B < 9.0$
weak	acid or base	$9.0 \leq \log K_A$ or $\log K_B < 15.74$
very weak	acid or base	$15.74 \leq \log K_A$ or $\log K_B$

Table 1.11. Selection of Several $pK_A = -\log K_A$ and $pK_B = -\log K_B$ Values of Corresponding Acid-Base Pairs

Acid Strength	pK_A	Acid	Base	pK_B
Strong	1.96	H_3PO_4	$H_2PO_4^-$	12.04
	1.91	HSO_4^-	SO_4^{2-}	12.09
	2.22	$(Fe(H_2O)_6)^{3+}$	$(Fe(OH)(H_2O)_5)^{2+}$	11.78
Medium	6.38	$CO_2 \cdot H_2O$	HCO_3^-	7.62
	7.12	$H_2PO_4^-$	HPO_4^{2-}	6.88
Weak	10.32	HCO_3^-	CO_3^{2-}	3.68
	12.32	HPO_4^{2-}	PO_4^{3-}	1.68
Very weak	15.74	H_2	OH^-	-1.74

To illustrate the extent of a protolytic reaction $HA + H_2O \rightleftarrows A^- + H_3O^+$, the protolytic degree in percent is used:

$$\alpha = \frac{[A^-]}{[HA] + [A^-]} \cdot 100\% \qquad (1.61)$$

Example:

Reaction   : $HCO_3^- + H_2O \rightleftarrows CO_3^{2-} + H_3^+O$
Measured data : $\{HCO_3^-\} = 1 \cdot 10^{-2}$ and $\{CO_3^{2-}\} = 1 \cdot 10^{-3}$
Result   : $\alpha = 100\% \cdot 1 \cdot 10^{-3}/(1 \cdot 10^{-2} + 1 \cdot 10^{-3}) = 9.1\%$

Aqueous solutions of a medium-strong acid and its corresponding base are capable of remaining at practically constant pH when limited amounts of strong acids or strong bases are added. Such solutions are called *buffers*. The degree to which an acid-base reaction is buffered is characterized by means of its buffer capacity β[1.1]:

$$\beta = dc/d(pH) \qquad (1.62a)$$

where dc is the amount of acid or base added in mol/L expressing the change of the extensive state variable. In the subsurface aqueous liquid phase, for instance in the soil solution, the buffer capacity is approximately given by

$$\beta = 2.30 \, ([H^+] + [OH^-] + [A^-][HA]/([A^-] + [HA])) \approx$$
$$2.3[HA][B]/([HA] + [B]) \qquad (1.62b)$$

Because $\{H^+\} = K_A\{HA\}/\{B\}$, if a strong acid or base is added to the aqueous phase [1.1]:

$$pH \approx -\log K_A - \log \frac{[HA] + c_A}{[B] - c_A}$$

or   (1.62c)

$$pH \approx -\log K_A - \log \frac{[HA] - c_B}{[B] + c_B}$$

where c_A and c_B are the concentrations of the strong acid and base, respectively. This buffer capacity is only of practical importance in the range of pH = $-\log K_A \pm 1$ = pH* ± 1. Examples for buffer levels of aqueous equimolar acid-base solutions are

HCO_3^-/CO_3^{2-}	pH* = 10.3	$H_2PO_4^-/HPO_4^{2-}$	pH* = 7.12
NH_4^+/NH_3	pH* = 9.25	$CO_2 \cdot H_2O/HCO_3^-$	pH* = 6.38

The buffer capacity is closely related to acidity and alkalinity. By definition, acidity and alkalinity reflect the addition and subtraction of protons required to get a definite pH change, where subtraction of protons is equivalent to addition of base. Hence, acidity and alkalinity are integral quantities of β over d(pH): ac $= +\int \beta$d(pH) and equivalent alk $= -\int \beta$d(pH), i.e., acidity and alkalinity are extensive state variables, and pH is the linked intensive state variable of the protolytic system (see Figure 1.51 in a later chapter).

As shown above, buffer systems occurring in subsurface water consist primarily of weak acids and their anion bases ($CO_2 \cdot H_2O/HCO_3^-$, HCO_3^-/CO_3^{2-}, and $H_2PO_4^-/HPO_4^{2-}$) and weak bases and their conjugate cation acids (NH_3/NH_4^+). However, the decisive buffering systems in the subsurface are the transphase exchange processes, and not these phase-internal reactions proceeding in the soil solution or in the groundwater. Most important are dissolution\precipitation and ion exchange for the protolytic state of the multiphase system subsurface (e.g., for the acidity of soils — see Figure 1.38).

1.4.4 OXIDATION-REDUCTION PROCESSES

Whereas acid-base reactions (protolysis) are characterized by the transfer of protons H^+ from an acid to a base (see Equation 1.56), oxidation-reduction processes (redox processes) are characterized by transfer of electrons e^- from a reducing agent (Red) to an oxidizing agent (Ox). Hence, the fundamental equation of a redox pair corresponds to Equation 1.56:

$$\nu\text{Red} \rightleftarrows \nu\text{Ox} + n e^- \qquad (1.63)$$

The reducing agent (reductant) acts as an electron donor and the oxidizing agent (oxidant) as the electron acceptor. The term *oxidation* refers to the process of electron donation (Red is oxidized), and the term *reduction* refers to the process of electron acceptance (Ox is reduced). Therefore, during a redox process, the oxidation state of an oxidant is decreased and the oxidation state of a reductant is increased, where the oxidation state (OS) is the charge of an atom in an assumed dissociated state (see §1.4.2). Atoms like C, S, N and Fe exhibit different oxidation states in different molecules. For instance, C has OS $= +4$ in HCO_3^-, 0 in CH_2O, and -4 in CH_4.

Similar to complete proton transfer, the complete electron transfer requires linking of two corresponding redox pairs (see also Equation 1.57). This is of special importance because unlike protons, free electrons cannot exist in aqueous solutions:

e⁻ - donation by Red_1 (oxidation)	$\nu_1\ Red_1$	$\rightleftarrows \nu_1 Ox_1 + ne^-$	(1.64a)
e⁻ - acceptance by Ox_2 (reduction)	$\nu_2\ Ox_2 + ne^-$	$\rightleftarrows \nu_2 Red_2$	(1.64b)

Redox reaction (Σ):	$\nu_1\ Red_1 + \nu_2 Ox_2 \rightleftarrows \nu_1 Ox_1 + \nu_2 Red_2$	
		(1.64c)

Determination of the equilibrium constant K of a redox process follows the same procedure used for protolytic reactions:

Fe^{2+}	$\rightleftarrows Fe^{3+} + e^-$	$\Delta_r g° = +74.40$ kJ/mol
$\frac{1}{4} O_2 + H^+ + e^-$	$\rightleftarrows \frac{1}{2} H_2O$	$\Delta_r g° = -118.65$ kJ/mol

$\frac{1}{4} O_2 + Fe^{2+} + H^+$	$\rightleftarrows \frac{1}{2} H_2O + Fe^{3+}$	$\Delta_r g° = -44.25$ kJ/mol
log $K_{Ox} = -13.03$	log $K_{Red} = 20.79$	log $K_{Redox} = 7.75$

In this example, the oxidation number of Fe increases from 2 to 3, and the oxidation number of oxygen decreases from 0 to –2.

The *electron activity* is characterized by means of pE, analogous to pH, the measure of proton activity. Because free electrons cannot exist in an aqueous solution, pE only reflects the tendency to donate electrons. pE is defined as (see Equation 1.50)

$$pE = -\log \{e^-\}. \qquad (1.65)$$

For a redox process $\Delta_r g$ is given according to Equation 1.43a by

$$\Delta_r g = \Delta_r g° + 2.3\ RT \log (\{Ox\}^\nu \{e^-\}^n / \{Red\}^\nu)$$

and in the thermodynamic equilibrium state where $\Delta_r g = 0$, it follows:

$$pE = pE° + (\nu/n) \log (\{Ox\}/\{Red\}) \qquad (1.66)$$

with $pE° = \Delta_r g°/(n\ 2.3\ RT)$. Instead of pE often the redox potential Eh is used:

$$Eh = (2.3\ RT/F)\ pE \qquad (1.67a)$$

Using Faraday's constant $F = 9.65 \cdot 10^4$ As/mol of electrons, and $RT_{25°C} = 2.479 \cdot 10^3$ Ws/mol of electrons, the following expression is obtained from Equation 1.67a (see Appendix 1 for SI units):

$$Eh = 0.059\ pE \text{ in volts}$$

Equations 1.66 and 1.67a lead to the Nernst equation:

$$Eh = Eh° + (\nu 2.3\ RT/(nF)) \log (\{Ox\}/\{Red\}) \qquad (1.67b)$$

with $Eh^\circ = \Delta_r g^\circ/(nF)$. Using again the example, $Fe^{2+} \rightleftarrows Fe^{3+} + e^-$ where $\Delta_r g^\circ = 74.4$ kWs/mol, pE° and Eh° may be calculated as $pE^\circ = 13.05$ and $Eh^\circ = +0.77$ V. An increase in the reducing potential or milieu always corresponds to a decrease of the pE° and Eh° values.

1.4.4.1 The pE-pH Diagram

pE-pH diagrams are frequently designated as stability diagrams and are widely used in hydrochemistry to characterize thermodynamic equilibrium states. For better understanding of these diagrams, their construction is illustrated with the simple example of iron as a solute in water.

As a first step, the stability of water itself is characterized:

$$\tfrac{1}{4} O_2(g) + H^+ + e^- \rightleftarrows \tfrac{1}{2} H_2O \qquad \Delta_r g^\circ = -118.65 \quad \log K = 20.8$$
$$H_2O + e^- \qquad\qquad \rightleftarrows \tfrac{1}{2} H_2(g) + OH^- \quad \Delta_r g^\circ = \quad 79.9 \quad \log K = -14.0$$

and according to Equation 1.45a and using Equation 1.39 ($\Delta_r g^\circ$ in kJ/mol), it follows:

$$1/(x_{O_2(g)}^{1/4} \{H^+\}\{e^-\}) = 10^{20.8}$$
$$pE = 20.8 - pH + \tfrac{1}{4} \log x_{O_2(g)} \tag{I}$$

$$x_{H_2(g)}^{1/2}\{OH^-\}/\{e^-\} = 10^{-14.0}$$
$$pE = -pH - 1/2 \log x_{H_2(g)} \tag{II}$$

Assuming $x_{O_2} = 1$ and $x_{H_2} = 1$ (i.e., partial pressures of 1 bar), Equations I and II correspond to lines I and II in Figure 1.42a. In the domain above line I, H_2O is converted into gaseous oxygen and H^+ by oxidation, and in the domain below line II H_2O is decomposed into gaseous hydrogen and OH^- [1.64].

The stability line between $Fe(OH)_3$ and $Fe(OH)_2$ as solids — designated by (c) — results from (see line III in Figure 1.42b)

$$Fe(OH)_3(c) + H^+ + e^- \rightleftarrows Fe(OH)_2(c) + H_2O$$
$$\Delta_r g^\circ = -26.2, \log K = 4.59$$
$$1/(\{H^+\}\{e^-\}) = 10^{4.59} \text{ and } pE = 4.59 - pH \tag{III}$$

$Fe(OH)_2$ is stable below this line, and $Fe(OH)_3$ is stable above it.

The stability line for dissolved (dissociated) Fe^{3+} ions in relation to $Fe(OH)_3(c)$ results from

$$Fe(OH)_3(c) + 3H^+ \rightleftarrows Fe^{3+} + 3H_2O \qquad \Delta_r g^\circ = -27.5, \log K = 4.82$$
$$\{Fe^{3+}\}/\{H^+\}^3 = 10^{4.82} \text{ and } pH = 1.61 - \tfrac{1}{3} \log \{Fe^{3+}\} \tag{IV}$$

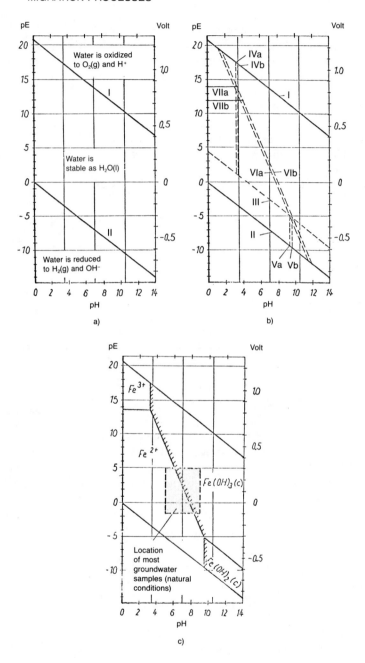

Figure 1.42. pE-pH diagram of iron in aqueous solution: *(a)* stability domain of water; *(b)* stability lines of the iron-water system; *(c)* stability regions of the dissolved cations of iron and their corresponding crystalline phases. *Sources:* Freeze and Cherry [1.27] and Davis and De Wiest [1.20].

Assuming an activity of $\{Fe^{3+}\}$ = 10^{-5} and 10^{-6}, lines IVa and IVb, respectively, are obtained in Figure 1.42b.

With the corresponding assumption of $\{Fe^{2+}\}$ = 10^{-5} and 10^{-6}, the reaction

$$Fe(OH)_2(c) + 2H^+ \rightleftarrows Fe^{2+} + 2H_2O$$
$$\Delta_r g^\circ = -75.7, \log K = 13.26$$
$$\{Fe^{2+}\}/\{H^+\}^2 = 10^{13.26} \text{ and pH} = 6.63 - \tfrac{1}{2} \log \{Fe^{2+}\} \qquad (V)$$

provides the lines Va and Vb.

Finally, the lines separating Fe^{2+} and $Fe(OH)_3(c)$

$$Fe(OH)_3(c) + 3H^+ + e^- \rightleftarrows Fe^{2+} + 3H_2O$$
$$\Delta_r g^\circ = -101.9, \log K = 17.85$$
$$\{Fe^{2+}\}/(\{H^+\}^3\{e^-\}) = 10^{17.85} \text{ and}$$
$$pH = \tfrac{1}{3}(-\log \{Fe^{2+}\} + 17.85 - pE) \qquad (VI)$$

and Fe^{3+} and Fe^{2+} are estimated:

$$Fe^{3+} + e^- \rightleftarrows Fe^{2+} \qquad \Delta_r g^\circ = -74.4 \text{ kJ/mol} \qquad \log K = 13.0$$
$$\{Fe^{2+}\}/(\{Fe^{3+}\}\{e^-\}) = 10^{13.0} \text{ and}$$
$$pE = 13.0 + \log\{Fe^{3+}\} - \log\{Fe^{2+}\} \qquad (VII)$$

With $\{Fe^{2+}\}$ and $\{Fe^{3+}\}$ equal to 10^{-5} and 10^{-6}, respectively, Equation VI yields lines VIa and VIb, and Equation VII yields lines VIIa and VIIb in Figure 1.42b. Figure 1.42c shows the completed pE-pH diagram.

In using stability diagrams, it is important to keep in mind that lines separating the stability domains of solid phases and dissolved ions are only valid for assumed activities of these ions. pE-pH diagrams are based on standard potentials μ° which, in turn, may vary according to the crystal structure of the solid phase, as is the case with $Fe(OH)_3(c)$, for instance. Nevertheless, these diagrams are useful tools in assessing the subsurface systems state and, hence, measured soil- and groundwater quality data.

1.4.4.2 Typical Redox Processes in the Soil- and Groundwater Zone

In the soil- and groundwater zone, two typical sequences of redox processes may be distinguished. One is given when molecular oxygen is available in sufficient quantities by continuous supply via the soil air (open systems); the other, when there is no additional oxygen available (closed systems).

After recharging water has reached groundwater, the soil air is practically no longer able to deliver the required molecular oxygen. Therefore, the groundwater zone behaves as a closed redox system. The oxygen dissolved in infiltrating water is consumed by oxidizing organic substances. If the oxygen O_2 dissolved in groundwater has been consumed, nitrate takes over the function as electron acceptor and becomes reduced. The sequence of redox pairs proceeds towards decreasing pH and Eh. Following the consumption of nitrate, manganese (IV) is reduced to manganese (II), iron (III) to iron (II), and, finally also, sulfates to sulfides, carbon dioxide to methane, and nitrogen to ammonium. This succession of typical redox processes is shown in Table 1.12, where CH_2O represents the organic substances (the electron donors). These "sugar units" occur, for instance, as mono-, di-, oligo-, and polysaccharoses, fatty acids, or phenols.

Table 1.12. Redox Processes in a Closed Thermodynamic System and Experimentally Determined Redox Potentials

Process	Reaction
(1) Aerobic respiration	$CH_2O + O_2(aq) \rightarrow CO_2 + H_2O$
(2) Denitrification	$CH_2O + \frac{4}{5} NO_3^- + \frac{4}{5}H^+ \rightarrow CO_2 + \frac{4}{5} N_2(g) + \frac{7}{5} H_2O$
(3) Mn-(IV) reduction	$CH_2O + 2MnO_2(c) + 4H^+ \rightarrow 2Mn^{2+} + 3H_2O + CO_2$
(4) Fe(III) reduction	$CH_2O + 8H^+ + 4Fe(OH)_3 \rightarrow 4Fe^{2+} + 11H_2O + CO_2$
(5) Sulfate reduction	$CH_2O + \frac{1}{2} SO_4^{2-} + \frac{1}{2} H^+ \rightarrow \frac{1}{2} HS^- + H_2O + CO_2$
(6) Methane fermentation	$CH_2O + \frac{1}{2} CO_2 \rightarrow \frac{1}{2} CH_4 + CO_2$
(7) Nitrogen fixation	$CH_2O + H_2O + \frac{2}{3} N_2 + \frac{4}{3} H^+ \rightarrow \frac{4}{3} NH_4 + CO_2$

	Eh	
• Beginning of NO_3^- reduction	0.45 to 0.55 V	(pH = 7)
• Beginning of Mn^{2+} formation	0.35 to 0.45 V	(pH = 7)
• Detection limit for solved oxygen	0.33 V	
• Detection limit for NO_3^-	0.22 V	
• Beginning of Fe^{2+} formation	0.15 V	
• Beginning of SO_4^{2-} and S^{2-} formation	−0.03 V	
• Detection limit for SO_4^{2-}	−0.18 V	

Source: Jackson and Inch [1.41] and Reissig [1.64].
Note: (c) denotes crystalline or solid phase.

Table 1.13, on the other hand, lists the sequence of typical redox processes occurring in the aeration zone. In this zone, oxygen is continuously and sufficiently supplied by molecular diffusion (see §1.3.1, Equation 1.20, and Figure 1.29).

The redox processes listed in Table 1.12 and 1.13 require the catalytic effect of enzymes produced by microorganisms in order to proceed at significant reaction rates. The most important organisms involved in this

Table 1.13. Redox Processes in an Open Thermodynamic System

Process	Reaction
(1) Aerobic respiration	$O_2(aq) + CH_2O \rightarrow CO_2(aq) + H_2O$
(2) Sulfate reduction	$O_2(aq) + \frac{1}{2}HS^- \rightarrow \frac{1}{2}SO_4^{2-} + \frac{1}{2}H^+$
(3) Fe-(II) oxidation	$O_2(aq) + 4Fe^{2+} + 10H_2O \rightarrow$ $4Fe(OH)_3 + 8H^+$
(4) Nitrification	$O_2(aq) + \frac{1}{2}NH_4^+ \rightarrow \frac{1}{2}NO_3^- + H^+ \rightarrow \frac{1}{2}H_2O$
(5) Mn-(II) oxidation	$O_2(aq) + 2Mn^{2+} + 2H_2O \rightarrow 2MnO_2(c) + 4H^+$

Source: Jackson and Inch [1.41].
Note: (c) denotes crystalline or solid phase.

biocatalysis are bacteria in the groundwater zone, and in the top soil also fungi, algae, yeasts, and protozoans. These bacteria have the size of large colloids and small suspended particles (see Table 1.2). As discussed before, enzymes lower the required activation energy of redox reactions decisively (see Figure 1.39). It should also be noted that the local micromilieu surrounding these microorganisms differs considerably from the statistical average in the REV (see §1.2.3).

Redox processes deliver bacteria the energy required for their metabolism. A significant reaction rate of a biologically catalyzed redox process is only given if reaction enthalpies of at least $\Delta_r h \geq 60$ kJ/mol substrate are released [1.27]. Bacteria are classified as aerobic, anaerobic, and facultative bacteria according to their oxygen requirements [1.88]. Although bacteria play such an important role, the investigation of microbial processes in the soil- and groundwater zone is still in the beginning stages of development.

The systems state of the subsurface (see fundamental Equation 1.1) is also for redox processes controlled by transphase reactions, as discussed for protolytic processes, where the extensive state variable Z symbolizes the acidity/alkalinity or the capacity of proton donation and acceptance. The equivalent extensive state variable characterizing redox processes is the supply of reductants or oxidants to the considered system, or more precisely the supply of electron acceptors or donors. In multiphase systems such as sludges, this variable is commonly designated as *biological* or *chemical oxygen demand* (BOD or COD). The systems state function BOD = f(pE), where pE is the intensive state variable (see Figure 1.51 in a later chapter), reflects one of the most important subsurface properties. This function must be estimated very carefully by laboratory and field tests. Without appropriate knowledge of this systems state function EAD = f(pE), where EAD stands for electron acceptor demand, effective bioremediation or subsurface water treatment for drinking water supply cannot be executed.

1.5

Phase Exchange

In addition to conversion of migrants in single mixphases of the multiphase system subsurface, attention must be given to exchange of migrants between mixphases. Transphase exchange processes (often termed *heterogeneous reactions* by chemists) and external exchange processes (the exchange with the environment) are of particular importance for the migrants' fate in the subsurface. This chapter discusses the dynamics of these processes (§1.5.1), the exchange processes in the literal sense (§1.5.2), dissolution and precipitation processes (§1.5.3), and biological processes (§1.5.4).

1.5.1 DYNAMICS OF EXCHANGE PROCESSES

The proper estimation of the time constants of exchange processes in relation to the time constant of the bulk migration process in the multiphase system subsurface is a key question in characterizing the dynamics of transphase exchange processes. The transphase exchange rate between two fluid mixphases in the subsurface — for instance, between mobile and immobile water — is controlled by diffusion. Molecular diffusion is also usually the controlling factor of exchange rates between a fluid and the solid phase. Accordingly, the exchange rate between two fluid mixphases in the subsurface f_E per volume of the multiphase system may be obtained from

$$f_E = (D_M/\overline{\Delta l})(\omega/V)^*\Delta c \qquad \text{in mol/(sm}_v^3) \qquad (1.68)$$

where $\overline{\Delta l}$ = average exchange path length
ω = inner surface area
$\omega/V = O_{sp}$ = specific surface area; $O_{sp} \approx 3/d_{10\%}$ [1.11, p. 120]
$(\omega/V)*$ = specific surface actually affected by the exchange process
D_M = molecular diffusion coefficient in the mixphase
Δc = concentration difference between the neighboring fluids as a measure of the process driving force (see Equations 1.38 and 1.46a)

Hence, the characteristic time constant t*, the product of resistance and corresponding capacitance, is given for the transphase exchange process by

$$t_i^* = [(D_M/\overline{\Delta l})(\omega/V)*]^{-1} \cdot ca \qquad (1.69a)$$

where ca is the specific storage density of the considered migrant in the two phases between which the exchange takes place (e.g., ca = $\sigma_{w,im}$ + $\sigma_{w,m}$ for immobile and mobile water).

The characteristic time constant of the bulk migration process in the multiphase system may be estimated by

$$t_b^* = \Delta\bar{x}/u \qquad (1.69b)$$

where $\Delta\bar{x} = \Delta L$ is the macroscopic distance traveled by the migrants.

Example:

For $(\omega/V)* = 10^3$ m^{-1}, $D_M = 10^{-9}$ m^2/s, $\overline{\Delta l} = 10^{-3}$ m, and ca = θ_m + θ_{im} = 0.4, we obtain $t_i^* = 400$ s, and for $\Delta L = 1$ m and u = 10^{-6} m/s, t_b^* = 10^6 s results, and hence $t_b^*:t_i^* = 2500:1$ is the time ratio characterizing the process dynamics. For the given example, the inner kinetics would be negligible, and an approximate equilibrium state between the exchanging phases could be assumed. But for u > 10^{-5} m/s and $\Delta L < 0.1$ m, the formulation of a nonequilibrium process model would be required.

It was also shown that the exchange rate f_E can frequently be approximated by a linear function of the hydraulic gradient (see Figure 1.34):

$$f_E = \alpha v \Delta c \qquad (1.69c)$$

This means that in such cases the exchange rate is considered as primarily convection controlled and not diffusion controlled.

1.5.2 EXCHANGE PROCESSES

Usually, the exchange processes are classified as filtering (sieving), sorption, ion exchange, and external exchange. Filtering, sorption, and ion exchange are overlapping categories which are often not easy to distinguish.

1.5.2.1 Filtering (Sieving)

In soil and rock, solutes, particularly suspended particles (see Table 1.2), are retained by filtering. This effect is primarily caused by geometric restrictions of pore channels. The tearing off of particles and retention of particles in the porous medium is again a process which proceeds towards an equilibrium state. In geohydraulics, the two partial processes of this two-way process are called *internal suffusion* and *internal clogging* [1.11]. Two criteria are important for characterizing the internal suffusion and clogging processes: a geometric and a hydraulic criterion.

The geometric filter criterion expresses the relation between the diameter d of the migrating particles and the minimum effective diameter $d_{c,min}$ of flow channels (bottlenecks) which particles will pass through:

$$d \leq F \, d_{c,min} \qquad (1.70)$$

where $0 < F < 1.0$. The magnitude of the slippage factor F depends on the manner in which d and $d_{c,min}$ are determined, the size of the transported particles (F increases with increased d), the particle shape, the bridging factor, and other quantities [1.11].

The hydraulic filter criterion assesses the tractive force acting upon particles in pore channels. This force is expressed by the hydraulic gradient (grad h) of Darcian flow. For grad $h > i_{crit}$, the particles of a diameter $d < F d_{c,min}$ are dragged; at grad $h < i_{crit}$, they remain in their position. The magnitude of i_{crit} is a function of particle size, shape, and density. Furthermore, the angle between the flow direction and the gravitational force affects i_{crit} [1.11].

In the internal clogging process, larger particles become lodged first in pore channels at points with minimal cross-sectional areas, which results in further narrowing of these bottlenecks. Particles retained become successively smaller, and the porous medium becomes progressively obstructed in the direction opposite to flow. This process is usually designated as *self-sealing*. Under aerobic conditions, small animalcules such as rotifers, tardigrades, zilitates, threadworms, and others (see Figure 1.48) are primarily responsible for establishment of an equilibrium state between channel clogging and restoration. The actions of these animalcules result in a definite permeability for most river beds. The clogged material consists of suspended organic and inorganic particles, colloids, and innumerable microorganisms, used by these animalcules as nutrients. In this way they clean up the pore channels or dig new channels.

Filtering is often simply described according to Equation 1.46b as [1.16]

$$f_{E,F} = kc \text{ in } g/(sm_v^3) \qquad (1.71)$$

where $f_{E,F}$ = mass rate of substance exchanged by filtering per volume soil
  c = concentration of the considered suspended substance in the flowing soil- or groundwater in g per m^3 water
  k = rate constant in $(m^3_w/m^3_v)\ s^{-1}$

The sieving effect (filtering) is called *dialysis* if macromolecules are investigated (see Table 1.2). Special natural or artificial semipermeable membranes allow passing of molecular solutes but retain colloids. Clogged soil layers, accumulation zones in soils, and silts may cause such dialytic effects.

Finally, ions may also be impacted by sieving effects. Thus, silts and clays are selective ion sieves. Due to the excess negative charge of clay minerals (see §1.1.4), migrating anions may be repulsed and hindered from passing due to their large size (see Figure 1.6). Clays may therefore act as a chloride barrier [1.35]. The small neutral water molecules and cations can pass such a barrier, although the cations are subjected to retardation. Even relatively thin layers of clay can act as such a semipermeable membrane. Therefore, water quality (e.g., salinity) may be quite different on either side of such a layer. This difference does not necessarily characterize a lack of water exchange or the existence of an aquiclude.

If two aquifers separated by a clay layer are considered as a closed thermodynamic system (no water and solute exchange with the environment), the pressure in aquifer I with higher mineralization (ion content) would be higher than that in aquifer II with lower mineralization. This results from the difference in water molecule activities. This activity is higher in the aquifer of lower mineralization ($\{H_2O\}^{II} > \{H_2O\}^I$). Therefore, the H_2O molecules tend to diffuse from the aquifer of lower mineralization to the aquifer of higher mineralization, according to their chemical potential μ_{H_2O}. The pressure difference (in Pa) which establishes at thermodynamic equilibrium is designated as osmotic pressure (see Equation 1.43c):

$$\Delta p_o = -(RT/V_{m,H_2O})\ \ln\ (\{H_2O\}^I/\{H_2O\}^{II}) \qquad (1.72)$$

where $\{H_2O\} \approx 1 - 0.017\ \Sigma b_i$ (for exact values, see, for example, Truesdell and Jones [1.86]).

In some cases these pressures may build up to very high values. Assuming, for example, a NaCl content of 12% expressed as mass fraction (w_{NaCl}) in aquifer I and 3% in aquifer II, the following results are obtained (see also Appendix 1 and §1.1.1):

$$b_{NaCl}^{II} = w/M = 0.03/0.0585 = 0.513 \qquad \text{and}\ b_{NaCl}^I = 2.05\ \text{mol/kg}$$

$$\{H_2O\}^{II} = 1 - 0.017\ ([Na^+] + [Cl^-]) = 1 - 0.017 \cdot 1.026 = 0.983$$

$\{H_2O\}^l = 1 - 0.017 \cdot 4.10 = 0.930$

$$\Delta p_o = -((2479 \text{ J/mol}) \cdot (55.56 \cdot 10^3 \text{mol/m}^3) \ln(0.930/0.983))$$
$$= 76.3 \text{ bar}$$

These pressures correspond to a water column height of about 750 m. Osmosis also plays an important role in root uptake of water and nutrients, or in nutrient transport in plants and living creatures.

1.5.2.2 Sorption

Sorption is an interactive process consisting of adsorption and desorption. Sorption includes attachment and release of gaseous, liquid, or solid migrants on surfaces of solids which serve as hosts. Migrant and host remain chemically unchanged during sorption with the exception of chemisorption, which is mostly irreversible. The migrants are fixed onto the solid surfaces by attraction forces and lose a part of their enthalpy during adsorption. The remaining enthalpy of adsorbed molecules is in the range typical of solids.

The molar enthalpy of adsorption $\Delta_r h$ is governed by the type of bond between adsorbed particles and their hosts. Sorption may be classified according to the bond type, which may be due to

- van der Waals attraction ($|\Delta_r h|$ usually < 50 kJ/mol)
- Coulombic attraction (opposite charges of adsorbed particles and absorbent), $K = (e_1 e_2)/(4\pi D a^2)$
- chemical bonding (ionic or atomic bonds: $|\Delta_r h| > 500$ kJ/mol)

Figure 1.43 illustrates the terms commonly used in chemistry to describe sorption. According to this scheme, an adsorbable species (migrant) is called *adsorpt* in the adsorbed state and *adsorptive* in the nonadsorbed state. The adsorbed particles are locally fixed within the

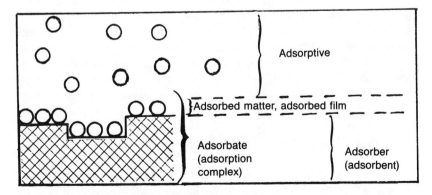

Figure 1.43. Sorption terminology. *Source:* Möbius and Dürschen [1.58].

adsorbed film. This film usually consists of one or a few layers if the film does not adopt the properties of the adsorber (see also Figure 1.41). The adsorber and the adsorbed film together form the *adsorbate* or *adsorption complex*.

1.5.2.3 Ion Exchange

If the particles (migrants) subjected to sorption are ions, the adsorption-desorption process is called *ion exchange* and the adsorbent is called the *ion exchanger* (i.e., cation or anion exchanger). The ion exchange capacity (cation exchange capacity CEC or anion exchange capacity AEC) is the amount of ions which the adsorbent can bind multiplied by their charges. The binding sites or potential bonds of the exchanger in the soil- and groundwater zone are always occupied by ions; i.e., independent of the ion concentration in the soil- and groundwater, there are no vacant bonds. Ions are hence only exchanged at the exchanger. Exchange always takes place in electrical equivalents according to the law of electrical charge balance. The exchange capacity is usually given as equivalent amount of substance n_{eq} per 100 g of adsorbate (i.e., solid matter) in mol/100 g (see Table 1.6). As mentioned previously, 1 mole of equivalents is equal to $6.023 \cdot 10^{23}$ charges or bonds of the exchanger.

Important ion exchangers in the soil- and groundwater zone include clay minerals; zeolites; iron, manganese, and aluminum hydroxides; and oxidohydrates. Organic matter, such as humic substances and metabolites of microorganisms (e.g., bioslime), has a remarkably high ion exchange capacity (see Table 1.6). The ion exchange capacity of the rock-forming minerals mica, feldspars, augites, and hornblendes may also be of importance in investigations of migration processes (see §1.1.4). All of these ion exchangers show a more or less pronounced ion selectivity.

Sorptivity generally decreases with increasing diameters of migrants. In ionic sorption, bonding forces also increase with increasing ion charges. Exceptions to this fundamental rule occur when the ions become incorporated into the crystal lattice, i.e., if they geometrically fit into interstitial voids. In this case they are not readily exchangeable. Apart from such staricstructural effects occurring between adsorbed matter and adsorbent, the following basic rule of binding intensities can be given for cations:

$$H^+ > Rb^+ > Ba^{2+} > Sr^{2+} > Ca^{2+} > Mg^{2+} > K^+ > Na^+ > Li^+ \quad (1.73)$$

Exchange capacities are strongly affected by pH. An exchanger is said to be amphoteric if its ability to exchange ions, whether cations or anions, changes at a definite pH. This pH value is called the *isoelectric point.* Amphoteric properties are typical not only of clay minerals, but also of silicic acid, aluminum and iron oxidohydrates, and manganese and iron oxides as well, particularly if they are not completely crystallized. These substances have functional groups which cause the amphoteric properties on their surfaces (see Figure 1.44).

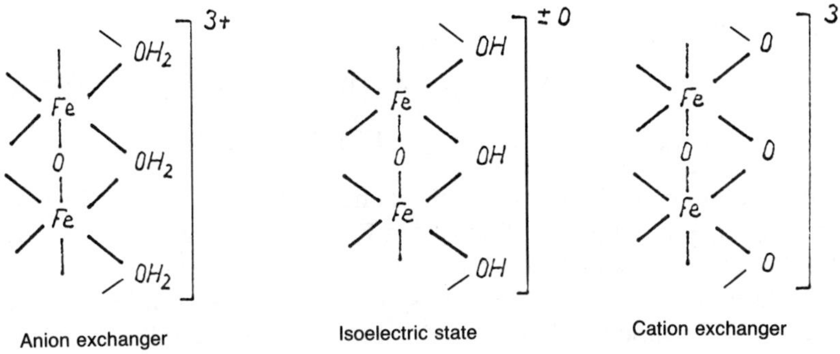

Figure 1.44. Amphoteric properties of iron oxidohydrate and iron oxide.

As an example, the exchange of the cations Na^+ and Mg^{2+} with the exchanger EX^- is considered here. According to Equation 1.42, this exchange may be reflected by

$$z_A B^{Z_B+} + z_B AEX_{Z_A} \rightleftarrows z_B A^{Z_A+} + z_A BEX_{Z_B} \qquad (1.74)$$

$$2Na^+ + MgEX_2 \rightleftarrows Mg^{2+} + 2NaEX$$

where A and B are the ions Mg and Na, respectively, and z_A and z_B their valencies.

In practice, of course, a considerably larger number of different cations compete for the exchange sites, and Equation 1.74 consists of a system of $\Sigma(i-1)$ equations, where i goes from 1 to n, and n is the number of the competing cations. Each of these equations describes the competition of two of the cations as shown in Equation 1.74. Additional equations balance the exchange sites (ΣEX_{z_i} = CEC) and the amount of substance of each species contained in the multiphase system.

The thermodynamic equilibrium constant follows from Equation 1.74 as

$$K_{A,B} = \{A^{Z_A+}\}^{Z_B}\{BEX_{Z_B}\}^{Z_A}/(\{B^{Z_B+}\}^{Z_A}\{AEX_{Z_A}\}^{Z_B}) \qquad (1.75a)$$

$$K_{Mg,Na} = \{Mg^{2+}\}\{NaEX\}^2/(\{Na^+\}^2\{MgEX_2\})$$

Considering the adsorbed film of the cation exchanger (see Figure 1.43) as a binary solid solution results in the common expressions of ion activities in the sorbed state [1.47]:

$$\{MgEX_2\} = \gamma_{Mg}x_{Mg(ad)} = \gamma_{Mg}n_{Mg}/(n_{Na} + n_{Mg})$$

$$\{NaEX\} = \gamma_{Na}x_{Na(ad)} = \gamma_{Na}n_{Na}/(n_{Na} + n_{Mg})$$

and hence in

$$K_{A,B} = \{A^{Z_A+}\}^{Z_B}(\gamma_B x_B)^{Z_A}/(\{B^{Z_B+}\}^{Z_A}(\gamma_A x_A)^{Z_B}) \qquad (1.75b)$$

$$K_{Mg,Na} = \{Mg^{2+}\}(\gamma_{Na}x_{Na})^2/(\{Na^{+2}\}\ (\gamma_{Mg}x_{Mg}))$$

A compilation of $K_{A,B}$ values may be found in Benson [1.8] and Benson and Teagne [1.9]. Typical values of log K are, for instance, Ca $\rightleftarrows$ Mg: –0.1; Na $\rightleftarrows$ K:0.4; and Na $\rightleftarrows$ NH$_4$:0.7.

The selectivity coefficient K^S of this exchange process is defined as the K value of Equation 1.75b for $\gamma = 1$. But other definitions of the selectivity coefficient, such as

$$K_{A,B}^S = \frac{[A^{Z_A+}]x_B}{[B^{Z_B+}]x_A} \quad \text{and} \quad K_{Mg,Na} = \frac{[Mg^{2+}]x_{Na}}{[Na^+]x_{Mg}} \qquad (1.75c)$$

are also in use. In these terms the activities are replaced by mass concentrations and their exponents are neglected.

An example given in Table 1.14 shows the cation distribution in the aqueous solution and in the adsorbed films of a soil sample. This example demonstrates that, for instance, Al^{3+} ions dominate in the adsorbed films, and Na$^+$ and K$^+$ ions in the soil liquid (see Equation 1.73).

Table 1.14. Composition of Cation Films on the Surface of Soil Solids and the Cation Composition in the Liquid Soilwater at Equilibrium State (in mmol charges per 100 g soil and in mmol charges per liter water)

		Al³⁺	Ca²⁺	Mg²⁺	Na⁺	K⁺	H⁺	Σ
Soilwater	n_c /L	0.06	0.47	0.11	0.30	0.33	0.24	1.51 mmol$_c$ /L
	x·100%	4.00	31.00	7.00	20.00	22.00	16.00	100.00%
Solid phase	n_c /100 g	10.20	2.18	0.26	0.13	0.42	1.05	14.2 mmol$_c$ /100 g
	x·100%	72.00	15.00	2.00	1.00	3.00	7.00	100.00%

Source: Reissig [1.64].

The distribution coefficient K_d is also frequently used to characterize the distribution of a cation A between the aqueous and the solid phase (in mL/g$_R$):

$$K_{d,A} = (n_s/m_s)/(n_w/V_w) = n_s\sigma_w/(\rho_b n_w) \qquad (1.76)$$

where n_s = amount of cations A sorbed onto the solids (in mol)
m_s = mass of the solids in g (g_{Rock})
n_w = amount of cations A solved in the aqueous solution (in mol)
V_w = volume of the aqueous solution in mL (cm_w^3)
ρ_b = bulk density of the dry soil or rock sample ($\rho_b = m_s/V$) in g/cm_V^3
σ_w = water-filled porosity or volumetric content ($\sigma_w = V_w/V$)

The amount of substance A adsorbed onto the solid phase related to the volume of the multiphase system can be expressed as

$$n_{A,s}/V = K_{d,A} \, \rho_b \, c \qquad (1.77a)$$

where $c_A = n_{A,w}/(\sigma_w V)$ and $\sigma_w = \theta_w$. Generally, the distribution coefficient $K_d = K_{d,A}$ reflects the equilibrium state. The simplest way to formulate a time-dependent K_d-value in characterizing the nonequilibrium state is given by [1.24]

$$K_d(t) = K_d(1 - \exp(-kt)) \qquad (1.77b)$$

where k is a rate constant. The value R, with

$$R = K_d \, \rho_b/\sigma_w + 1 = (K_d\rho_b + \sigma_w)/\sigma_w \qquad (1.78a)$$

is called the *retardation coefficient*. The travel speed of a considered migrant A is 1/R of the water molecules' speed ($u_w = Ru_A$) under the assumptions made for Equation 1.24.

If Equation 1.78a is used in connection with storage considerations (see also Equation 1.101a), σ_w represents the storage coefficient of the aqueous phase per unit volume of the multiphase system subsurface, and $K_d\rho_b = \alpha$, the storage coefficient of the solid phase for the same unit volume. Occasionally, the sum of both quantities is termed the *fictitious porosity*:

$$\sigma_f = \sigma_w + K_d\rho_b = \sigma_w + \alpha \qquad (1.78b)$$

1.5.2.4 External Exchange Processes

Mass exchange processes proceeding between the considered system in the soil-and groundwater zone and its environment should be regarded as external exchange processes. External exchange processes include the extraction of solutes by roots, and the extraction or injection of solutes in engineering tasks. Such natural and anthropogenically caused external exchange processes must be formulated as boundary conditions in mathematical modeling.

1.5.3 DISSOLUTION AND PRECIPITATION PROCESSES

Soil- and groundwater quality is also largely attributed (1) to dissolution of solids such as salts and minerals, of mobile or immobile liquids such as residually distributed chlorinated hydrocarbons or oil compounds, and of gases in an aqueous solution or in another fluid mixphase, and (2) to precipitation of dissolved solid, liquid, or gaseous species from an aqueous solution or another mixphase and their transfer onto the surface of solids in the multiphase system subsurface.

The outstanding power of water to dissolve matter is due to its large dielectric constant. Thus, for instance, the attractive force in a salt crystal with monovalent ions is equal to: $C^+ \cdot A^-/a^2$ where C^+ is the charge of the cation, A^- is the charge of the anion, and a is the distance between charge centers. In the aqueous environment, this force is reduced to $C^+ \cdot A^-/(a^2D)$, where the dimensionless dielectric constant of the solvent is D ($D_{water} = 81.5$). The ions are now able to overcome this reduced attractive force due to their enthalpy and to detach from the crystal surface. Upon arrival in the aqueous solution, they hydrate at once with water dipoles and form the hydro- or aquocomplexes (see §1.4.2). The hydration energy released during this process must be about the same as the lattice energy to ensure a continuous dissolution of crystalline matter.

The solubility of many salts and minerals in water increases with increasing temperature (see Table 1.15).

Table 1.15. Temperature Dependence of Solubility for Several Substances in Pure Water in g/kg Water (Activity Coefficient $\gamma = 1$)

Substance	0°C	10°C	20°C	Substance	0°C	10°C	20°C
$CaCl_2 \cdot 6H_2O$	603	560	745	KNO_3	133	215	315
$CaSO_4 \cdot 2H_2O$	1.76	1.93	2.04	$MgSO_4 \cdot 7H_2O$	301	356	408
$Ca(HCO_3)_2$	161.5	—	166	NaCl	356	357	358
KCl	282	313	344	$Na_2CO_3 \cdot 10H_2O$	68.6	120	216

Source: Matthess [1.55].

This corresponds to an endothermal process where the lattice energy is greater than the hydration energy, as shown in Figure 1.39. Exothermal dissolution processes in which the solubility decreases with increasing temperature (lattice energy < hydration energy) are rare in the subsurface [1.64].

The dissolution/precipitation process, as an exchange process between the solid (crystalline) and the aqueous mixphase, can be described by

$$A_mB_n(c) \rightleftarrows mA^{n+}(aq) + nB^{m-}(aq) \qquad (1.79)$$

For this process, the equilibrium constant according to Equation 1.45a is designated as the *solubility constant* K_S^*. When the activity of the solids $\{A_mB_n(c)\}$ is equal to one, the solubility constant is given by

$$K_S^* = \{A^{n+}(aq)\}^m\{B^{m-}(aq)\}^n \qquad (1.80a)$$

Example:

$$CaF_2(c) \rightleftarrows Ca^{2+} + 2F^-$$

$$\Delta_r g^\circ = -553.4 - 2 \cdot 276.7 - (-1162.7) = 55.9 \text{ kJ/mol} \quad \log K = -9.8$$

$$K_S^* = \gamma_{Ca}^{2+} \cdot \gamma_F^2[Ca^{2+}][F^-]^2 = 1.61 \cdot 10^{-10} \qquad (1.80b)$$

$$[Ca^{2+}] = 2[F^-] \text{ (charge balance)} \qquad (1.80c)$$

To get the concentrations $[Ca^{2+}]$ and $[F^-]$ from these two conditional equations, Equation 1.80c may be substituted into Equation 1.80b:

$$y = [Ca^{2+}] = \sqrt[3]{\frac{1.61 \cdot 10^{-10}}{4\gamma_{Ca}^{2+} \cdot \gamma_F^2}} = 3.4 \cdot 10^{-4/3} \sqrt[3]{\gamma_{Ca}^{2+} \cdot \gamma_F^2} \text{ in mol/L} \qquad (1.80d)$$

Equation 1.80d may now be solved iteratively using the initial value $\gamma = 1$:

According to Fig. 1.37	and Eq. 1.80d		Eq. 1.40	
γ_{Ca}^{2+}	γ_F	y	$I = 3y$	
1.00	1.00	$3.4 \cdot 10^{-4}$	$1.02 \cdot 10^{-3}$	
0.87	0.97	$3.7 \cdot 10^{-4}$	$1.09 \cdot 10^{-3}$	
0.86	0.97	$3.7 \cdot 10^{-4}$		

The final result is therefore given as

$$[CaF_2(aq)] = [Ca^{2+}] = 3.7 \cdot 10^{-4} \text{ mol/L}$$

and

$$[F^-] = 7.4 \cdot 10^{-4} \text{ mol/L}$$

Using mass concentrations, the solutes follow from

$$c_{CaF_2} = yM_{Ca} + 2yM_F = 3.7 \cdot 10^{-4}(40 + 2 \cdot 19)$$
$$= 28.8 \cdot 10^{-3} \text{ mg/L}$$

Considering the dissolution of less soluble substances, the solubility product K_S

$$K_S = [A^{n+}(aq)]^m[B^{m-}(aq)]^n \qquad (1.80e)$$

can be used to describe the process as $\gamma_{A^{n+}}$ and $\gamma_{B^{m-}}$ approach one.

It is important to keep in mind that K_S characterizes the dissolution only for γ-values near 1. For the example considered, Equation 1.80e results in

$$K_S = [Ca^{2+}][F^-]^2 = 1.61 \cdot 10^{-10} \text{ and with } 2[Ca^{2+}] = [F^-]$$

$$[Ca^{2+}] = \sqrt[3]{1.61 \cdot 10^{-10} / 4} = 3.4 \cdot 10^{-4} \text{ mol/L (compare}$$
$$\text{Equation 1.80d)}$$

$$[F^-] = 2 \cdot 3.4 \cdot 10^{-4} \text{ mol/L} \quad \text{and} \quad [CaF_2(aq)] = 3.4 \cdot 10^{-4} \text{ mol/L}$$

instead of the exact concentration $CaF_2(aq) = 3.7 \cdot 10^{-4}$ mol/L. A general expression of the saturation concentration or solubility is obtained from Equation 1.80e as

$$c_{sat} = [A_m B_n(aq)] = S = \sqrt[m+n]{K_s/m^m \cdot n^n)} \quad \text{in mol/L} \quad (1.80f)$$

$$[A^{n+}] = mc_{sat} \text{ and } [B^{m-}] = nc_{sat} \quad (1.80g)$$

Further examples are given in Table 1.16.

Table 1.16. Solubility of Selected Mineral Salts in Water at 25°C and p = 1 bar

Substance	Process	log K_s^*	S at pH = 7 mg/L
Halite	$NaCl \rightleftarrows Na^+ \; Cl^-$	+ 1.6	360,000
Sylvite	$KCl \rightleftarrows K^+ + Cl^-$	+ 0.9	264,000
SiO$_2$, amorphous	$SiO_2 + H_2O \rightleftarrows Si(OH)_4$	− 2.7	120
Quartz	$SiO_2 + H_2O \rightleftarrows Si(OH)_4$	− 3.7	12
Gypsum	$CaSO_4 \cdot 2H_2O \rightleftarrows Ca^{2+} + SO_4^{2-} + 2H_2O$	− 4.5	2,100
Calcite	$CaCO_3 \rightleftarrows Ca^{2+} \; CO_3^{2-}$	− 8.4	100[a], 500[b]
Fluorite	$CaF_2 \rightleftarrows Ca^{2+} + 2F^-$	− 9.8	160
Dolomite	$CaMg(CO_3)_2 \rightleftarrows Ca^{2+} + Mg^{2+} + 2CO_3^{2-}$	−17.0	90[a], 480[b]

Source: Freeze and Cherry [1.27].
[a]p_{CO_2} = 0.001 bar.
[b]p_{CO_2} = 0.1 bar.

In addition to electrolytic dissociation, solubility is also influenced by other reactions such as complex formation and hydrolysis. Frequently, the presence of other ions than the ions considered in a specific dissolution process may also impact their solubility. The presence of such foreign ions increases the ionic strength in the solution (see Equation 1.40), and hence lower the activity constants γ of the considered dissolved ions. If, for instance, in addition to $CaSO_4$

$$K^*_{CaSO_4} = \gamma_{Ca^{2+}} \cdot \gamma_{SO_4^{2-}} [Ca^{2+}][SO_4^{2-}] = 10^{-4.5} = const.$$

(see Equation 1.80a and Table 1.16)

NaCl is also dissolved, the foreign ions Na^+ and Cl^- lower the ionic strength I (Equation 1.40) and consequently the $\gamma_{Ca^{2+}}$ and $\gamma_{SO_4^{2-}}$. Hence, the concentrations $[Ca^{2+}]$ and $[SO_4^{2-}]$ must increase in order to maintain $K_S^* = 10^{-4.5}$. This effect, called the *ionic strength effect,* is illustrated in Figure 1.45.

If, instead of NaCl, the salt CaF_2 is added, the concentration of Ca^{2+} as one of the considered ions in the dissolution process increases in the solution, and the solubility of gypsum decreases. This effect is called the *common ion effect.* Similar to the previous example, the three unknown concentrations $[Ca^{2+}]$, $[SO_4^{2-}]$, and $[F^-]$ may be determined from the following three equations:

$$K^*_{CaSO_4} = \gamma_{CA^{2+}} \cdot \gamma_{SO_4^{2-}} [Ca^{2+}][SO_4^{2-}] = 10^{-4.5} \qquad (1.81a)$$

$$K^*_{CaF_2} = \gamma_{CA^{2+}} \cdot \gamma_{F^-} [Ca^{2+}][F^-]^2 = 10^{-9.8} \qquad (1.81b)$$

$$2[Ca^{2+}] = 2[SO_4^{2-}] + [F^-] \text{ (charge balance)} \qquad (1.81c)$$

If thermodynamic equilibrium does not exist, the quotient of Equation 1.80a, the saturation index, differs from unity:

$$SI = \{A^{n+}(aq)\}^m \{B^{m-}(aq)\}^n / K_S^* \qquad (1.82)$$

where SI > 1 corresponds to a supersaturated solution, SI < 1 to an unsaturated solution, and SI = 1 to a saturated solution.

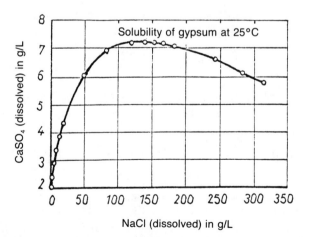

Figure 1.45. Solubility of gypsum in aqueous solution at different concentrations of NaCl. *Source:* Freeze and Cherry [1.27].

The quantity $|1\text{-SI}|$ is frequently considered as the driving force of dissolution/precipitation processes. Consequently, according to Equation 1.46a, the reaction rate results in

$$f_{solution} = k_S(1 - SI); \quad f_{precipitation} = k_P(SI - 1) \qquad (1.83)$$

In dissolving oxides and hydroxides in soil- and groundwater, pH is of particular importance.

Example:

The dissolution of metal hydroxides in acidic aqueous solutions may be described by

$$Me(OH)_n(c) \quad \rightleftarrows \quad Me^{n+} + nOH^- \quad \log K_s^* \qquad (1.84a)$$

$$nH^+ + nOH^- \quad \rightleftarrows \quad nH_2O \quad -n \log K_W \qquad (1.84b)$$

$$\Sigma: Me(OH)_n(c) + nH^+ \rightleftarrows Me^{n+} + n H_2O \quad \log K \qquad (1.84c)$$

For $\tau_i = 1$, Equation 1.84c is given in the general form (see also Equation 1.45a):

$$\log K = \log K_S^* -n \log K_W = \log [Me^{n+}] - n \log [H^+]$$

$$\log [Me^{n+}] = \log K_S^* -n \log K_W - npH \qquad (1.84d)$$

Hence, the dissolution of $Fe(OH)_2(c)$ in the acidic milieu is given at thermodynamic equilibrium between the solid and aqueous mixphase by

$$\log [Fe^{2+}]_{aq} = -14.7 - 2(-14) - 2 \text{ pH} = 13.3 - 2 \text{ pH}$$

A change of pH from 8 to 7, therefore, results in an increase of $[Fe^{2+}]_{aq}$ from 2 mmol/L to 200 mmol/L; that is, per pH unit the concentration increases by a factor 100. For trivalent metal ions, the multiplication factor of concentration change is 1000 per pH unit, and for monovalent ions 10. Equation 1.84d is plotted for several metals in Figure 1.46.

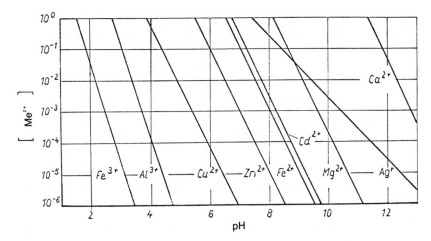

Figure 1.46. Concentration of dissolved metal ions at equilibrium state with solid oxides or hydroxides. *Source:* Stumm and Morgan [1.83].

In precipitating oxides, hydroxides, and oxidohydrates, microcomponents such as the heavy metals Cu, Co, Ni, Pb, and Zn are attached onto the microflocs' surface or are enclosed in the interior of these associates. Such a form of precipitation is called *coprecipitation*.

Contrary to real and potential electrolytes, nonelectrolytes do not dissociate into cations and anions in aqueous solutions. They dissolve as neutral species. Nonelectrolytes are called *polar* when their asymmetric charge distribution generates dipoles (see, for example, Figure 1.2c). Nonelectrolytes usually have functional groups such as hydroxyl, carboxyl, or primary amino groups [1.64], which associate with H_2O molecules by hydrogen bridging (see §1.4.2). The solubility of polar nonelectrolytes in water is therefore high. Typical polar nonelectrolytes are, for instance, glucose, alcohols, amines, phenols, aldehydes, and halogenated hydrocarbons.

During the solution process the nonpolar nonelectrolytes are preferably built into the interstitial voids of the water molecules [1.64] (see Figure 1.6 and 1.7). Their solubility in water is therefore relatively low. The solubility of nonpolar substances in water (e.g., petroleum products) is characterized by their substance/water ratio. In Reissig [1.64] the following data are given:

- Benzene   1700 mg/L
- Light gasoline   60 mg/L
- Gasoline   50 . . . 500 mg/L
- Fuel oil   10 . . . 50 mg/L
- Petroleum   0.1 . . . 5 mg/L
- Diesel fuel   10 . . . 50 mg/L

Finally, the dissolution and release of gaseous species i, (i.e., the exchange process between the gaseous and the aqueous mixphase) is considered:

$$i(aq) \rightleftarrows i(g) \qquad (1.85)$$

The driving force of this process is given as $F_{Ch} = \mu_i(g) - \mu_i(aq)$ according to Equation 1.42b for $\nu_i = 1$. Hence, the substance exchange rate in mol/(sm_V^3) follows according to Equation 1.46a by

$$f_{ex} = k_{ex}\Delta\mu = k_{ex}\Delta_r g \qquad (1.86a)$$

According to Equations 1.85, 1.45a, and 1.45b, the thermodynamic equilibrium state is attained at

$$K = \{i(g)\}/\{i(aq)\} = \exp(-\Delta_r g^\circ/(RT)) \qquad (1.86b)$$

Hence, with Equation 1.39, it follows at standard conditions:

$$p_i/p = \{i(aq)\} \exp(-\Delta_r g^\circ/(RT)) \qquad (1.87a)$$

and for pure water or other pure liquids with the activity $\{i(liq)\} = 1$:

$$p_i/p = x_i = \exp(-\Delta_r g^\circ/(RT)) \qquad (1.87b)$$

Consequently, the dissolved amount of gas in g/L results in

$$\lambda = x_i\rho_i = (p_i/p)\rho_i = \rho_i \exp(-\Delta_r g^\circ/(RT)) \qquad (1.88a)$$

$$\lambda = x_iK_i \text{ or } \lambda = p_iK_p \qquad (1.88b)$$

Equation 1.88b is called the *Henry-Dalton law*. It does not apply to gases chemically reacting with water such as NH_3 (see the example according to Equation 1.57 and 1.58). But for CO_2 the deviation can be neglected. Thus, for example, the water vapor pressure at 25°C and $p = 1$ bar is given according to Equation 1.87b by (see also Equation 1.43b, Table 1.1)

$$\Delta_r g^\circ = \mu_{H_2O}^\circ(g) - \mu_{H_2O}^\circ(l) = -228.77 - (-237.35) = 8.58 \text{ kJ/mol}$$

$$p_{H_2O} = 1 \text{ bar} \exp\left(\frac{-8.58 \text{ kJ/mol}}{2.479 \text{ kJ/mol}}\right) = 0.0314 \text{ bar} = 3140 \text{ Pa}$$

The solubility of gases in water decreases with increasing temperature. Solubility data for several gases at 0°C are given in Table 1.17; solubility data for the common constituents of air are presented in Table 1.18.

The solubility of the carbonates $CaCO_3$ and $CaMg(CO_3)_2$ strongly depends on the amount of CO_2 dissolved in water. At constant partial pressure p_{CO_2}, the solubility of CO_2 decreases with increasing tempera-

Table 1.17. Solubility of Several Gases at 0°C and at a Partial Pressure of 1.013 bar in Water

Gas	mL/L	mg/kg	Gas	mL/L	mg/kg
He	9.70	1.70	Ar	57.8	100.7
H_2	21.48	1.92	NH_3	1300	1000
N_2	23.59	28.80	CO_2	1713	3346
O_2	49.22	69.45	H_2S	4690	7100
CH_4	55.63	39.59			

Source: Matthess [1.55].

ture; therefore, the solubility of carbonate minerals also decreases with increasing temperature, in contrast to most other minerals (see Table 1.15).

Consider the example of the carbonate–carbonic acid equilibrium, which at a given partial pressure of CO_2 or x_{CO_2} is described by the following system of six equations with the six unknown concentrations $[CO_2 \cdot H_2O] = [H_2CO_3]$, $[HCO_3^-]$, $[CO_3^{2-}]$, $[Ca^{2+}]$, $[H^+]$, and $[OH^-]$:

$$CO_2(g) + H_2O \rightleftarrows H_2CO_3(aq) \qquad K_1 = \{H_2CO_3\}/x_{CO_2}$$
$$H_2CO_3 \rightleftarrows HCO_3^- + H^+ \qquad K_2 = \{H^+\}\{HCO_3^-\}/\{H_2CO_3\}$$
$$HCO_3^- \rightleftarrows H^+ + CO_3^{2-} \qquad K_2 = \{H^+\}\{CO_3^{2-}\}/\{HCO_3^-\}$$
$$CaCO_3 \rightleftarrows Ca^{2+} + CO_3^{2-} \qquad K_4 = \{Ca^{2+}\}\{CO_3^{2-}\}/1$$
$$H_2O \rightleftarrows H^+ + OH^- \qquad K_5 = \{H^+\}\{OH^-\}/1 = K_w$$
$$[H^+] + 2[Ca^{2+}] = [HCO_3^-] + 2[CO_3^{2-}] + [OH^-] \qquad (1.89)$$

The first four equations allow the following statement for pH:

$$\{H^+\} = (K_1 K_2 K_3/K_4)\, x_{CO_2}\, \{Ca^{2+}\})^{1/2} \qquad (1.90)$$

As with the previous examples, the equilibrium constants $K_1, \ldots, K_5$ are calculated according to Equation 1.45b. A compilation of equilibrium constants K at different temperatures is given in Table 1.19. Commonly, the system of Equations 1.89 is solved by computer methods. A simple iterative procedure which avoids the use of computers is described in Freeze and Cherry [1.27].

Table 1.18. Solubility of Air Constituents in Water at Their Natural Atmospheric Partial Pressure in mg/L

Gas	0°C	5°C	10°C	15°C	20°C	25°C
O_2	14.46	12.68	11.24	10.10	9.18	8.38
N_2	22.88	20.25	18.09	16.37	15.10	14.11
CO_2	1.00	0.83	0.69	0.59	0.51	0.44

Source: Rösler and Lange [1.67]

Table 1.19. Equilibrium Constants for the Determination of the Carbonate/Acid Equilibrium in Pure Water for p = 1 bar

°C	log K_{CO_2}	log $K_{H_2CO_3}$	log $K_{HCO_3^-}$	log K_{CaCO_3}	log K_{H_2O}
0	−1.12	−6.58	−10.62	−8.340	−14.94
5	−1.20	−6.52	−10.56	−8.345	−14.73
10	−1.27	−6.47	−10.49	−8.355	−14.53
15	−1.34	−6.42	−10.43	−8.370	−14.35
20	−1.41	−6.38	−10.38	−8.385	−14.17
25	−1.47	−6.35	−10.33	−8.400	−14.00

Source: Freeze and Cherry [1.27].

A portion of the dissolved ions may react with water molecules to form undissociated acids or undissociated bases. This process is called *hydrolysis* when occurring with water, and *solvolysis* when occurring with other solvents. Consequently, aqueous solutions resulting from such processes are no longer neutral. The basic hydrolytical process equations are given by

$$Me^+ + H_2O \rightleftarrows MeOH + H^+ \quad \text{(cation hydrolysis)} \quad (1.91a)$$

$$A^- + H_2O \rightleftarrows HA + OH^- \quad \text{(anion hydrolysis)} \quad (1.91b)$$

In general, anions and cations of strong acids and bases do not undergo hydrolysis. Their salts yield neutral aqueous solutions. However, this does not hold for other types of salts:

- salts of weak acids and strong bases which undergo anion hydrolysis, e.g., $C_2H_3O_2^- + H_2O \rightleftarrows HC_2H_3 + OH^- + O_2$
- salts of strong acids and weak bases which undergo cation hydrolysis, e.g., $NH_4^+ + H_2O \rightleftarrows H_3O^+ + NH_3$
- salts of weak acids and weak bases which undergo both cation and anion hydrolysis, e.g., $NH_4C_2H_3O_2$ or NH_4CHO_2
- salts of polyprotic acids which undergo polyprotic hydrolysis, e.g., Na_2S

1.5.4 BIOLOGICAL METABOLISM OF MIGRANTS

Migrants in the subsurface are also exchanged between living and dead matter and are thereby transformed. This exchange process can be reflected by two subprocesses: immobilization $\rightleftarrows$ mineralization. Immobilization or fixation reflects the incorporation of inorganic substances such as NH_4^+ or NO_3^- into proteins, nucleic acids, and other organic matter (N fixation), while mineralization describes the conversion of organic matter into inorganic substances such as CO_2, H_2O, and NH_4^+, and into relatively stable, high-molecular weight substances including humic acid and fulvic acid. Both mineralization and fixation are caused by microbial energetic and structural (biosynthetic) metabolisms.

This section presents an ecological approach (§1.5.4.1), which emphasizes vital activity (§1.5.4.2), and substance and energy conversion (§1.5.4.3) as inseparable entities. Finally, some examples of microbial substance exchange and transformation are shown (§1.5.4.4).

1.5.4.1 Ecological Fundamentals

The distribution of organisms in the biosphere is not accidental. Particular constellations of environmental factors determine their spatial and temporal pattern of occurrence. In characterizing such a pattern, the terms *living space, biotope, location,* and *habitat* are often used as synonyms. Usually they indicate a limited space in which an individual organism lives, including all environmental factors.

Environmental impacts are divided into abiotic (physical and chemical) and biotic factors. Depending upon the constellation of the environmental factors, organisms will occupy certain living spaces (ecological niches) and form biological communities (biocenoses). Within a biocenosis there are different populations (systems of individuals of one species). The spatial and temporal integration of vital activities and environment, as well as their functional unity, constitutes an ecosystem.

The intensity with which living organisms manifest themselves is closely related to a distinct pattern of environmental factors. The functional dependence reflects the reaction of an individual or a population to variations of environmental factors and is characterized by an optimum (preferendum), a minimum, and a maximum (pessimum). This tolerance band of a species in pure culture is called its *physiological amplitude,* and when in association with other species, its *ecological amplitude.* A divergence between these two amplitudes is an expression of rival behavior at a particular location.

The following matrix of interactions is often used in characterizing the rivalry between two microbial populations:

Effects	j on i		
	+ +	+0	+ -
i on j	0+	00	0-
	-+	-0	- -

where   00 represents neutralism
            0+ and +0 represent commensalism (one benefits from the other)
            + + represents mutualism (mutual benefit)
            + - and - + represent parasitism

-0 and 0- represent amensalism (antithesis of commensalism)
- - represents competition

The development of biological communities is controlled by an integral set of environmental conditions, but may also be controlled by a single limiting factor, the intensity of which is above or below a specific threshold (Liebig's minimum law).

Within a functioning ecosystem, a state of approximate equilibrium between constructive and destructive forces is maintained. This dynamic state is called *flow equilibrium*. The term *ecological equilibrium* is used to characterize the structural and dynamic equilibrium of biological communities by continuous self-regulation, self-organization, and self-reproduction under given environmental conditions [1.12; 1.82]. The stability of an ecological system, expressed as its ability to tolerate changes of the environmental factors, depends on its variety and complexity and of the exchange of material and energy with its environment.

1.5.4.2 The Subsurface as a Living Space

Ecosystems are characterized by analyzing their environmental factors, the biological communities, and their interactions. The limiting, controlling, and impacting abiotic environmental factors may be divided into

- climatic factors (light, temperature, moisture)
- edaphic factors (soil conditions)
- chemical factors (electron donors, electron acceptors, nutrients, trace substances, toxic substances) [1.12; 1.82]

Environmental factors

Climatic factors

The depth to which light penetrates into the subsurface is minute. Thus, subsurface ecosystems have no chance for photoautotrophic production. Permanent darkness leads to the development of a blind, pigmentless fauna having no daily or seasonal rhythm [1.17]. The relatively constant low temperature of the groundwater zone in the middle northern latitudes produces a groundwater fauna of cryophilic organisms with a small temperature range (cold stenothermic organisms). They require a liquid environment for their movement as well as for their supply of nutrients and electron donors and acceptors. While algae and anaerobes are favored by abundant moisture, fungi and actinomycetes predominate in dry soils.

Edaphic factors

The properties of soil and rock are important factors for settlement of organisms. The size and shape of cavities limit the size, form, and locomotion of subsurface organisms. Flowing soil- and groundwater, the transport medium, is essential to supply organisms with nutrients and electron donors and acceptors. The mineral content of soils and rocks may support the organism's activity.

Chemical factors

Most microorganisms and all multicellular animals require oxygen as an efficient electron acceptor for maintaining their vital activities; that is, they need an environment characterized by positive Eh values. These organisms constitute the aerobic biota. On the other hand, anaerobic and facultative microbes do not require molecular oxygen as an electron acceptor.

Heterotrophic microbes derive their energy from the catabolism or degradation of hydrocarbons, while autotrophic microbes use reduced inorganic substances as an energy source or electron donor. The concentration of O_2 in subsurface water decreases with depth and may become zero if groundwater is loaded with efficient electron donors such as organic hydrocarbons. A sufficient supply of O_2 in groundwater is only ensured by recharge from infiltrating precipitation, irrigation, and surface water, as well as by molecular diffusion via the soil air [1.45].

The quality and quantity of nutrients supplied to a biological community play an important role in composition and activity of biocenoses [1.17]. Microbes require both macronutrients (C, H, O, N, P, S, K, Mg, and Na) and micronutrients (Ca, Fe, Mn, Cu, Zn, and others). For nutrients, simple organic carbon compounds, particularly simple sugars, and inorganic nitrogen (i.e., NH_4^+ and NO_3^-) are essential. When carbon and nitrogen are combined in large molecules such as lignin, cellulose, and proteins, they are not readily available for microbial metabolism. Trace metals such as copper and zinc are also required. They serve as cofactors in enzymes.

In groundwater which is anthropogenically unaffected, the subsurface fauna is mostly independent of the ionic composition. In water with a high salt or iron content, however, biocenoses have fewer species. Extreme anthropogenic pollution leads to a species impoverishment of the biocenosis or to the dying out of all of their organisms. Various species have a well-defined pH range in which they are active, and are most active at a well-defined optimum pH. Accordingly, they can be classified into acidophilic, neutrophilic, and alkalophilic microbes.

Toxic effects are caused by an excess of inorganic ions such as Zn or

Mn, by organic toxins such as antibiotics and pesticides, and by leachate from industrial wastes. Ammonia is also toxic to living cells, which use it as an energy source, including bacteria such as *Nitrobacter.*

Biological communities

The biological communities in the subsurface are determined by microorganisms and animalcules.

Microorganisms

The main microorganisms in the subsurface are bacteria, actinomycetes, and fungi including molds and yeasts (see Figure 1.47). The single cellular geomicrobes (protists) may be classified as

A.   Procaryotes
    1.   Bacteria (heterotrophic and autotrophic; aerobic, facultative and anaerobic; mostly mobile)
    2.   Actinomycetes or filamentous bacteria (heterotrophic; aerobic; nonmotile)
    3.   Blue-green bacteria, formerly referred to as algae (capable of photosynthesis)
B.   Eucaryotes
    1.   Fungi (including yeasts)
    2.   Algae (capable of photosynthesis)
    3.   Protozoa

The dimensions of geomicrobes vary from one nanometer in diameter, to several centimeters in length for fungal mycelia. Fungi, actinomycetes, and algae have many distinguishing morphological characteristics, whereas bacteria are less distinctive in structure. They are divided into two major taxonomic groups, the Gram positives and the Gram negatives.

Geomicrobes may be suspended in water, but most of them form biofilms covering the surfaces of solid particles. Microorganisms may be passively transported by soil- and groundwater, or they may propel themselves, resulting in hydrobiomigration ranging from a few mm up to a few cm per day.

Animalcules

Groundwater animalcules are thin, pigmentless, and blind. More than one hundred species have been identified in central European aquifers.

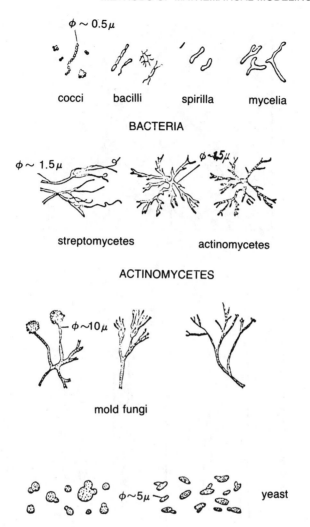

$\phi \sim 0.5\mu$

cocci   bacilli   spirilla   mycelia

BACTERIA

$\phi \sim 1.5\mu$ $\phi \sim 1.5\mu$

streptomycetes   actinomycetes

ACTINOMYCETES

$\phi \sim 10\mu$

mold fungi

$\phi \sim 5\mu$ yeast

FUNGI

Figure 1.47. Important microorganisms in the subsurface. *Sources:* Hölting [1.35], Uhlmann [1.88], and Davis [1.19].

These include crustacea (copepods, amphipods, octrapods), worms (eddyworms, bristle worms, threadworms, wheel animalcules), web spinners (mites), as well as tardigrades and ciliata (see Figure 1.48, Uhlmann [1.88], and Husmann [1.39]).

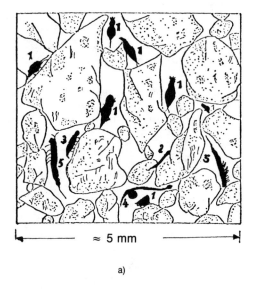

≈ 5 mm

a)

order of diameter 0.1 mm

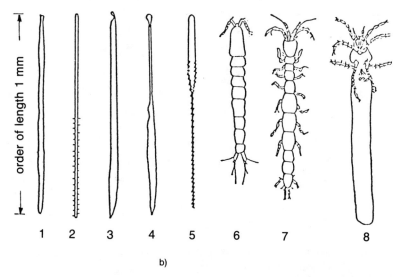

order of length 1 mm

1   2   3   4   5   6   7   8

b)

Figure 1.48. Groundwater animalcules. *(a)* Schematic section showing life forms in the subsurface, where bacteria are represented by points on the solid particles, and typical animalcules are labeled by numbers: *1,* rotifera; *2,* large ciliata; *3,* tardigrades; *4,* threadworms (nematodes); *5,* copepods (harpacticoids). *(b)* Examples of worm-shaped groundwater animalcules: *1,* eddyworm (turbellaria); *2,* bristle worm (oligochaeta); *3* and *4,* ciliata; *5,* gastrotrich; *6,* copepod; *7,* water hog louse; *8,* water mite. *Source:* Uhlmann [1.88].

1.5.4.3 Biogenic Substance and Energy Conversion

Biosynthetic metabolism and energy metabolism are inseparable; they must proceed simultaneously. The process of biomass production, continuous replacement of biomass, and transport of internal substances, as well as locomotion or biomigration, requires a continuous supply of free energy. This requirement may be met either by physical sources (radiation-driven photosynthesis) or by chemical sources (redox processes). Organisms transform this radiation or chemoenergy into the biologically useful compound adenosine triphosphate (ATP), the universal energy carrier of living matter.

Basic structure

Within an ecological system, the simultaneous conversion of biomass and energy takes place on three biogenic levels. These are the levels of producers, consumers, and decomposers. The following activities are characteristic of these three groups (see Figure 1.49):

- Producers make organic matter out of CO_2 and other inorganic substances by using radiation energy, called *photoautotrophic production,*

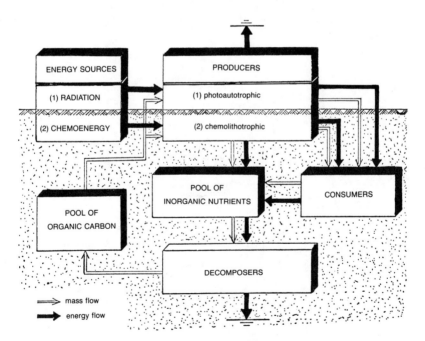

Figure 1.49. Material and energy flow in a hypothetical groundwater ecosystem. The system is open and connected with surface ecosystems by energy and material exchange.

or chemical energy, called *chemolithotrophic production*. The producers are carbon autotrophic; i.e., they can directly assimilate CO_2 to cover their C demand for their biosynthetic metabolism. Hence, only the chemolithotrophic bacteria act as subsurface producers. However, the productivity of chemolithotrophic bacteria is low, and the substance conversion rates of their constructive and energetic metabolism are relatively slow.

- Consumers attack organic substrates of living organisms. If they rely on living hosts they are called *parasites*. Consumers may be microorganisms or animalcules.

- Decomposers (saprophytes) attack organic matter of the environment and break it down. Only microorganisms are decomposers, and not animalcules. Both consumers and decomposers meet their energy demand through the catabolism of organic substrates. They are also carbon heterotrophic in their constructive metabolism; i.e., they take carbon out of organic sources for production of their own biomass.

As a rough generalization of the subsurface mass flow of biosynthesis, one may assume that

- about 20 to 60% of the organic substrate is used for biomass production
- about 40 to 80% is "burned" (respired) for energy production
- about 10 to 30% of the total substrates is nonutilizable, i.e., waste (see Figure 1.50)

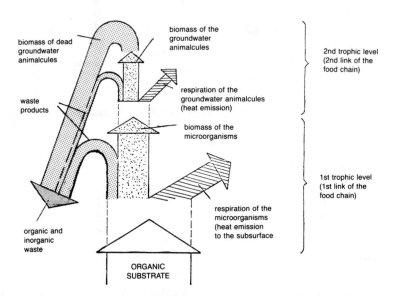

Figure 1.50. Diagram of a two-level food chain in the soil- and groundwater zone. *Source:* Uhlmann [1.88].

Therefore, out of 1 g of organic matter, only 0.2 to 0.6 g biomass can be produced; the other part is "lost" in the form of heat and waste [1.12].

Energy Production

Photosynthetic activity in the subsurface is insignificant. Instead the redox processes, which yield energy by transferring electrons from a donor (reductant) to an acceptor (oxidant), are important (see §1.4.4):

Energy gain (respiration)

Electron donor	—— e^- ——>	*Electron acceptor*
* *Hydrogen*		* *Oxygen*
■ simple organic compounds		Aerobic respiration
■ proteins, fats, . . .		* *Nitrate* NO_3^-
—living organic matter		Nitrate respiration
—dead organic matter		(denitrification)
* *Reduced substances*		* *Sulfate* SO_4^{2-}
such as H_2S, FeS_2, $S°$, NH_4^+,		Sulfate respiration
NO_2^-, Fe^{++}, Mn^{++}, etc.		* *other oxidized substances*

The energy yield or electron flux is high if both the electron donor and the electron acceptor are effective and available in sufficient amounts. An efficient donor transferring electrons to an inefficient acceptor causes an electron blockage and a high electron activity, and therefore a low pE, or a negative Eh as discussed in §1.4.4. Ideal living conditions for microbes are characterized by high redox potentials or low electron activities, which require an efficient electron acceptor and a sufficient rate of electron transfer.

The most effective "fuel" (electron donor) for biological energy production is the hydrogen of carbohydrates, which is usually present as $C_6H_{12}O_6$ (glucose), other carbohydrates (CH_2O_x sugars), or hydrogen H_2 itself. Table 1.12 shows a sequence of redox processes with such electron donors and with oxygen (O_2), nitrate (NO_3^-), sulfate (SO_4^{2-}), and other oxidizing agents acting as electron acceptors. According to the electron acceptor involved, these energy-yielding heterotrophic processes may be designated as

- aerobic respiration (respiration)
- nitrate respiration (denitrification)
- sulfate respiration (sulfate reduction)

The electron acceptor which is most biologically suitable is molecular oxygen O_2. Table 1.13 shows a sequence of redox processes with O_2 dissolved in subsurface water as the electron acceptor and carbohydrate (CH_2O), HS^-, Fe^{++}, NH_4^+, and Mn^{++} as electron donors. Other combinations of electron donors and electron acceptors are also possible, provided the energy yield is sufficient to meet the microbes' requirements. Thus, Fe^{++} can serve as an electron donor and NO_3^- as an acceptor to form a combination which is often used in subsurface water treatment for drinking water supply.

The energy yield is highest in the case of aerobic respiration, since both the donor and the acceptor are efficient, but it is nevertheless only about 5% of the energy yielded by methane fermentation (see Table 1.12). Such low outputs result from energy consumption during the generation of the reduced substances N_2, Mn^{2+}, Fe^{2+}, and HS^-, produced by reduction of NO_3^-, Mn^{4+}, Fe^{3+}, SO_4^{2-}, etc. (see Table 1.12). However, the O_2 demand of aerobic respiration is high (see Table 1.20), and thus O_2 which is dissolved in groundwater may be consumed quickly.

Table 1.20. Specific Oxygen Demand for Selected Substrates

Substrate	g O_2/g substr.	Substrate	g O_2/g substr.
fat	2.85	starch	1.18
phenol	2.38	cellulose	1.18
lignin	1.89	glucose	1.06
humic substances	1.5	acetic acid	0.94
protein	1.46	oxalic acid	0.18

Source: Uhlmann [1.88].

If endogenous biomass is "burned" instead of external substrate such as carbohydrates or exogenous biomass, the energy yielding process is called *endogenous respiration*. Pure endogenous respiration takes place only if no inorganic nutrients are available. The following equation characterizes this process:

$$C_{106}H_{180}O_{45}N_{16}P_1 + 718.5\ O_2 \rightarrow$$
$$106CO_2 + 66H_2O + 16NH_3 + PO_4^{3-} \qquad (1.92)$$

This equation describes the complete mineralization of biomass. Not all of the organic matter in the soil- and groundwater zone can be completely mineralized. A part remains stabilized, for example, as humic substances which are biologically difficult or impossible to decompose.

Biomass production

The process by which nutrients are changed into living tissue is known as *assimilation*. It is a constructive metabolism, i.e., a biosynthetic process. The basic stoichiometric formula for biomass is given by $C_{106}H_{180}O_{45}N_{16}P_1$.

- The carbon in biomass comes from CO_2. Chemolithotrophic microbes such as sulfur, nitrogen, and iron bacteria are able to assimilate C into organic compounds:

- —from carbon dioxide and water, a process termed *chemosynthesis*
- —from imported carbohydrates because there is no photosynthetic glucose production in the subsurface
- —from organic matter decomposed by heterotrophic microbes
- The hydrogen in biomass originates from hydrocarbons, or in the case of CO_2 conversion from water, a process termed *chemosynthesis.*
- The oxygen in biomass comes from O_2 after electron acceptance and from other compounds, e.g., PO_4^{3-}.
- The nitrogen of biomass originates from NH_3, NH_4^+, and NO_3^-.
- The phosphorus in biomass comes from orthophosphate, PO_4^{3-}.

Nutrients for microbes should contain C, H, O, N, and P according to the composition of biomass, i.e., in a mole relation of about 106:180:45:16:1.

Production of biomass by microorganisms is tightly related to the population increase. Typical biomass production curves have four time phases:

1. a lag phase (necessary adaptation)
2. a logarithmic phase (geometrical or exponential growth)
3. a stationary phase (equilibrium between growth and dying rate)
4. a death phase with decreasing population

If a special substrate, for instance a special pesticide, is repeatedly applied to a soil, the lag phase is reduced.

Kinetics of bioconversion

Bioconversion of matter involves enzymes. Enzymes are proteins produced by living cells. They are highly specific; i.e., they usually catalyze only one type of reaction, giving the lock-and-key hypothesis of enzyme catalyzation. Enzymes are classified according to the type of reaction they catalyze, so that oxidoreductases catalyze redox processes, hydrolases catalyze hydrolysis, and transferases catalyze the transfer of molecule clusters. Enzymes are not consumed in reactions and are needed only in minute quantities. Often they require additional small amounts of cofactors, usually trace metals, to perform their catalytic function. The simplest case of a single substrate going to a single product is given by the following model:

$$\text{Enzyme + Substrate} \rightleftarrows \text{Enzyme Substrate Complex} \rightleftarrows$$
$$\text{Enzyme + Product}$$

where the activated enzyme substrate complex has a higher free energy than the reactants (see Figure 1.39). The Michaelis-Menten kinetic model is the simplest model of an enzyme catalyzed reaction, where the enzyme concentration is thought to be constant. For estimating the reaction rate r, a limiting rate r_{max}, the concentration of substrate [S], and the Michaelis-Menten constant $[k_m]$ must be known (compare also the model Equation 1.46g)

$$r = \frac{r_{max} [S]}{[S] + [k_m]} \tag{1.93}$$

Inhibition of enzyme activity is one of the major self-regulatory tools of living cells.

The kinetics of conversion of biological substance is reflected in its simplest form by a first order reaction model, which follows from Equation 1.93 for $[S] << [k_m]$:

• degradation rate of the substrate [S]

$$d[S]/dt = k[S] \tag{1.94}$$

More sophisticated examples of models are given in the following:

• biomass production rate, $d[M]/dt$

$$d[M]/dt = \mu[M] \tag{1.95a}$$

or with an inhibiting term INH

$$d[M]/dt = \mu[M] \cdot INH = \mu[M]([M_m] - [M])/ [M_m] \tag{1.95b}$$

• development of biofilms

$$d[M]/dt = \mu[M] - k[M] \tag{1.95c}$$

where [M] is the biomass concentration, μ is the growth rate, and k is the dying or loss rate factor. This factor k depends upon the hydraulic friction, the disaggregation of the biofilm caused by dying or dissolution processes, animalcule activities, and gas bubbles caused by endogenous respiration of bacteria and fungi [1.88].

Further models may be derived, and by combining these basic models, models for the linked constructive and energetic metabolism may be developed:

$$\frac{d[S]}{dt} = -[M]k_{max} \left(\frac{[S] - [S_{min}]}{[k_s] + [S]} \right) \left(\frac{O}{[k_o] + [O]} \right) \quad \text{for } [O] > [O_{min}] \tag{1.96a}$$

$$\frac{d[O]}{dt} = f_1 d[S]/dt - f_2 k[M][O]/([k_o] + [O]) \quad \text{for } [O] > [O_{min}] \tag{1.96b}$$

$$\frac{d[M]}{dt} = -f_3 d[S]/dt - k[M] \quad \text{for } [O] > [O_{min}] \tag{1.96c}$$

where [S], [0] and [M] are the concentrations of substrate, oxygen, and biomass in g per m_v^3

k_{max} is the maximum substrate utilization rate per g of microorganisms in s^{-1}

$[k_s]$ and $[k_o]$ are the substrate and oxygen half-saturation constants,
i.e., concentration of S or O yielding 50% of the maximum decay rate
in g per m_v^3

$[S_{min}]$ and $[O_{min}]$ are thresholds, i.e., minimum concentrations of S and
O which are still usable in an aerobic biodegradation process

f_1 is the oxygen/substrate consumption ratio (see Table 1.20)

f_2 is the oxygen/biomass consumption ratio of dead cell material

f_3 is the yield coefficient of biomass production

k is the microbial decay rate in s^{-1}

For anoxic bioconversion processes, the oxygen terms are replaced by
those of the electron acceptor used, such as nitrate, sulfate, or others.

Assessment of Applicability

In general, chemoautotrophic bacteria should not be regarded as the
base of the subsurface food chain because their growth rates are relatively low. Nevertheless, they play an important role in self-purification
processes and cleanup procedures in the soil- and groundwater zone
[1.12].

Productive groundwater ecosystems generally require constant import
of organic matter and dissolved oxygen. Dissolved organic substances
are the main food sources for bacteria, while subsurface animalcules
preferentially consume particulate organic substances with adhered bacteria. The excrements of such animalcules are oxidized completely to CO_2
and H_2O by bacterial decomposers. The animalcules constantly remove a
large fraction of organic particles, such as detritus and bacteria which
obstruct flow channels, and thereby make room for new inflowing substances [1.40].

In summary, organisms' metabolism in the subsurface has the following impacts on migration processes:

- Inorganic substances are extracted from the soil- and groundwater by
  chemoautotrophic biomass production.
- Organic substances in water, including hydrocarbons, proteins, fats, and
  phenols, are decomposed or mineralized by respiration. Bacteria and
  fungi are primarily responsible for biochemical decomposition of dissolved substances, while subsurface animalcules and some bacteria eliminate particulate organic substances (see Figure 1.49). Thus, organic substances are extracted from the soil- and groundwater, but organic and
  inorganic components such as CO_2, NH_4^+, PO_4^{3-}, and SO_4^{2-} are produced.
  The biological formation of CO_2 primarily increases the solubility of
  carbonates (see §1.5.3).

- Organic acids, such as lactic, tartaric, butyric, and salicylic acids, are produced in these metabolic processes and lead to the formation of heavy metal complexes, which are considerably more mobile in the subsurface than the free heavy metal ions (see §1.4.2).
- Redox reactions can be catalyzed by enzymes produced by microorganisms. The rates of such reactions may be several orders of magnitude higher than the rates of purely chemical, nonenzymatic reactions. Microorganisms compete with these chemical reactions for the available nutrients (see §1.4.2).
- Suspended particles are effectively retained due to the sieving effect of biofilms (see §1.5.2).
- The "grazing" activity of groundwater animalcules causes the hydraulic permeability to tend toward a dynamic equilibrium state and to prevent total clogging of pore channels.
- Biofilms also have a high cation exchange capacity as a result of the generally negative charge of microorganisms' cell walls (see §1.5.2).

1.6

Storage Process Models

Each mixphase of the multiphase system subsurface is capable of storing migrants. The fundamental nonhysteretic systems state function, Equation 1.1, links the extensive state variable storage with the intensive state variables temperature, pressure, and composition, where the bracket terms in Equation 1.1b represent storage capacities. The graphic symbol of a lumped storage element is shown in Figure 1.14 and next to Equation 1.97, where the flux arrows symbolize the change of stored substance or heat.

In the simplest case only, one intensive state variable has a significant effect on storage (i.e., stored energy, ions, or others). In solute migration this occurs in the ideal case of isothermal and isobaric conditions when composition changes are caused by one component only. Figure 1.51 shows some important practical examples where the systems state function is governed by a single intensive state variable. Storage function plots are generally characterized by two kinds of relatively linear segments: buffer segments with a high storage capacity and steep slope, where $s = f(P)$ is poorly defined, and normal segments with a well-defined functional relationship of $s = f(P)$. Buffer segments usually reflect conversion processes such as evaporation/condensation, freezing/thawing, dissolution/precipitation, ion exchange, and others (discussed in §1.4.3., §1.4.4., §1.5.2., §1.5.3., §1.5.4., and in Chapter 1.7).

For normal segments, Equations 1.1a and 1.1b yield in their simplest form:

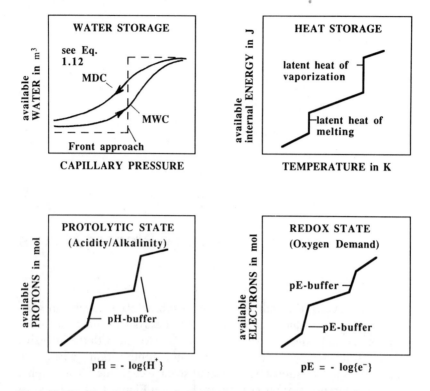

Figure 1.51. Storage functions S = f(P) caused by addition or extraction of water, heat, protons, or electrons to or from a considered REV or lumped system. One point of the storage function S_o = f(P_o) serves as the arbitrary reference point. Addition or extraction of protons proceeds by adding H^+-donors (e.g., strong acids) or H^+-acceptors (e.g., strong bases). Addition or extraction of electrons proceeds by adding an electron donor (e.g., CH_2O) or an electron acceptor (e.g., O_2).

$$s = ca \cdot P \tag{1.97a}$$

$$ds/dt = ca \cdot dP/dt \tag{1.97b}$$

where   s = stored substance or heat per unit volume
ca = storage coefficient or capacity per unit volume
ds/dt = storage rate per unit volume
P = effective intensive state variable of the considered process (e.g., temperature, activity, fugacity)

With $P_i = \{i\}$, the specific stored amount of migrant i in the subsurface can be written according to Equation 1.97a (in mole or g per m^3 subsurface volume) as

$$s_i = \sum_j (ca_j \{i\}_j) \qquad (1.98)$$

where $\{i\}_j$ is the activity of the migrant i in the mixphase j, and ca_j the storage coefficient resulting from the volumetric content of the mixphase j in the subsurface multiplied by the storage density of this phase. This basic equation should be now considered in more detail.

For mobile fluid phases only, the terms of Equation 1.98 are relatively easy to describe using mathematical models; thus, for instance,

for mobile water

$$(ca \cdot \{i\})_{w,m} = \sigma_{w,m} (1/\gamma)\{i\}_{w,m} = \theta_{w,m}[i]_{w,m} \qquad (1.99)$$

$$\sigma_{w,m} = \theta_{w,m} \approx \phi(1 - U/12)$$

with $U = d_{60}/d_{10}$ for sands and gravels, and

for soil air

$$(ca \cdot \{i\})_a = \sigma_a(1/\gamma)f_i = \theta_a(RT)^{-1}px_i \qquad \text{for ideal gas} \qquad (1.100)$$

For immobile fluid and solid phases in the subsurface, the terms of Equation 1.98 are not as easily describable with mathematical models. Stored matter at the pore scale is heterogeneously distributed — for example, in the form of films covering solid particles — and the activity of the considered stored species is not directly measurable. Using nondestructive measuring techniques, the only practically measurable subsurface state variable in solute migration is the activity or concentration of a migrant in a mobile fluid mixphase after its extraction. However, at thermodynamic equilibrium the activity $\{i\}_{m,fl} \approx [i]_{m,fl}$ can also be used to characterize the activity $\{i\}$ of the immobile fluid and solid phases because at equilibrium the intensive state variable activity $\{i\}$ must be equal in all mixphases of a REV.

The following nonhysteretic equilibrium storage models have proven particularly useful for the two-phase model concept mobile/immobile (see §1.2.2):

$$s_i = (\theta_o + \alpha)[i] \qquad (1.101a)$$

$$ds_i/dt = (\theta_o + \alpha) \, d[i]/dt, \; \theta_o + \alpha = \text{const.} \qquad (1.101b)$$

$$s_i = \theta_o[i] + K[i]^q \qquad (1.102a)$$

$$ds_i/dt = (\theta_o + Kq[i]^{q-1}) \, d[i]/dt, \; K, q = \text{constants} \qquad (1.102b)$$

$$s_i = \theta_o[i] + K'_{i,max}[i]/(1 + K'[i])$$ (1.103a)

$$\frac{ds_i}{dt} = \left(\theta_o + \frac{K's_{i,max}}{1 + K'[i])^2} \right) \frac{d[i]}{dt}$$ (1.103b)

where $[i] = c$ represents the intensive and s the extensive state variable
$s_{i,max}$ = maximum storage amount for migrant i (e.g., CEC)
K' = a constant, e.g., $K' = K_d \rho_b / s_{i,max}$ in ion exchange processes

If only storage processes are considered to occur in the immobile mixphases, the storage fraction of the mobile mixphase θ_o drops out of Equations 1.101 to 1.103, and the adsorption isotherms are obtained:

- Henry's adsorption isotherm (Equation 1.101)
- Freundlich's adsorption isotherm (Equation 1.102)
- Langmuir's adsorption isotherm (Equation 1.103)

A literature review [1.85] and our own experience lead to the following recommendations of the model applicabilities:

- Equation 1.101: applicable to storage processes at relatively low migrant concentrations
- Equation 1.102: especially applicable to the sorption of sulfate, cadmium, and herbicides
- Equation 1.103: especially applicable to the sorption of gases, and phosphorus. However, the best results for phosphate are obtained using

$$s_i = \theta_o[i] + \frac{K'_1 s_{1,i,max}[i]}{1 + K'_1[i]} + \frac{K'_2 s_{2,i,max}[i]}{1 + K'_2[i]}$$ (1.103c)

A hysteretic storage model approach is often required, for example for heavy metal adsorption/desorption processes. Such storage models should be formulated and fitted in a similar manner as discussed for the hydraulic retention model $\theta_w = f(p_c, \partial p_c / \partial t)$ of Equation 1.12.

Storage and exchange process models must be formulated consistently, so that they do not contradict each other at the equilibrium state. This can be guaranteed if the equilibrium storage process model is derived from a nonequilibrium linked storage and exchange process model. The following models fulfill these requirements. They reflect the storage of the considered migrant i in the mixphase II, which exchanges this migrant with the mixphase I, where c = [i] again represents the intensive and s the extensive state variable:

Reversible linear kinetic storage model of first order

$$ds_{II}/dt = k_I c_I - k_{II} s_{II}$$ (1.104)

For $ds_{II}/dt \to 0$, it follows that

$$s_{II} = (k_I/k_{II})c_I = \alpha c_I$$

See Equation 1.101a and 1.46d.

Reversible nonlinear kinetic storage model

$$ds_{II}/dt = k_I c_I^q - k_{II}s_{II} \tag{1.105}$$

For $ds_{II}/dt \to 0$, it follows that

$$s_{II} = (k_I/k_{II})c_I^q = Kc_I^q$$

See Equation 1.102a.

Bilinear kinetic storage process model

$$ds_{II}/dt = k_I c_I (s_{II,max} - s_{II}) - k_{II}s_{II} \tag{1.106}$$

For $ds_{II}/dt \to 0$, using $(k_I/k_{II}) = K'$, it again follows that

$$s_{II} = K's_{II,max} \, c_I/(1 + K'c_I)$$

See Equation 1.103a.

Practical application for the preceding equations are recommended as follows:

- Equation 1.104: applicable to the sorption of herbicides, organic solutes, nitrite, nitrate, and phosphate
- Equation 1.105: applicable to the sorption of herbicides and phosphorus with $q < 1$
- Equation 1.106: applicable at relatively high concentrations of migrants — for example, to reflect the ion exchange

A nonhysteretic model approach is also often required for these models.

Similar to the derivation and the synchronization of Equations 1.101 through 1.103 (the storage models) and 1.104 through 1.106 (the non-equilibrium linked storage and exchange process models), a variety of additional storage and exchange process models may be derived, synchronized, and made hysteretic.

1.7

Heat Migration

Heat is internal energy consisting of a potential component, embodied in the cohesion of molecules, and a kinetic component, the vibrational, translational, and rotational energy of molecules. Sensible heat increases proportional to temperature. Latent heat change is caused by a qualitative change of potential and kinetic energy during aggregate state changes. According to Equation 1.1, heat is the extensive state variable and temperature the intensive state variable, as shown in Figure 1.51.

Heat transport, storage, transformation, and exchange in the multiphase system subsurface may be reflected by process models which are very similar to solute migration models with respect to their mathematical structure. Therefore, heat transport, storage, transformation, and exchange processes together are designated as *heat migration,* regardless of their phenomenological differences. Consequently, the following paragraphs are closely related to the explanations in Chapters 1.3 to 1.6.

1.7.1 HEAT TRANSPORT IN THE SUBSURFACE

Heat transport in the subsurface takes place by heat conduction, heat convection, and hydrodynamic heat dispersion.

1.7.1.1 Heat Conduction

In heat conduction, particles with higher energy, and therefore higher temperature, transfer energy to neighboring particles having lower temperatures. This type of heat flow obeys Fourier's law:

Table 1.21. Heat Conductivity of Selected Solids, Liquids, and Gases at 0°C and 1 bar

Substance	λ W/(mK)	Substance		λ W/(mK)	Substance	λ W/(mK)
Silver	419	H_2O	0°C	0.54	Feldspars	about 2.5
			12°C	0.58		
SiO_2 ‖ to axis	13.6		30°C	0.63	Quartzite	about 6.0
_ to axis	7.2					
$\sqrt{\_\cdot\|}$	≈ 10	Paraffin		0.25	Granite	2.2 to 3.4
Potash feldspar	4.2	Petroleum		0.15	Basalt	1.4 to 2.0
Muscovite	0.4	O_2, N_2		0.024	Limestone	2.0 to 3.0
FeS_2	38	H_2		0.18	Marble	2.8 to 3.2
$CaCO_3$ _ to axis	4.3	Air	0°C	0.024	Sandstone	1.5 to 5
			20°C	0.026		
$CaSO_4$	5.2				Clay	0.8 to 1.2
NaCl	9.6	CO_2		0.014	Coal	about 5

Note: 1 cal/(msgrad) = 4.19 J/(msK) = 4.19 W/(mK). For a more comprehensive data compilation see Schön [1.73].

$$\vec{q}_M = -\lambda \text{ grad } T \tag{1.107}$$

where $\vec{q}_M$ = sensible heat flux due to heat conduction in W/m²
λ = heat conductivity in W/(mK) = J/(smK)
T = temperature in K

Selected heat conductivity data for solids, liquids, and gases important to heat migration in the soil- and groundwater zone are shown in Table 1.21. However, the heat conductivity λ_o of the multiphase system subsurface cannot be directly estimated from these data. Heat conductivity values span a range whose limits may be estimated by the arithmetic and geometric mean:

$$\lambda_{o,max} = (1 - \phi) \lambda_s + \theta_w\lambda_w + \theta_a\lambda_a \tag{1.108a}$$

$$1/\lambda_{o,min} = (1 - \phi)/\lambda_s + \theta_w/\lambda_w + \theta_a/\lambda_a \tag{1.108b}$$

These models correspond to systems where the mixphases solids, water, and air are arranged parallel and perpendicular to the heat flux (see Figure 1.52). For completely fluid-filled soil or rock, λ_o may be estimated by Russel's parametric model [1.33]:

$$\lambda_o = \lambda_{fl} \frac{(1 - \phi)^{2/3} + (\lambda_{fl}/\lambda_s)(1 - (1 - \phi)^{2/3})}{(1 - \phi)^{2/3} - (1 - \phi) + (\lambda_{fl}/\lambda_s)(2 - \phi - (1 - \phi)^{2/3})} \tag{1.108c}$$

This model yields two useful limits of λ_o for $\lambda_{fl} = \lambda_{air}$ and $\lambda_{fl} = \lambda_{water}$. For rock partially saturated with water and air, an estimate can be obtained which connects both values and takes into account the convex shape of this function (see Figure 1.53a). This approach may also be extended to frozen soils where a third limit point λ_o for $\lambda_{fl} = \lambda_{ice}$ is

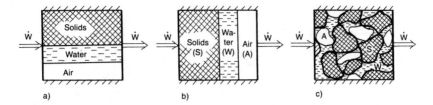

Figure 1.52.   Models of different phase distributions for the multiphase system "solids, water, and air" to derive the heat conductivity λ_o of the mixphase "subsurface."

considered. Nevertheless, more accurate calculations or simulations require soil and rock tests to estimate $\lambda_o = f(\phi, \theta_w)$ for each soil type and the corresponding bulk density.

Figures 1.53c and 1.53d show functional dependencies $\lambda_o = f(\rho_b, w)$ according to Balstrup [1.3]. These functional plots can be converted to the graphical forms of Figures 1.53a and 1.53b if the following relationships are taken into account where rock, soil, and subsurface are used synonymously:

$$\rho_b = (1 - \phi)\rho_s$$

where ρ_b is the dry bulk density of rock (= partial density of solids = mass concentration of solids in the multiphase system subsurface) and ρ_s is the density of solids (see Table 1.5),

$$\rho_{b,w} = \rho_b + \theta_w\rho_w$$

where $\rho_{b,w}$ is the wet bulk density of rock and ρ_w is the density of water, and

$$w = \rho_{b,w}/\rho_b - 1$$
$$= \theta_w\rho_w/\rho_b$$

where w is the water content of rock (= mass fraction of water in rocks).

Example:

For $\rho_s = 2.65$ g/cm^3 and $\theta_w = \phi = 0.36$ (complete saturation of pore spaces by water) we obtain $\rho_b = 1.70$ g/cm^3, $\theta_w = 0.36$, $\rho_{b,w} = 2.06$ g/cm^3, and w = 0.21.

The preceding discussion, and a lot of measured data, allow the following general statements:

- λ_o does not increase significantly with increasing temperature (about a $1/4\%$ per K or °C).

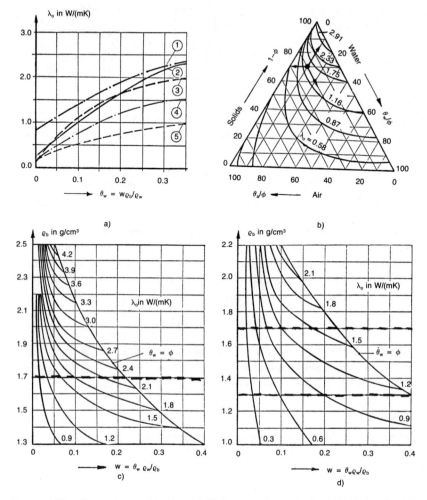

Figure 1.53. Dependency of heat conductivity on water content and bulk density. *(a)* Heat conductivity as a function of volumetric water content: *1,* coarse-grained soil with ρ_b = 1.7 g/cm³ according to *(c)*; *2* and *3*, uniform sands, relatively equal-sized grains; *4* and *5*, fine-grained soils with ρ_b = 1.7 g/cm³ and ρ_b = 1.3 g/cm³ according to *(d)*. *(b)* Heat conductivity as a function of water and air content. *(c)* Heat conductivity as a function of dry bulk density and water content of coarse-grained soils. *(d)* As for *(c)* but for fine-grained soils according to Balstrup [1.3].

- λ_o increases with increasing water content at constant dry bulk density ρ_b in unconsolidated and consolidated rocks.
- λ_o increases with increasing dry bulk density ρ_b at w = constant.
- λ_o increases with increasing particle size, i.e., with decreasing number of transfer points at w = constant and ρ_b = constant.

- λ_o depends on the mineral content of rocks: e.g., sands having a high fraction of quartz grains have larger λ_o values than those with high feldspar content. Clay minerals lower the λ_o values for soils.

1.7.1.2 Heat Convection and Hydrodynamic Dispersion

According to Equation 1.22, convective heat transport in the subsurface may be written as

$$\vec{q}_c = \sum_j (W/V)_j \vec{v}_j \tag{1.109a}$$

where $W = f(T)$ according to Figure 1.51b
         $\vec{q}_c$ = heat flux rate due to convection in W/m^2
   $(W/V)_j$ = specific heat of the phase j in J/cm^3
        $\vec{v}_j$ = flow rate (= Darcy's velocity) of the phase j in m/s

The latent heat is often neglected in this equation (see Figures 1.51 and 1.54). Such an approach is particularly useful when investigations are made in the temperature range of 0 to 50°C. Equation 1.109a then takes on the following form:

$$\vec{q}_c = \sum_j (\rho c T)_j \vec{v}_j \tag{1.109b}$$

where c is the storage coefficient or specific heat capacity in $J/(gK)$ and ρ is the density of the considered phase. First estimates for ρ_j and c_j are shown in Table 1.22. For air and oil, the following estimates may be used:

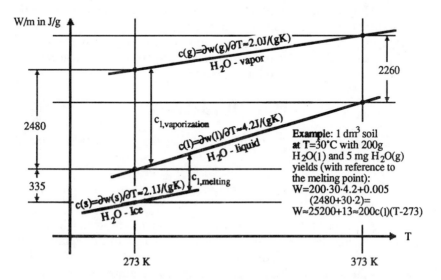

Figure 1.54. Specific heat content in J per g H_2O approximated for the natural occurring range of heat migration in the soil- and groundwater zone.

Table 1.22. Selected Parameters of Heat Migration

Substance	c J/(gK)	ρc J/(cm³K)	λ_o W/(mK)
Air	1.00	0.0013	0.02
Water	4.19	4.19	0.60
Ice	2.14	1.88	2.30
Snow, fresh	2.14	0.13	0.08
Granite	0.84	2.18	4.61
Clay	1.38	2.47	0.92
Humus	1.84	2.39	1.26
Sand, dry	0.84	1.26	0.17
Sand, wet	1.26	1.67	1.68

Source: Rethali [1.65]. For a more comprehensive data compilation see Schön [1.73].

Air (g)

$$\rho_a = 1.29 \cdot 10^{-3} \text{ g/cm}^3 \qquad c_a = 1.0 \text{ J/(gK)}$$

at 0°C and p = $1 \cdot 10^5$ Pa = 1 bar;

Oil (l)

$$\rho_{oil} = 0.85 \text{ g/cm}^3 \qquad c_{oil} = 2.0 \text{ J/(gK)}$$

For more accurate calculations and simulations, the state function W/ V = f(T) according to Figure 1.51 must be determined by calorimetric laboratory tests. This is especially important if heat migrations should be investigated near the freezing point, and in warm, relatively dry soils.

The heat transport due to hydrodynamic dispersion is given according to Equation 1.25 by the following model:

$$\vec{q}_{D,j} = -\overline{\overline{D}}_{D,j} \text{ grad } (W/V)_j \qquad (1.109c)$$

which upon replacing W/V by $(\rho c)T$ using the previously mentioned assumptions yields:

$$\vec{q}_{D,j} = -\overline{\overline{D}}_{D,j} (\rho c)_j \text{ grad } T_j \qquad (1.109d)$$

where $\vec{q}_{D,j}$ is the heat flux rate due to hydrodynamic dispersion in W/m², and $\overline{\overline{D}}_{D,j}$ is the heat dispersion, which is assumed to be equal to the coefficient $\overline{\overline{D}}_{D,j}$ of solute dispersion (see §1.3.2). Consequently, it has been assumed that δ-values obtained from heat migration tests may also be used in solute migration simulation. However, this is often not a very good approach.

1.7.1.3 Total Transport Process

In the multiphase system subsurface, the total heat transport process, i.e., the sum of all transport phenomena according to Equation 1.33, is given by

$$\vec{q} = -\lambda_o \, \overline{\mathrm{grad} \, T} + \sum_j [(W/V)_j v_j - \overline{\overline{D}}_{D,j} \, \mathrm{grad} \, (W/V)_j] \quad (1.109e)$$

and approximated under the aforementioned conditions by

$$\vec{q} = -\lambda_o \, \overline{\mathrm{grad} \, T} + \sum_j [(\rho c T)_j v_j - \overline{\overline{D}}_{D,j} \, (\rho c)_j \, \mathrm{grad} \, T_j] \quad (1.109f)$$

As the average temperature $\overline{T}$ in the multiphase system subsurface deviates only slightly from T_j of the mobile fluid phases, the quantity $\overline{\overline{D}}_j = \overline{\overline{D}}_{D,j} + \lambda_o/(\rho c)_j$ may again be introduced (see Equation 1.34). Thus, the following model is obtained:

$$\vec{q} = \sum_j [(\rho c T)_j v_j - \overline{\overline{D}}_j (\rho c)_j \, \mathrm{grad} \, T_j] \quad (1.109g)$$

1.7.2 PHASE EXCHANGE, STORAGE, AND TRANSFORMATION PROCESSES

Heat conduction plays an important role in the subsurface not only for heat transport but also for the exchange processes between the phases of the REV. In heat migration the characteristic time constants in establishing thermodynamic equilibrium in the REV are given according to Equation 1.69a by

$$t^* = ((\lambda/\overline{\Delta l})(\omega/V))^{-1} \cdot ca \quad (1.110)$$

For parameters comparable to those used in the example in §1.5.1, exchange between the mobile and immobile water, with $(\omega/V) \approx 20(\omega V)^* = 2 \cdot 10^4 \, m^{-1}$, $\lambda_w = 0.6 \, J/(smK)$, $\overline{\Delta l} = 10^{-3} \, m$, and $ca = (\theta_m + \theta_{im})(\rho c)_w \approx 1.7 \cdot 10^6 \, J/(m^3 K)$, is characterized by the time constant: $t^* = 0.14 \, s$, which is 3000 times less than the time constant corresponding to solute exchange. Thus, heat exchange in the REV is several hundred times more rapid than solute exchange. The internal kinetics of heat exchange may nearly always be neglected in unconsolidated rocks, and the average temperature $\overline{T}$ in the REV is practically equal to the temperature of each single phase T_j in the multiphase system subsurface.

The heat storage equation in the subsurface is derived from Equation 1.98:

$$s_Q = \sum_j [\sigma_j(W/V)_j] = f(T) \quad \text{in } J/cm_v^3 \quad (1.111)$$

For temperatures ranging from 0 to about 50°C, $(\sigma W/V)_{gas}$ is usually small compared to $(\sigma W/V)_{water}$ and $(\sigma W/V)_{solids}$ in soils, and the following approximate equation may be recommended for heat storage (see also example calculations in Figure 1.54):

$$s_{Q,s} \approx [\theta_w(\rho c)_w + (1 - \phi)(\rho c)_s]\bar{T} = f^*(\bar{T}) \qquad (1.112)$$

These storage process equations may be linked with heat transport and transformation equations without difficulty because the same intensive state variable $T \approx \bar{T}$ is used. This average temperature of the REV is a state variable which can be monitored in practice by relatively simple measuring techniques.

Finally, particular attention must be given to internal heat sources and sinks. In the subsurface, biological and chemical exchange and conversion of substances simultaneously absorb heat through endothermal reactions and release heat through exothermal reactions (see Chapters 1.4 and 1.5). However, these phenomena can usually be neglected in investigations of heat migration problems in the subsurface. Heat sources and sinks caused by the changes of the aggregate state of subsurface water expressed as latent heat cannot otherwise be ignored (see Figure 1.54):

Latent heat of vaporization: liquid $\rightleftarrows$ gaseous

$$c_{l,v} \approx 2500 - 2.4 \, (\bar{T} - 273) \text{ in J/g}$$

Latent heat of melting: solid $\rightleftarrows$ liquid

$$c_{l,m} = 335 + 2.1 \, (\bar{T} - 273) \text{ in J/g}$$

Thus, the process of freezing and melting in the subsurface can be modeled as a first-order kinetic transformation reaction:

$$\dot{w}_T = k_T(\bar{T} - 273.15)\delta' \qquad (w_T > 0 \text{ corresponds to a source}) \qquad (1.113)$$

$$\text{where} \quad \delta' = \begin{cases} 1 \text{ at } \bar{T} > 273.15 \text{ K and } \sigma_{ice} > 0 \\ 1 \text{ at } \bar{T} < 273.15 \text{ K and } \sigma_{water} > 0 \\ 0 \text{ in all other cases} \end{cases}$$

The amount of ice or water formed can be determined from the integral $\int \dot{w}_T dt$ and the transformation rates $335\rho_{ice}\partial\theta_{ice}/\partial t = k_T(\bar{T} - 273.15)\delta'$. In a similar manner, we can model the evaporation and condensation process in the subsurface, which may be very important in dry and hot soils.

1.8

Mathematical Description of the Total Migration Process

A mathematical migration process model should always be developed based upon a qualitative process assessment (see §1.8.1). Our approach will be demonstrated with three examples in §1.8.2. The mathematical process model is finally to be completed with initial and boundary conditions, as discussed in §1.8.3.

1.8.1 PROCESS ASSESSMENT

1.8.1.1 Qualitative Assessment

A qualitative migration assessment in the soil-and groundwater zone is based on the

(1) flow analysis of the mobile fluid phases and

(2) assessment of the storage, transformation, and exchange processes to which the migrants are subjected during their transport with the flowing fluid phases.

The flow analysis of mobile fluid phases in the geological milieu is studied in geofiltration [1.52] and geohydraulics [1.11]. Flow analysis is an indispensable requirement for any estimation of the migrants' fate. Apart from bulk fluid flow, the net motion of migrants is controlled by internal reactions and exchange processes. These processes should be assessed on the basis of the migrants' mobility and degradability in the subsurface environment. Migrants are frequently classified as (1) conservative (or persevering), if they are subject to neither exchange nor biochemical transformation (e.g., Cl⁻), or (2) persistent (or refractory), if they do not undergo biochemical transformation (e.g., heavy metals) [1.16]. Additionally it is useful to assess whether the precipitated or adsorbed migrants are remobilizable or whether mobile migrants dispersed in the fluid phase support remobilization (i.e., dissolution or desorption) of immobile species. Finally, effort should be made to assess the extent of storage, transformation, and exchange (involved quantities) to which the migrants are subjected. Summarizing, a synopsis is given in Figure 1.55.

	Conservative	Persistent	Remobilizable	Remobilizing	Electrol. dissociation	Complex formation	Coagulation	Acid-base reaction	Redox process	Filtration	Sorption	Ion exchange	Dissolution-precipitation	Biological accumulation	Aerobic decomposition	Anaerobic decomposition	Autolysis
Susp. inorg. solids	○	○					○	○		◉	◉						
Ammonium			◉					◉	◉			◉			◉	◉	
Toxic metals	◉	◉				◉	○	◉		◉	◉	◉	◉	◉			
Iron	◉	◉				○	○	◉		◉	◉	◉	◉	◉	○		
Manganese	◉	◉				○	○	◉		◉	◉	◉	◉	◉	○		
Chloride	◉				◉	◉	◉	◉									
Fluoride					◉	◉		◉					◉				
Sulfate	○				○	◉	◉	◉								◉	
Nitrate	○				○											◉	
Hydrocarbons			○			◉	○		○	◉	○			◉	◉	◉	
Phenols			◉	◉		◉			○		◉			◉	◉	◉	
Organic Cl comp.	○	◉	◉			○			○	○	◉			◉			
Organic F comp.	◉	◉				○			○	○	◉						
Bacteria			○	◉			○			◉	◉			◉	◉	◉	◉
Viruses											◉			◉			◉

◉ dominant   ○ partial   ☐ insignificant influence

Figure 1.55. Schematic overview of the fate of migrants in the soil- and groundwater zone. *Source:* adapted from Damroth et al. [1.16, Table 2.2].

1.8.1.2 Formation of the Mathematical Model

The integral mathematical formulation of subsurface quality models includes four stages:

1. In the first stage, the three-step succession of figurative models according to Figure 1.14 should be followed. In this stage, the number of phases and migrants to be modeled must be identified (see Chapter 1.2). This succession should be performed for each migrant, and Figure 1.14c must be designed for each. This figure must reflect all subprocesses to which the considered migrant is subjected, including transport, storage, phase-internal reactions, and transphase and external exchange or approximations thereof.
2. The second stage must reflect the coupling structure of the considered migrants. Typical models of this structure are the stoichiometric or chemical balance equations for phase-internal reactions, and balance equations of ion capacities for description of transphase exchanges.
3. In the third stage, the integral mathematical process equations must be set up. In addition to the coupling equations, the mathematical process description consists of a system of partial differential equations with one subsystem for each migrant. The number of equations for such a subsystem is equal to the number of nodal elements or considered phases. Each equation expresses a balance for a migrant in one of the mixphases of the REV. Hence, such a solute or heat balance equation has the following structure:

$$
\boxed{\begin{array}{c}\text{Trans-}\\\text{port}\\\text{TR}\end{array}} = \boxed{\begin{array}{c}\text{Storage}\\ \\ S\end{array}} + \boxed{\begin{array}{c}\text{Exchange}\\ \\ EX\end{array}} + \boxed{\begin{array}{c}\text{Internal}\\\text{reactions}\\ IR\end{array}} + \boxed{\begin{array}{c}\text{External}\\\text{sources/sinks}\\ ESS\end{array}} \quad (1.114)
$$

Interlinkage of these equations within each subsystem is due to transphase exchanges. Interlinkage between different subsystems arises from internal reactions (stoichiometric relations). However, interlinkage may also be caused by source-sink terms.
4. Finally, in the fourth stage, submodels derived in Chapters 1.4, 1.5, and 1.6 are incorporated into the balance equation system 1.114.

Summarizing, Figure 1.56 shows the entire four-stage approach.

1.8.2 METHODOLOGICAL EXAMPLES

The following three examples are considered:

1. heat migration in the vadose subsurface
2. oxygen migration in the aerated subsurface
3. deironization of water in the subsurface

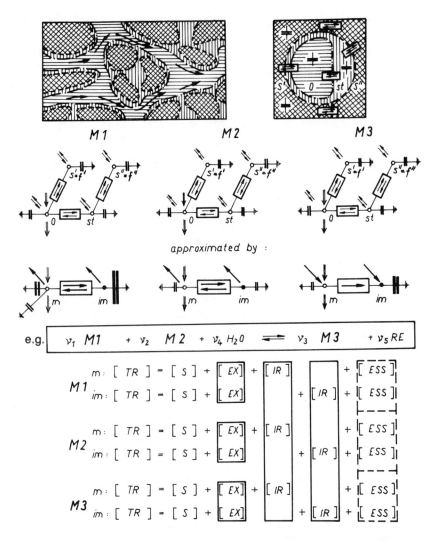

Figure 1.56. Setting up the integral mathematical migration model with three interlinked migrants: *M1*, migrant 1; *0*, mobile fluid phase; *st*, stagnant fluid phase; *R*, rock phase; *m* and *im*, mobile and immobile phase (see §1.2.2); *v*, stoichiometric numbers; *TR*, transport; *S*, storage; *EX*, exchange; *IR*, internal reactions; *ESS*, external source-sink term.

1.8.2.1 Heat Migration in the Vadose Subsurface

Let us first consider the heat migration in the vadose zone caused by vertical one-dimensional infiltration of water.

Flow Model

Provided both soilwater and soil air can be regarded as incompressible, the flow model derived from the general Equation 1.14 is given by

$$\frac{\partial}{\partial z}\ (K_w\ \frac{\partial}{\partial z}\ (\frac{p_w}{\rho_w g}\ +\ z))\ =\ \frac{\partial \theta_w}{\partial t}\ -\ w_w \qquad (1.115a)$$

$$\frac{\partial}{\partial z}\ (K_a\ \frac{\partial}{\partial z}\ (\frac{p_a}{\rho_a g}\ +\ z))\ =\ \frac{\partial \theta_a}{\partial t} \qquad (1.115b)$$

where

K_w, K_a = permeability or conductivity coefficients of soil for water and air in m/s (see also Figure 1.22)

z = vertical spatial coordinate in m (upward is positive)

p_w, p_a = pressure of soilwater and soil air in Pa

ρ_w, ρ_a = soilwater and soil air density in kg/m^3

g = gravitational acceleration in m/s^2

θ_w, θ_a = volumetric water and air content, dimensionless

p_c = capillary pressure; $p_c = p_a - p_w$ in the subsurface (see Equation 1.9)

w_w = soilwater source-sink term, e.g., caused by melting of soil ice and freezing of soilwater, or the water uptake by plant roots, in s^{-1} ($w_w > 0$ is source)

The outputs of this model are the pressures p_w and p_a. Inserting them into Equation 1.14, the flow rates v_w and v_a may be calculated, and making use of Equation 1.12, the volumetric contents θ_w and θ_a, as well as their changes $\partial\theta_w/\partial t$ and $\partial\theta_a/\partial t$, may be determined. The permeabilities K_w and K_a and the source-sink term w_w are nonlinear functions which must already be known. They depend on p_a, p_w, θ_a, and θ_w (see Luckner et al. [1.54]).

Quality Model

Heat migration in the subsurface can be simulated with a single-migrant model. The four-stage abstraction sequence of figurative models is shown in Figure 1.57, based on Figure 1.14.

Exchange Process Dynamics

Temperatures of solids, stagnant water, flowing water, and soil air are in most cases in equilibrium, yielding: $\bar{T} = T_o = T_{st} = T_s = T_a$. Consequently, the four nodal points representing the four lumped mixphases in Figure 1.57c can be superimposed and treated as one node symbolizing the REV as a whole (see Figure 1.57d). Therefore, heat migration is usually reflected by a single-migrant and single-phase model.

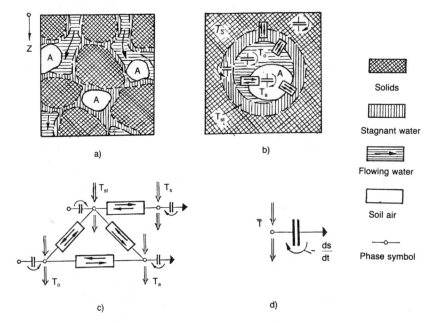

Figure 1.57. Multistage abstraction sequence of figurative models for heat migration in the soil zone: T_o, temperature of the flowing soilwater; T_{st}, temperature of the stagnant soilwater; T_s, temperature of the solids; T_a, temperature of the soil air; $\bar{T}$, temperature of the mixphase subsurface.

Transport Processes

Approximating all exchange processes at equilibrium state, the following equation represents the integral transport process of the mixphase system subsurface (see also Equation 1.109e and 1.109g):

$$\vec{q}_\Sigma = -[\lambda_o + D_{D,w}(\rho c)_w + D_{D,a}(\rho c)_a]$$
$$\text{grad } \bar{T} + [(\rho c)_w v_w + (\rho c)_a v_a]\bar{T} \qquad (1.116a)$$

Heat transport by vapor diffusion is neglected or included in λ_o, and $D_{D,a}(\rho c)_a$ is usually small in comparison with $D_{D,w}(\rho c)_w$. For $D_{D,a} = 0$ and $v_a = 0$, it follows then immediately:

$$\vec{q}_\Sigma = -[\lambda_o + D_{D,w}(\rho c)_w] \text{ grad } \bar{T} + (\rho c)_w v_w \bar{T} \qquad (1.116b)$$

Storage and Transformation Processes

Heat storage in the subsurface is approximated as follows (see Figure 1.51):

$$s_Q = (W/V)_{soil} \approx \Sigma \sigma_j (W/V)_j = f(\bar{T}) \qquad (1.117a)$$

Data for this function are obtained experimentally, after which the derivative ds_Q/dt as a function of temperature must be determined.

Latent heat and sensible heat are frequently considered separately (see Figure 1.51): $s_Q = s_{Q,l} + s_{Q,s}$. The term $ds_{Q,l}/dt$ is then approached by a source-sink term of ice formation and melting, or of evaporation and condensation (see Equation 1.113), and the sensible heat change $ds_{Q,s}/dt$ can be expressed as

$$ds_{Q,s}/dt = [\theta_w(\rho c)_w + \theta_i(\rho c)_i + (1 - \phi)(\rho c)_s]\partial\bar{T}/\partial t$$
$$+ (\rho c)_w\bar{T}\partial\theta_w/\partial t + (\rho c)_i\bar{T}\partial\theta_i\partial t \qquad (1.117b)$$

Total Balance

Substituting the established submodels Equations 1.116b and 1.117b in the total balance model according to Equation 1.114 results in

$$\frac{\partial}{\partial z}\ ([\lambda_o + D_{D,w}(\rho c)_w]\ \partial\bar{T}/\partial z - (\rho c)_w v_w\bar{T})$$

Transport process

$$= [\theta_w(\rho c)_w + \theta_i(\rho c)_i + (1-\phi)(\rho c)_s]\partial\bar{T}/\partial t + (\rho c)_w\bar{T}\partial\theta_w/\partial t + (\rho c)_i\bar{T}\partial\theta_i/\partial t$$

Storage process

$$-\rho_iC_{l,m}\partial\theta_i/\partial t \qquad + \qquad \theta_aC_{l,v}\partial\rho_v/\partial t \quad - \quad wc_{ex} \qquad (1.118a)$$

Transformation processes   external
"icewater"   "water vapor"   exchange

This model may now be transformed to avoid terms containing $\bar{T}$ (zero point problem). Such terms must be replaced by terms which contain derivatives or differences of $\bar{T}$ only. This can be achieved by using

- Equation 1.115a in the form of $-(\rho c)_w\bar{T}\partial v_w/\partial z = (\rho c)_w\bar{T}\partial\theta_w/\partial t - (\rho c)_w\bar{T}w_w$
- the total derivative as $\dfrac{\partial}{\partial z}\ ((\rho c)_w v_w T) = (\rho c)_w\bar{T}\ \dfrac{\partial v_w}{\partial z} + (\rho c)_w v_w\ \dfrac{\partial\bar{T}}{\partial z}$
- and the external exchange term in the form of

$$wc = \begin{cases} w_w(\rho c)_w\bar{T} & \text{if } w_w \text{ is a sink} \\ \\ w_w(\rho c)_w\bar{T}_{ex} & \text{if } w_w \text{ is a source} \end{cases}$$

Thus,

$$\frac{\partial}{\partial z} \left([\lambda_o + D_{D,w} (\rho c)_w] \frac{\partial \bar{T}}{\partial z} \right) - (\rho c) v \frac{\partial \bar{T}}{\partial z}$$

$$\underbrace{\hspace{8cm}}_{\text{Transport process}}$$

$$= \underbrace{[\theta_w(\rho c)_w + \theta_i(\rho c)_i + (1 - \phi)(\rho c)_s] \frac{\partial \bar{T}}{\partial t}}_{\text{Storage process}} \underbrace{-\rho_i c_{l,m} \partial \theta_i / \partial t + \theta_a c_{l,v} \partial \rho_v / \partial t}_{\substack{\text{Transformation processes} \\ \text{"ice-water" "water-vapor"}}}$$

$$\underbrace{(\rho c)_w w_w (\bar{T} - \bar{T}_{ex})}_{\text{Source = sink term}} \tag{1.118b}$$

where $\bar{T}_{ex} = \bar{T}$ if w_w is a sink term.

Kinetic models of first order are commonly applied to ice transformation into liquid water, liquid water into vapor, and to transformations in the opposite direction. These models use deviations from the equilibrium state as driving forces (see also Equation 1.113). The symbols in Equation 1.118 are defined as

$$\lambda_o = \text{thermal soil conductivity in W/m}^2$$
$$D_{D,w} = \text{hydrodynamic dispersion coefficient in the flowing soilwater;}$$
$$\quad\quad D_{D,w} = \delta v_w \text{ in m}^2\text{/s, where } \delta \text{ is the dispersity}$$
$$D_{D,a} = \text{hydrodynamic dispersion coefficient in the flowing soil air;}$$
$$\quad\quad D_{D,a} = \delta v_a \text{ in m}^2\text{/s; where } v_a \text{ is the bulk velocity of soil air}$$
$$\rho = \text{density in kg/m}^3$$
$$c = \text{specific heat capacity in kJ/(kgK)}$$
$$z = \text{vertical space coordinate in m}$$
$$\bar{T} = \text{temperature of the subsurface in K}$$
$$v = \text{Darcy flux in z-direction in m/s}$$
$$\theta_w, \theta_i, \theta_a = \text{volumetric water, ice, and air content}$$

The quality model 1.118 results in $\bar{T} = \bar{T}(z,t)$, time-dependent changes of θ_w and θ_i due to formation and melting of ice, and ρ_v due to evaporation/condensation. The soil temperature $\bar{T}(z,t)$ and the volumetric ice content θ_i impact the parameters K_w and K_a and hence the flow model. The heat migration model is, therefore, with the partial differential Equations 1.115 and 1.118, rather complex.

1.8.2.2 Migration of Oxygen in the Aerated Subsurface

Let us consider the vertical one-dimensional case.

Flow model

Compared to Equation 1.115, the flow model of this example is simplified by assuming that soil air pressure p_a is nearly equal to atmospheric pressure. Consequently, Equation 1.115b reduces to $p_a \approx p_{atm}$. This does not mean that v_a must be zero, because K_a is relatively large, resulting in finite air flow velocities v_a even at very small gradients $\partial p_a / \partial z$. With the above assumptions, and $h_w = p_w/(\rho_w g) + z$, Equation 1.115 yields

$$\frac{\partial}{\partial z}\left(K_w \frac{\partial h_w}{\partial z}\right) = \frac{\partial \theta_w}{\partial t} - w_w \qquad (1.119)$$

Quality Model

This example again develops the quality model based on the multistage abstraction sequence by means of figurative models according to Figure 1.58.

Transport

In the present example, oxygen transport is assumed to take place in the subsurface air in gaseous state by molecular diffusion (see §1.3.1 and Figure 1.29; v_a equals zero), and in flowing subsurface water in the dissolved state by convection and hydrodynamic dispersion (see §1.3.2). Under these conditions, transport processes are reflected as follows (see also Equation 1.34) for subsurface air:

$$\vec{fc}_{m,aO_2} = -\theta_a D_{a,O_2} \, \text{grad} \, c_{a,O_2} = -D_{o,O_2} \, \text{grad} \, c_{a,O_2} \qquad (1.120a)$$

and for subsurface water:

$$\vec{fc}_{o,O_2} = -D_{o,O_2} \, \text{grad} \, c_{o,O_2} + v_o c_{o,O_2} \qquad (1.120b)$$

Storage Processes

O_2 is stored in the flowing soilwater, in the stagnant soilwater, and in the soil air (see Figure 1.58 and Equation 1.101):

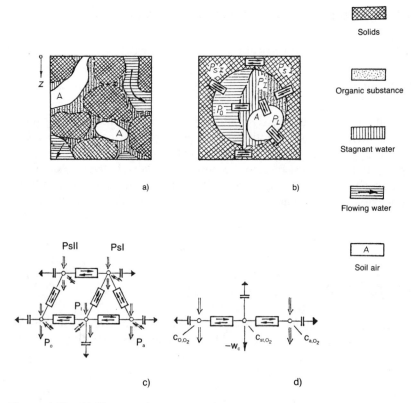

Figure 1.58. Multistage abstraction sequence of figurative models for oxygen migration in the aerated subsurface; P activity potential of the migrating species. Indices: *s*, solids; *a*, air; *o*, flowing soilwater; O_2, oxygen; *st*, stagnant water.

$$s_{o,O_2} = \theta_o c_{o,O_2} \; ; \quad \frac{\partial s_{o,O_2}}{\partial t} = \theta_o \frac{\partial c_{o,O_2}}{\partial t} + c_{o,O_2} \frac{\partial \theta_o}{\partial t} \qquad (1.121a)$$

$$s_{st,O_2} = \theta_{st} c_{st,O_2} ; \quad \frac{\partial s_{st,O_2}}{\partial t} = \theta_{st} \frac{\partial c_{st,O_2}}{\partial t} + c_{st,O_2} \frac{\partial \theta_{st}}{\partial t} \qquad (1.121b)$$

$$s_{a,O_2} = \theta_a c_{a,O_2} \; ; \quad \frac{\partial s_{a,O_2}}{\partial t} = \theta_a \frac{\partial c_{a,O_2}}{\partial t} + c_{a,O_2} \frac{\partial \theta_a}{\partial t} \qquad (1.121c)$$

Conversion

It is assumed that oxygen is only consumed by reactions involving organic substances which are suspended in stagnant water, and that this consumption is independent of the O_2 concentration as long as it is greater than zero:

$$\text{wc} = \text{constant for } c_{st,O_2} > 0 \qquad (1.122)$$

Transphase exchange

For the present example it is assumed that O_2 is only exchanged

(A) between the flowing and stagnant subsurface water by molecular diffusion (see §1.3.1 and §1.5.1)
(B) between the stagnant subsurface water and the soil air by dissolution and degassing of oxygen (see §1.5.3)

These exchanges are considered to be nonequilibrium processes. When related to volume, they are given in $kg/(sm_v^3)$ as follows (see Equations 1.68, 1.86a, and 1.83):

$$\text{(A)} \quad f_{ex,D} = (D_M/\overline{\Delta l})\,(c_{o,O_2} - c_{st,O_2})O_{Sp}^* \qquad (1.123a)$$

$$\text{(B)} \quad f_{ex} = k_{ex}\Delta\mu \approx k'_{ex}(SI - 1)\,O_{Sp}^* \qquad (1.123b)$$

where k'_{ex} is in $kg/(sm_v^2)$ and the dimensionless saturation index is given by

$$SI = \frac{c_{st,O_2}/@}{x_{a,O_2}/0.2015} \qquad (1.123c)$$

where x_{a,O_2} = O_2 mole fraction in the subsurface air; x_{a,O_2} = $p_{a,O_2}/p_a$
  @ = value characterizing the equilibrium state according to Table 1.18 (c_{st,O_2} and @ in mg/L)
  O_{Sp}^* = effective specific surface area; O_{Sp}^* = $(\omega/V)^*$ in m^2 water interface per m^3 volume of the subsurface (see also Equation 1.68)

Total Balance

Contrary to the first example, the total quality model consists of a system of three partial differential equations with three dependent variables. The structure of this model is illustrated in Figure 1.58d. The dependent variables are the oxygen concentrations in the three mixphases under consideration c_{o,O_2}, c_{st,O_2}, and c_{a,O_2} in mol/L:

■ *Flowing subsurface water (index o):*

$$\underbrace{\frac{\partial}{\partial z}(D_d\,\partial c_{o,O_2}/\partial z - v_o c_{o,O_2})}_{\text{Transport Process}} = \underbrace{\theta_o\,\partial c_{o,O_2}/\partial t + c_{o,O_2}\partial\theta_o/\partial t}_{\text{Storage process}} \qquad (1.124a)$$

$$+ \underbrace{(D_o/\overline{\Delta l})(c_{o,O_2} - c_{st,O_2})O_{sp}^*}_{\text{Exchange process "mobile/immobile water"}}$$

■ *Stagnant subsurface water (index st):*

$$\underbrace{0}_{\text{Transport}} = \underbrace{\theta_{st}\partial c_{st,O_2}/\partial t + c_{st,O_2}\partial\theta_{st}/\partial t}_{\text{Storage process}} + \tag{1.124b}$$

$$\underbrace{(D_o\overline{\Delta l})(c_{o,O_2} - c_{st,O_2})O_{Sp}^* + k_{ex}'O_{Sp}^* \left(\frac{0.2015\, c_{st,O_2}}{@x_{a,O_2}} - 1\right)}_{\text{Exchange processes}} + \underbrace{wc\delta'}_{\text{Consumption}}$$

■ *Subsurface air (index a):*

$$\underbrace{\partial(D_{m,aO_2}\,\partial c_{o,O_2}/\partial z)/\partial z}_{\text{Transport process}} = \underbrace{\theta_a\,\partial c_{a,O_2}/\partial t + c_{a,O_2}\,\partial\theta_a/\partial t}_{\text{Storage process}} +$$

$$\underbrace{-k_{ex}'O_{sp}^* \left(\frac{0.2015\, c_{st,O_2}}{@x_{a,O_2}} - 1\right)}_{\text{Exchange process}} \tag{1.124c}$$

where c_{o_2} = oxygen concentration in mg per L water
  $v_o = v_w$ = Darcy flux of the subsurface water in m/s
  θ_o = volumetric content of flowing water
  θ_{st} = volumetric content of stagnant water
  θ_a = volumetric content of subsurface air
  $\overline{\Delta l}$ = average diffusion path in m
  k_{xx}' = exchange rate constant in kg per s and m^2
  $@$ = oxygen saturation content according to Table 1.18
  wc = oxygen consumption in mg/(sm$_v^3$), usually equal to $w_w c_{o,O_2}$
  $\delta' = \delta' = 1$ at $c_{st,O_2} > 0$, and $\delta' = 0$ at $c_{st,O_2} = 0$
$D_d = D$ = hydrodynamic dispersion coefficient in m^2/s
  D_o = molecular diffusion coefficient of O_2 in subsurface water
D_{m,a,O_2} = diffusion coefficient of O_2 in soil air according to Figure 1.29
  O_{sp}^* = effective specific surface area; $O_{sp}^* = (\omega/V)^*$

The *flow model* Equation 1.119 provides $v_w = K_w\,\partial h_w/\partial z$, θ_w, $\partial\theta_w/\partial t$, and w_w. The values of θ_o, θ_{st}, and θ_a, and their temporal derivatives, are estimated from these values. These quantities act as parameters or as internal and boundary conditions of Equation 1.124.

1.8.2.3 Deironization of Water in the Subsurface

In the third example, we will consider a scenario in which groundwater is characterized by a pH of about 6.5, a pE of approximately zero, and an iron concentration of about 20 mg/L (see also Figure 1.42). We consider further that groundwater flow in the aquifer is driven by a small natural gradient and pumped by a single well. In order to avoid well-clogging by iron hydroxides and to minimize traditional water treatment, it is desirable to eliminate iron (deironization) in the subsurface. For this purpose, pumping must be performed discontinuously. Between pumping intervals, water with a dissolved oxygen concentration of $c_{O_2,max}$ is injected into this well, i.e., with a concentration of approximately 10 mg/L for air-treated water and approximately 50 mg/L for O_2 treated water.

The model objective is to determine the Fe (II), Fe (III), and O_2 concentrations in groundwater as functions of space (r) and time (t) for any pumping and injection regime. An example would be 5 days of pumping with a discharge rate Q and 2 days of injection with a rate 2/3 Q.

Flow Model

The flow model is assumed to be steady state and given by

$$v(r) = -\frac{Q}{2\pi M\,r} \qquad (1.125)$$

where   Q = volume flux in m^3/s (Q>0: discharge)
           M = thickness of the aquifer in m
            r = cylinder coordinate in m (r = 0 corresponds to the well axis)
       v(r) = v_r = Darcy flux in m/s

This model is an appropriate approximation for confined flow conditions near the well (r < 10 to 20 m), small natural flow gradients, and permeabilities greater than 10^{-4} m/s.

Quality Model

In designing the quality model, we once again start from the abstraction chain of figurative models shown in Figure 1.59. It is assumed that no transport takes place in solids or along their surfaces or cation films, respectively (no coupling of s' and s" or f' and f", see Figure 1.59c). The following three migrants are coupled according to their redox process equation, given in Table 1.13:

- the cation Fe^{2+} = Fe (II)
- the gas molecule O_2 = O_2(aq) dissolved in groundwater
- the $Fe(OH)_3$ floccules which are still colloidally dissolved or suspended

Change in pH due to formation of protons H^+ is not taken into consideration because the system is assumed to be well buffered.

In this example, concentrations are used related to the volume of the subsurface. They are designated as $\bar{c}$ and given in $mmol/dm_v^3$ or mg/dm_v^3:

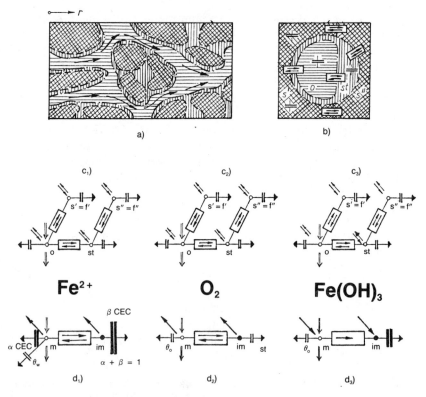

Figure 1.59. Multistage abstraction sequence of figurative models for migration of iron (II), oxygen, and iron hydroxide in the water-saturated subsurface (the same symbols as in Figures 1.57 and 1.58 are used).

Concentration $\bar{c}$ used:

Fe^{2+} : $\bar{c}_{m,Fe} = \theta_o c_{m,Fe}$; $\bar{c}_{im,Fe} = s_{im,Fe}$; $\bar{c}_{c,m,Fe} = z\,\bar{c}_{m,Fe}$;
$\bar{c}_{c,im,Fe} = z\,\bar{c}_{im,Fe} = s_{c,im,Fe}$

O_2 : $\bar{c}_{m,O_2} = \theta_o c_{m,O_2}$; $\bar{c}_{im,O_2} = \theta_{st} c_{im,O_2} = s_{im,O_2}$

$Fe(OH)$: $\bar{c}_{m,Fe(OH)_3} = \theta_o c_{m,Fe(OH)_3}$; $\bar{c}_{im,Fe(OH)_3} = s_{im,Fe(OH)_3}$

Migrant Fe^{2+}

■ *Transport processes* Fe (II) is transported with the flowing groundwater by convection and dispersion (see §1.3.2, Equation 1.34):

$$\vec{fc}_{Fe} = -\theta_o D\,\frac{\partial c_{m,Fe}}{\partial r} + v_r c_{m,Fe} = \frac{1}{\theta_o}\left(-\theta_o D\,\frac{\partial \bar{c}_{m,Fe}}{\partial r} + v_r \bar{c}_{m,Fe}\right) \quad (1.126)$$

■ *Storage and exchange* Fe^{2+} is exchanged between groundwater and solids (cation exchange). This exchange is regarded as an equilibrium process between the flowing aqueous phase (o) with part of the solid surface $\alpha s = s'$ or film area $\alpha f = f'$, and between the stagnant aqueous phase (st) with the remaining part of the solid surface $\beta s = s''$ or film area $\beta f = f''$ (where $\alpha + \beta = 1$). Consequently, only two phases need be considered: the mobile (m) phase in equilibrium with f' and the immobile phase (im) in equilibrium with f'' (see Figure 1.59d and §1.2.2).

The storage capacity of the immobile phase is attributable to the film portion $f'' = \beta O_{sp}^{*}$ and the stagnant water content θ_{st}. Frequently the storage capacity of the stagnant water is small in comparison to the film storage capacity. Hence, θ_{st} can be added to the storage capacity of flowing groundwater θ_{o}. The storage capacity of the mobile phase then results from the film portion $f' = \alpha O_{sp}^{*}$ and from the volumetric water content $\theta_{w} = \theta_{o} + \theta_{st}$ (see Figure 1.59d$_1$).

The cation exchange capacity CEC of solids increases due to precipitation of iron hydroxide formed during deironization. This phenomenon is frequently enhanced by increasing biomass content in dead-end pores during the biologically catalyzed process.

Now let us consider the example given in Table 1.14 in more detail, and assume that CEC has a constant value of 14.2 mmol charges per 100 g solids (see also Table 1.6). With a dry bulk density of 1.7 g/cm^3, the volume-related cation exchange capacity $\overline{\text{CEC}}$ amounts to 240 mmol$_c$/dm$_V^3$. At thermodynamic equilibrium, the cation-equivalent concentration in groundwater is assumed to be 1.5 mmol$_c$/L (see Table 1.14). At a volumetric water content of $\theta_{w} = 0.36$, for example, this corresponds to a volume-related positive charge concentration of about 0.5 mmol$_c$ per dm$_V^3$. In other words, the overwhelming majority of cations is resident in the films.

If 20 mg Fe^{2+} per liter are now added, the concentration in groundwater would be changed by $(2 \cdot 20 \text{ mg/L})/(56 \text{ mg/mmol}) \approx 0.7 \text{ mmol}_c/$L, or by $0.36 \cdot 0.7 \approx 0.25 \text{ mmol}_c/\text{dm}_V^3$), and a Fe^{2+} cation ratio in groundwater of about $0.7/(1.5 + 0.7) \approx 1/3$ would result. When compared to the divalent cations in Table 1.14, such Fe^{2+} concentration in groundwater could likely cause an occupation of approximately 1/6 of the total CEC at thermodynamic equilibrium, and cause a change of the charge concentration of about $240/6 = 40 \text{ mmol}_c/\text{dm}_V^3$ in the films.

As Figure 1.60 shows, the Henry isotherm describes the cation distribution between the immobile phase (film solution) and the mobile groundwater (aqueous phase) over the considered range with sufficient accuracy. But this figure also makes it clear that Henry's isotherm does not hold for iron concentrations in groundwater > 30 to 40 mg/L.

The nonequilibrium models of cation exchange which correspond to Henry's and Langmuir's isotherm are obtained in volume-related units as follows (see also §1.5.5):

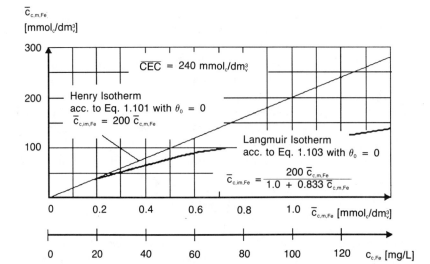

Figure 1.60. Henry's and Langmuir's adsorption isotherms for Fe^{2+} equivalents in the mobile and immobile phase per unit volume of the aquifer estimated according to Table 1.14 for the present example.

Equilibrium	**Nonequilibrium**

Henry's isotherm:

$\bar{c}_{c,im,Fe} = K\bar{c}_{c,m,Fe}$

$K = 200$ according to Figure 1.60

$EX = (k_1\bar{c}_{c,m,Fe} - k_2\bar{c}_{c,im,Fe})O^*_{sp}$

$= k_2(K\bar{c}_{c,m,Fe} - \bar{c}_{c,im,Fe})O^*_{sp}$

with $k_1/k_2 = K = 200$ (see Figure 1.60 and Equations 1.46d and 1.68)

Langmuir's isotherm:

$$\bar{c}_{c,im,Fe} = \frac{200\bar{c}_{c,m,Fe}}{1.0 + 0.833\bar{c}_{c,m,Fe}}$$

$EX = k_1 O^*_{sp}[(\overline{CEC} - \bar{c}_{c,im,Fe})\bar{c}_{c,m,Fe}$

$- k_2\bar{c}_{c,im,Fe}]$

with $\overline{CEC} = 240$ mmol$_c$/dm$_v^3$ and $K/\overline{CEC} = 200/240 = 0.833$.

■ *Storage models* These models are given with *Henry's* storage characteristic as

$$S_{im,Fe} = \bar{c}_{im,Fe} = \beta\, 200\bar{c}_{m,Fe} \tag{1.127a}$$

$$S_{m,Fe} = \bar{c}_{m,Fe}(1 + \alpha 200) \tag{1.127b}$$

and with *Langmuir's* storage characteristic as

$$S_{im,Fe} = \bar{c}_{im,Fe}\,\beta\,200/(1 + 0.833\,\bar{c}_{c,m,Fe}) \tag{1.127c}$$

$$S_{m,Fe} = \bar{c}_{m,Fe} \left(1 + \frac{\alpha 200}{1 + 0.833\bar{c}_{c,m,Fe}}\right) \tag{1.127d}$$

■ *Exchange models*   These models are given with *Henry's* characteristic as

$$EX_{Fe} = k_2 O^*_{sp}(\beta\ 200\bar{c}_{m,Fe} - \bar{c}_{im,Fe}) \tag{1.128a}$$

and with *Langmuir's* characteristic as

$$EX_{Fe} = k_1 O^*_{sp}[0.833(\beta\ 240 - \bar{c}_{c,im,Fe})\ \bar{c}_{m,Fe} - \bar{c}_{im,Fe}] \tag{1.128b}$$

where $\bar{c}_{c,im}\bar{c}_m = \bar{c}_{im}\bar{c}_{cm}$

kO^*_{sp} = exchange rate constant in s^{-1} estimated from $D_m/\Delta l$ and ω/V according to Equation 1.68

α,β = partitioning parameters ($\alpha + \beta = 1$), dimensionless

K = k_1/k_2, ratio of rate coefficients; in the example $K = 200$

$\overline{CEC}$ = volume-related cation exchange capacity; in the example $\overline{CEC} = 240$ mmol$_c$/dm^3_V

■ *Conversion*   The conversion of Fe^{2+} by oxidation takes place in two stages [1.26]. The stoichiometric balance equations of both stages and of their sum are given by

$$4\ Fe^{2+} + O_2 + 4\ H^+ \xrightarrow[\text{chemical}]{\text{biological}} 4\ Fe^{3+} + 2\ H_2O + \text{energy}$$

$$4\ Fe^{3+} + 12\ H_2O \xrightarrow{\hspace{2cm}} 4\ Fe(OH)_3 + 12\ H^+ \tag{1.129}$$

$$4\ Fe^{2+} + O_2 + 10\ H_2O \xrightarrow[\text{chemical}]{\text{biological}} 4\ Fe(OH)_3 + 8\ H^+ + \text{energy}$$

Figure 1.61 illustrates these reactions which proceed in the adsorption films (see also Figure 1.43). The conversion rate of this total process can be derived from Equation 1.46b as:

$$r_{Fe} = kO^*_{sp}\ [Fe^{2+}][O_2] = k^*\ \bar{c}_{Fe}\ \bar{c}_{O_2} \tag{1.130a}$$

However, in Hummel [1.38] and Mreganga [1.56] the following reaction equation is suggested:

$$r_{Fe} = k'O^*_{sp}\ [Fe^{2+}][O_2]/[H^+]^2 = k''O^*_{sp}[Fe^{2+}][O_2][OH^-]^2 \tag{1.130b}$$

which corresponds to Equation 1.130a if pH, and hence $[H^+]$, are considered constant.

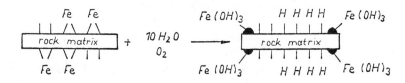

Figure 1.61.   Transformation processes of Fe^{2+} bound to solids according to Equation 1.129.

Migrant O_2

■ *Transport processes*   The dissolved molecular O_2 is transported by flowing groundwater in the same manner as Fe^{2+} (see Equation 1.126):

$$\vec{fc}_{O_2} = (1/\theta_o)(-\theta_o D \partial \bar{c}_{m,O_2}/\partial r + v_r \bar{c}_{m,O_2})$$

■ *Storage and exchange*   The dissolved molecular oxygen O_2 is assumed to be stored only in the flowing and stagnant water, and exchanged between both phases. Under these conditions the following partial process models are obtained:

Storage:

$$s_{m,O_2} = \theta_o c_{m,O_2} = \bar{c}_{m,O_2} \tag{1.131a}$$

$$s_{im,O_2} = \theta_{st} c_{im,O_2} = \bar{c}_{im,O_2} \tag{1.131b}$$

Exchange:

At equilibrium $c_{im,O_2} = c_{m,O_2}$, thus at *nonequilibrium*

$$EX_{O_2} = k_2 O_{sp}^* \left(\frac{\theta_{st}}{\theta_o} \bar{c}_{m,O_2} - \bar{c}_{im,O_2} \right) \tag{1.131c}$$

is valid if *Henry's* characteristic is used.

■ *Conversion*   The conversion rate of O_2 must correspond to the conversion rate of Fe^{2+}. Due to the stoichiometric balance according to Equation 1.129, 4 mol Fe^{2+} react with 1 mol O_2. Thus, using the model Equation 1.130a, the conversion rate of O_2 is

$$r_{O_2} = (1/4)\, r_{Fe} = 0.25\, k^* \bar{c}_{Fe}\, \bar{c}_{O_2} \qquad (1.132a)$$

Furthermore, a consumption of O_2 due to oxidation of other substances including biomass should be expected in the mobile and immobile phase. Considering this as only a bulk approach, the following two limiting cases may be distinguished:

1. The groundwater influx does not contain any other oxygen-consuming solutes besides Fe (II). It is assumed that all oxygen-consuming immobile substances were previously oxidized. If the oxygen-enriched infiltrate (injected water) now penetrates once again into the same part of the aquifer, Equation 1.132a can be used.
2. The oxidizable immobile substance stored in the aquifer is regarded to be large enough to neglect its consumption due to oxidation by dissolved O_2 in the injected water. The additional consumption of oxygen in the immobile phase may then be reflected by a first-order reaction rate:

$$r_{O_2} = k^{**} \bar{c}_{O_2} \qquad (1.132b)$$

However, for the example considered, the following simplified model is used:

$$r_{O_2} = (\phi/4)\, r_{Fe} \qquad (1.132c)$$

with $\phi > 1$. The largest value of ϕ has to be chosen for the first push-pull cycle. Subsequently, with increasing cycle numbers, ϕ tends asymptotically to a limit.

Migrant Fe(OH)₃

■ *Transport processes*   The colloidally dissolved or suspended $Fe(OH)_3$ particles are transported with the flowing groundwater by convection and dispersion according to Equation 1.126.

■ *Storage and exchange*   The $Fe(OH)_3$ microfloccules formed in the adsorption films are assumed to remain attached or stored (see Figure 1.61). The microfloccules formed in the stagnant water should also become immediately adsorbed; thus, their exchange with the flowing groundwater can be neglected. Hence, only the microfloccules formed in the flowing water are considered to be subjected to exchange with the immobile phase. This process should be reflected by a model according to Henry's exchange characteristic.

The $Fe(OH)_3$ microfloccules covering the rock matrix age and become dehydrated. This aging process may be simply described in two stages:

$$
\begin{array}{c}
\begin{array}{c}
Fe{\Large\langle}\!\!\!\!\begin{array}{l}OH\\-OH\\OH\end{array}\\[1em]
Fe{\Large\langle}\!\!\!\!\begin{array}{l}OH\\-OH\\OH\end{array}
\end{array}
\xrightarrow[\text{First stage}]{\text{"Aging"}}
\begin{array}{c}
Fe=O\\\;\;\;\;\searrow OH\\[0.5em]
\;\;\;\;\nearrow OH\\Fe=O
\end{array}
+\;H_2O
\xrightarrow[\text{Second stage}]{\text{"Aging"}}
\begin{array}{c}
Fe{\Large\langle}\!\!\!\!\begin{array}{l}O\\O\end{array}\\[0.5em]
Fe{\Large\langle}\!\!\!\!\begin{array}{l}O\\O\end{array}
\end{array}
+\;3H_2O
\end{array}
\qquad (1.133)
$$

$Fe(OH)_3$ (limonide)	α–FeOOH (geothite)	Fe_2O_3 (amorphous)
iron gel	β–FeOOH (lepidocrocite)	Fe_2O_3 (crystalline)

During aging, the density of the iron precipitate increases significantly, and its volume decreases correspondingly. In the present example, therefore, it is assumed that the permeability coefficient of the aquifer is not affected by precipitation of $Fe(OH)_3$, $FeOOH$, and Fe_2O_3. Hence, the quality model does not feed back to the flow model (see Figure 3 in the Introduction). Nevertheless, for real cases, this assumption must be checked by calculations of the precipitated volume.

Taking these considerations into account, the following submodels result. For storage:

$$S_{m,Fe(OH)_3} = \theta_o c_{m,Fe(OH)_3} = \bar{c}_{m,Fe(OH)_3} \qquad (1.134a)$$

$$S_{im,Fe(OH)_3} = \bar{c}_{im,Fe(OH)_3} \qquad (1.134b)$$

For exchange:

$$EX_{Fe(OH)_3} = k_2 O^*_{sp} \bar{c}_{m,Fe(OH)_3} \qquad (1.134c)$$

■ *Conversion*   The conversion rate of $Fe(OH)_3$ is linked with the conversion rate of Fe^{2+}. According to Equation 1.129, 4 moles Fe^{2+} react with 4 moles $Fe(OH)_3$. Thus, using Equation 1.130a it follows

$$r_{Fe(OH)_3} = r_{Fe} = k^* \bar{c}_{Fe} \bar{c}_{O_2} \qquad (1.135)$$

■ *Total balance*   The overall mathematical quality model, the structure of which is illustrated in Figure 1.59d_1 to d_3, now consists of a system with six partial differential equations (see also Figure 1.56). The following volume-related concentrations are used as the six dependent variables:

1. $\bar{c}_{m,Fe}$	3. $\bar{c}_{m,O_2}$	5. $\bar{c}_{m,Fe(OH)_3}$
2. $\bar{c}_{im,Fe}$	4. $\bar{c}_{im,O_2}$	6. $\bar{c}_{im,Fe(OH)_3}$

This system is coupled by the exchange and conversion processes and has the following structure:

$$\text{div} \left(\frac{1}{\theta_o} \left(-\theta_o D \frac{\partial \bar{c}_{m,Fe}}{\partial r} + v_r \bar{c}_{m,Fe} \right) \right) = (\phi/\theta_o + \alpha K) \frac{\partial \bar{c}_{m,Fe}}{\partial t}$$
$$+ k_2 O_{sp}^* (\beta K \bar{c}_{m,Fe} - \bar{c}_{im,Fe}) + k' \bar{c}_{m,Fe} \bar{c}_{m,O_2} - \overline{wc}_{m,Fe} \quad (1.136a)$$

$$0 = \frac{\partial \bar{c}_{im,Fe}}{\partial t} - k_2 O_{sp}^* (\beta K \bar{c}_{mFe} - \bar{c}_{im,Fe})$$
$$+ k'' \bar{c}_{im,Fe} \bar{c}_{im,O_2} - \overline{wc}_{im,Fe} \quad (1.136b)$$

$$\text{div} \left(\frac{1}{\theta_o} \left(-\theta_o D \frac{\partial \bar{c}_{m,O_2}}{\partial r} + v_r \bar{c}_{m,O_2} \right) \right) = \frac{\partial \bar{c}_{m,O_2}}{\partial t}$$
$$+ k_2 O_{sp}^* (\bar{c}_{m,O_2} \theta_{st}/\theta_o - \bar{c}_{im,O_2}) + (\phi/4) k' \bar{c}_{m,Fe} \bar{c}_{m,O_2} - \overline{wc}_{m,O_2} \quad (1.136c)$$

$$0 = \frac{\partial \bar{c}_{im,O_2}}{\partial t} - k_2 O_{sp}^* (\bar{c}_{m,O_2} \theta_{st}/\theta_o - \bar{c}_{im,O_2})$$
$$+ (\phi/4) k'' \bar{c}_{im,Fe} \bar{c}_{im,O_2} - \overline{wc}_{im,O_2} \quad (1.136d)$$

$$\text{div} \left(\frac{1}{\theta_o} \left(-\theta_o D \frac{\partial \bar{c}_{m,Fe(OH)_3}}{\partial r} + v_r \bar{c}_{m,Fe(OH)_3} \right) \right) = \frac{\partial \bar{c}_{m,Fe(OH)_3}}{\partial t}$$
$$+ k_2 O_{sp}^* \bar{c}_{m,Fe(OH)_3} - k' \bar{c}_{m,Fe} \bar{c}_{m,O_2} - \overline{wc}_{im,Fe(OH)_3} \quad (1.136e)$$

$$0 = \frac{\partial \bar{c}_{im,Fe(OH)_3}}{\partial t} - k_2 O_{sp}^* \bar{c}_{m,Fe(OH)_3} - k'' \bar{c}_{im,Fe} \bar{c}_{im,O_2} - \overline{wc}_{im,Fe(OH)_3} \quad (1.136f)$$

where div $\vec{\chi}$ stands for $\partial \chi/\partial r + \chi/r$ of a radial flow and transport pattern and $\overline{wc}$ represents a source/sink term.

The migration model according to Equation 1.136 represents the tracer case.

1.8.3 INITIAL AND BOUNDARY CONDITIONS

Each migration process takes place in a limited space and a limited time. The spatial boundaries separate the considered system from its environment. The same applies to the time limits. Only in special cases, space and time are considered as unlimited.

All boundaries refer to the independent variables used, such as the space and time variables of the flow and quality model. For instance, in the second example of §1.8.2.2 (see Equations 1.119 and 1.124), the independent variables are z (space variable) and t (time variable).

In solving real problems, very careful consideration must be made to delineate the independent variables. This task of defining spatial and temporal limits for the object under investigation is often one of the most important steps in migration modeling. Delineation of the independent variables threatens the continuity between the investigated system and its surrounding environment, which is not evaluated in the model. In the second example of §1.8.2.2, for instance, the limits of the independent variables are $z_1 < z < z_2$ and $t_1 < t < t_2$.

The dependent variables of the flow and quality models are functions of the independent variables. Generally, they can be detected in the mathematical model by their derivatives with respect to the independent variables. For the example of Equations 1.119 and 1.124, the dependent variables are the following:

$$v_o = -(K_w \partial h_w / \partial z)h_w; \quad \theta_w;$$
$$fc_D = -(D_o \partial c_{o,O_2} / \partial z); \quad (v_o c_{o,O_2}); \quad \theta_o;$$
$$c_{st,O_2}; \quad \theta_{st}$$
$$fc_{m,a,O_2} = -(D_{M,a,O_2} \partial c_{a,O_2} / \partial z); \quad c_{a,O_2}; \quad \theta_a$$

The following first-order derivatives occur in Equations 1.119 and 1.124:

$$\frac{\partial \theta_w}{\partial t}; \frac{\partial (v_o c_{o,O_2})}{\partial z}; \frac{\partial c_{o,O_2}}{\partial t}; \frac{\partial \theta_o}{\partial t}; \frac{\partial c_{st,O_2}}{\partial t}; \frac{\partial \theta_{st}}{\partial t}; \frac{\partial c_{a,O_2}}{\partial t}; \frac{\partial \theta_a}{\partial t} \qquad (1.137a)$$

and the following terms can be written as first-or second-order derivatives:

$$\frac{\partial v_o}{\partial z} \text{ or } \frac{\partial}{\partial z} (K_w \frac{\partial h_w}{\partial z}); \quad \frac{\partial fc_D}{\partial z} \text{ or } \frac{\partial}{\partial z} (D_o \frac{\partial c_{o,O_2}}{\partial z});$$

$$\frac{\partial fc_{m,a,O_2}}{\partial z} \text{ or } \frac{\partial}{\partial z} (D_{m,a,O_2} \frac{\partial c_{a,O_2}}{\partial z}) \qquad (1.137b)$$

The values exhibited by the dependent variables at the limits of the independent variables are called the *external boundary conditions*. For each of the first-order derivatives in the mathematical model, the corresponding boundary condition must be formulated on one boundary; for the second-order derivatives, two conditions on two boundaries must be found.

For all time derivatives of first order, the preferred limit is the initial time $t_1 = 0$. The values describing the initial systems state at time $t_1 = 0$ are usually called the *initial conditions*. The initial conditions of the example considered are the dependent variables at z and $t = 0$:

$$c_{o,O_2}(z,0); \quad \theta_o(z,0); \quad c_{st,O_2}(z,0); \quad c_{a,O_2}(z,0); \quad \phi_a(z,0) \quad (1.138a)$$

Initial conditions must be known for each transient investigation.

Convection terms of the quality models must be treated in the same manner as the first-order time derivatives; i.e., they must be defined on one of the space boundaries, preferably at the inflow points of the considered model region. For the example this means

$$(v_o c_{o,O_2})|_{z_1 \text{ or } z_2, t} \qquad t_1 < t < t_2 \qquad (1.138b)$$

For the terms according to Equation 1.137b, either the dependent variables (h_w, c_{o,O_2}, and c_{a,O_2}) or their first-order derivatives, and hence the flux quantities, must be prescribed on both boundaries. The prescribing functions are termed *boundary conditions* of first and second order, respectively. Hence, for our example, the following boundary conditions must be specified:

First type:

$$h_w(z_1,t), \; h_w(z_2,t), \; c_{o,O_2}(z_1,t), \; c_{o,O_2}(z_2,t),$$
$$c_{a,O_2}(z_1,t), \; c_{a,O_2}(z_2,t) \qquad (1.138c)$$

Second type:

$$v_w(z_1,t), \; v_w(z_2,t) \; \vec{fc}_D(z_1,t), \; \vec{fc}_D(z_2,t),$$

$$\vec{fc}_{m,a,O_2}(z_1,t), \; \vec{fc}_{m,aO_2}(z_2,t) \qquad (1.138d)$$

Boundary conditions of the third type are combinations of the first and second type. For flow models, third-type boundary conditions are of particular interest when a hydraulic transition or resistance occurs on the boundary, such as a semipervious or clogged bottom layer of a lake, or a skin resistance around a well screen [1.11]. For quality models, third-type boundary conditions should always be formulated when a given heat or mass flux is fed into or withdrawn from the considered model region, which splits into the convective and the dispersive flux at the boundary. In the example considered, the following third-type boundary conditions could be prescribed for the flow and quality model, respectively:

$$h_{w,b} + a_o \, \partial h_{w,b}/\partial z = h_{ex} \qquad (1.139a)$$

where h_{ex} and a_o are prescribed (see, e.g., [1.11, p. 188 and 206]), and

$$\left(v_o c_{o,O_2} - D_o \frac{\partial c_{o,O_2}}{\partial z}\right)\Big|_{z_1 \text{ or } z_2, t} = (cv)_{ex} = \dot{m}_{ex}/A \qquad (1.139b)$$

where $\dot{m}_{ex}A$ and D_o are prescribed.

The environment additionally acts upon the interior of the delineated system. Conditions which become effective within the spatially delineated system are designated as *internal boundary conditions.* In this example, these are the boundary conditions which are effective in the region $z_1 < z < z_2$. Internal boundary conditions must be formulated as source-sink terms and are added to the right-hand side of the mathematical flow and quality model, respectively.

In specifying internal boundary conditions, one should take into account that in migration problems, the source-sink terms of the flow model nearly always become effective as source-sink terms in the quality model. In the example considered, w_w in Equation 1.119 is such a term, describing the impact of the external system "plant" onto the investigated system "soilwater flow." Presetting of this function $w_w = f(z,t,T,\theta_o)$ is indispensable in simulating soilwater flow problems. Plant water extraction also implies solute extraction $wc = w_w c$, which forms the internal boundary condition of the quantity model (Equation 1.124).

References for Part 1

1. Ackermann, G. Lehrwerk Chemie, Lehrbuch 5: Elektrochemie. Leipzig: Dt. Verl. f. Grundstoffindustrie, 1974.
2. Anderson, M. P. Using models to simulate the movement of contaminants through groundwater flow systems. Critical Reviews in Environmental Control (1979), pp. 97–156.
3. Balstrup, T. Varmepumpeanlaeg: Varmeovergangsforhold i jord, Bd. 1. Kobenhavn: Teknologisk inst. Forlog, Denmark, 1977.
4. Bear, J. Dynamics of fluids in porous media. New York: Elsevier, 1972.
5. Behrens, H., and K.-P. Seiler. Field tests on propagation of conservative tracers in fluvioglacial gravels of Upper Bavaria. In Quality of Groundwater. Amsterdam: Elsevier, 1981, pp. 649–657.
6. Behrens, H., and K.-P. Seiler. Beziehungen zwischen der Dispersivität und der Länge des Fließweges des Grundwassers nach Geländeversuchen in den fluvioglazialen Kiesen des oberen Loisachtales und der Dornach. GSF-Bericht; R 290 des Institutes für Radiohydrometrie, München, 1982, pp. 323–330.
7. Beims, U., and L. Luckner. Grundlagen der Ermittlung repräsentativer Durchlässigkeitsparameter. Berlin, Zeitschr. für angewandte Geologie 20 (1974) 7, pp. 304–313.
8. Benson, L. V. A tabulation and evaluation of ion exchange data on smectites, certain zeolites and basalt. Berkeley Laboratory, 1980 (Topical Report No. 2; LBL-10541).
9. Benson, L. V., and L. S. Teagne. A tabulation of thermodynamic data for chemical reactions involving 58 elements common to radioactive waste package systems. Berkeley: University of California, Lawrence Berkeley Laboratory, 1980 (Topical Report No. 4; LBL-11448).
10. Bloemen, G. W. Calculation of hydraulic conductivities of soils from texture and organic matter content. Z. f. Pflanzenernährung und Bodenkunde 143 (1980) 5, pp. 581–605.

11. Busch, K.-F., and L. Luckner. Geohydraulik. Leipzig: Dt. Verl. für Grund-stoffindustrie, 1972.

12. Busch, K.-F., et al. Ingenieurökologie. Jena: Fischer-Verl., 1983.

13. Cordes, J. F. Größen-und Einheitssysteme; SI-Einheiten. Analytiker Ta-schenbuch, Bd.2, Berlin: Akademie-Verlag, 1981, pp. 3–29.

14. Cushman, J. H. Volume averaging, probabilistic averaging, and ergodicity. Advances in Water Resources 6 (1983), pp. 182–184.

15. Czolbe, P., H.-J. Kretzschmar, and G. Kühnel. Reservoirmechanische Untersuchungen zur Gas-Wasser-Strömung in Gasspeichern, Teil I: Labora-tive Messungen der Phasenpermeabilität und der Gassorption im Spei-chergestein. Zeitschr. f. angew. Geologie 28 (1982) 2, pp. 66–72.

16. Damroth, H. et al. Wasserinhaltsstoffe im Grundwasser—Reaktionen, Tran-sportvorgänge und deren Simulation. Berlin: Schmidt-Verl., 1979 (Berichte/Umweltbundesamt, 1979, 4).

17. Danielopol, L. D. Introduction to groundwater ecology. Limnol, Inst. österr. Akademie der Wiss., 1979 (Lecture notes for the UNESCO Training Course in Limnology).

18. D'Ans, J., and E. Lax. Taschenbuch für Chemiker und Physiker. Berlin: Springer-Verl., 1949.

19. Davis, J. B. Petroleum microbiology. New York: Elsevier, 1967.

20. Davis, S. N., and R. J. M. De Wiest. Hydrogeology. New York: Wiley & Sons, 1966.

21. Diersch, H.-J. Modellierung und numerische Simulation geohydrodynamis-cher Transportprozesse. Berlin, Akad. der Wissensch. dehemmoiliger DDR, Diss. B, 1984.

22. Dyck, S., et al. Angewandte Hydrologie, Teil I: Berechnung und Regelung des Durchflusses der Flüsse. Berlin: Verl. für Bauwesen, 1976.

23. Everett, L. G. Groundwater monitoring. New York: General Electric Com-pany, Technology Marketing Operation, 1980.

24. Fenske, P. R. Time-dependent sorption on geological materials. J. of Hydrology 43 (1979), pp. 415–425.

25. Fischer, R. Laborexperimentelle Untersuchungen und Modellierung der Pyritverwitterung. Techn. Univ. Dresden, Abschlußarbeit im postgr. Stu-dium Grundwasser, 1983.

26. Fischer, R. Entwicklung und Anwendung eines mathematischen Modells zur Berechnung der Komplexbildung in natürlichen Wässern unter besonderer Berücksichtigung der Eisenmigration sowie der Pyrit-und Markasitverwit-terung im Grundwasserleiter. Techn. Univ. Dresden, Sektion Wasserwesen, Diss. A, 1983.

27. Freeze, A. R., and J. A. Cherry. Groundwater. Englewood Cliffs: Prentice-Hall, 1979.

28. Gelhar, L. W. Stochastic analysis of flow in aquifers. Advances in ground-water hydrology, 2. Ed., Bd. 57. Minneapolis: Am. Water Res. Association, 1976.

29. Gelhar, L. W., and C. L. Axness. Three-dimensional stochastic analysis of macrodispersion in aquifers. Water Res. Research 19 (1983) 1, pp. 161–180.

30. Globus, A. M., and V. M. Mogilevskij. Termoosmose i termosamo diffuzia zidkoj fazy v kapillarnoporistoj srede. Issled. processov abmena energiej i

vesestvom v sisteme pocva-rastenie-vozduh. Leningrad: Nauka, 1972, pp. 70–80.

31. Godavikov, A. A. Vedenie v mineralogiu. Novosibirks: Nauka, 1973.

32. Greenkorn, R. A. Flow phenomena in porous media—fundamentals and applications in petroleum, water, and food production. New York: Dekker, 1982.

33. Griethe, H. P. Beitrag zur Bestimmung der Wärmetransporteigenschaften von nichtbindigen Böden unter besonderer Berücksichtigung des teilgesättigten Zustandes. Aachen: Inst. für Wasserbau und Wasserwirtschaft, 1977 (Mitt./Inst. f. Wasserbau und Wasserwirtschaft, 21).

34. Grunert, S., and H. Schneider. Grundlagen der Ingenieurgeologie. Techn. Univ. Dresden, Sektion Bauwesen, Lehrbriefe.

35. Hölting, B. Hydrogeologie—Einführung in die Allgemeine und Angewandte Hydrogeologie. Stuttgart: Ferd. Enke Verl., 1980.

36. Hubbert, M. K. Entrapment of petroleum under hydrodynamic conditions. Bull. of the Am. Ass. of Petrol. Geologists 37 (1953), pp. 1954–2026.

37. Hubbert, M. K. The theory of ground water motion. Journal of Geology 48 (1940) 8, pp. 785–944.

38. Hummel, J. Aufbau eines Migrationsmodells der unterirdischen Enteisenung für eine eindimensionale Stromröhre. Techn. Univ. Dresden, Abschlußarbeit im postgr. Studium Grundwasser, 1983.

39. Husmann, S. Die gegenseitige Ergänzung theoretischer und angewandter Grundwasser-Limnologie, mit Ergebnissen der Wasserwerke Wiesbadens. Die Sicherstellung der Trinkwasserversorgung Wiesbadens. Wiesbaden: Stadtwerke, 1971, pp. 79–90.

40. Husmann, S. Die Bedeutung der Grundwasserfauna für die biologischen Reinigungsvorgänge im Interstitial von Lockergesteinen. Gas-u. Wasserfach/Wasser, Abwasser 119 (1978) 6, pp. 293–332.

41. Jackson, R. E., and K. J. Inch. Hydrogeochemical processes affecting the migration of radionuclides in a fluvial sand aquifer at the Chalk River nuclear laboratories. Ottawa: NHR, 1980, p. 58 (NHRI Paper No 7, Sci. Series, 104).

42. Kirkham, D., and W. L. Powers. Advanced soil physics. New York: Wiley & Sons, 1972.

43. Kittner, H., W. Starke, and D. Wissel. Wasserversorgung. Berlin: Verlag für Bauwesen, 1975.

44. Klotz, D., and H. Moser. Hydrodynamic dispersion as aquifer characteristic—Model experiments by means of radioactive tracers. In Isotope Techniques in Groundwater Hydr., Vienna, IAEAE-SM-182/42.

45. Köpke, U. Hydrobiologische Grundlagen und Probleme der Grundwasserbewirtschaftung. Berlin: Institut für Wasserwirtschaft, 1983.

46. Kortüm, G. Einführung in die chemische Thermodynamik, Weinheim: Verl. Chemie, 1963.

47. Krass, M., et al. Daten zur Geochemie der Elemente. Berlin: Zentrales Geol. Institut, 1969.

48. Kretzschmar, H.-J., and P. Czolbe. Reservoirmechanische Untersuchungen zur Gas-Wasser-Strömung in Gasspeichern, Teil 2: Reservoirmechanische

Berechnungskonzeption zu Phasendispersion und Feldbedingungen. Z. für angew. Geologie 28 (1982) 3, pp. 126–132.

49. Langguth, H.-R., and R. Voigt. Hydrogeologische Methoden. Berlin: Springer-Verl., 1980.

50. Luckner, L. Maßstabsprobleme der hydrodynamischen Dispersion. In 27. Mezdunarodnuj geologiceskij kongress, Moskva 1984, Tezisy, tom VII, Sek. 13–16. Moskva: Nauka, 1984, pp. 456–457.

51. Luckner, L., and C. Nitsche. Bildung systembeschreibender Modelle der Migrationsprozesse in der Aerationszone und ihre digitale Simulation. Geodät. und geophysikalische Veröffentlichungen. Berlin: Reihe IV (1980) 32, pp. 82–99.

52. Luckner, L., and G. Schreiber. Parametermodelle für die Saugspannungs-Sättigungs-Verteilung und den kapillaren k-Wert ungesättigter Böden. Geodätische und geophysikalische Veröffentlichungen. Berlin: Reihe IV (1980) 32, pp. 99–109.

53. Luckner, L., and K. Tiemer. Mathematische Modellbildung der Geofiltration und Migration. Techn. Univ. Dresden, Lehrbrief im postgr. Studium Grundwasser, 1981, H.1.

54. Luckner, L., M. T. van Genuchten, and D. R. Nielsen. A consistent set of parametric models for the flow of two immiscible fluids in the subsurface. Water Res. Research 25 (1989) 10, pp. 2187–2193.

55. Matthess, G. The properties of groundwater. New York: Wiley & Sons, 1982.

56. Mreganga, M. G., et al. Precipitation of iron in aerated ground waters. Proc. ASCE, Journal of the Sanitary Eng. Div. 92 (1966) Feb., pp. 199–213.

57. Mironenko, V. A., and V. M. Schestakow. Osnovy gidrogeomechaniki. Moskva: Nedra, 1974.

58. Möbius, H.-H., and W. Dürschen. Lehrwerk Chemie, Lehrbuch 5: Chemische Thermodynamik. Leipzig: Dt. Verl. f. Grundstoffindustrie, 1973.

59. Molz, F. J., et al. An examination of scale-dependent dispersion coefficients. Groundwater 21 (1983) 6, pp. 715–725.

60. Pickens, J. F., and G. E. Grisatz. Scale dependent dispersion in a stratified granular aquifer. Water Res. Research 17 (1981) 4, pp. 1191–1211.

61. Plummer, L. N. WATEQF-alpha FORTRAN IV version of WATEQ, a computer program for calculating chemical equilibrium of natural waters. U.S. Geological Survey, 1978.

62. Prickett, T. A. A random-walk solute transport model for selected groundwater quality evaluations. Illinois State Water Survey 65 (1981).

63. Rao, P. V., et al. A stochastic approach for describing convective dispersive solute transport in saturated porous media. Water Res. Research 17 (1981) 4, pp. 963–968.

64. Reissig, H. Grundlagen der Hydrogeochemie. Techn. Univ. Dresden, Lehrbrief im postgr. Studium Grundwasser, 1982.

65. Rethali, L. Groundwater in civil engineering. Budapest: Akademiai Kiado, 1983.

66. Rösler, H.-J. Lehrbuch der Mineralogie. Leipzig: Dt. Verl. f. Grundstoffindustrie, 1979.

67. Rösler, H.-J., and H. Lange. Geochemische Tabellen. Leipzig: Dt. Verl. f. Grundstoffindustrie, 1975.
68. Rösler, H.-J., and R. Starke. Einführung in die Tonmineralogie. Freiberg: Bergakademie Freiberg, 1967 (Lehrbriefe für das Hochschulfernstudium).
69. Rosal, A. A. Metody opredelenia migracionnyh parametrov. Obzor VUEMC, Moskva (1980), p. 63.
70. Schestakow, W. M. Dinamika podzemnyh vod. Izd. Moskovskogo Universiteta, 1979, p. 368.
71. Schestakow, W. M. Modeli perenosa v neodnorodnyh plastah i porodah. In Wiss. Konferenz der Techn. Univ. Dresden zur Simulation der Migrationsprozesse im Boden-und Grundwasser, Bd. II. Dresden: Techn. Universität, 1979, pp. 56–61.
72. Schestakow, W. M. Migrationsmodelle für heterogene Medien mit Blockstruktur. Berlin: Zeitschr. für angew. Geologie, Bd. 33 (1987) 12, pp. 314–320.
73. Schön, J. Petrophysik: Physikalische Eigenschaften von Gesteinen und Mineralen. Berlin: Akademie-Verlag, 1983.
74. Schwartz, F. W. Macroscopy in porous media: the controlling factors. Water Res. Research 13 (1977) 4, pp. 743–752.
75. Simmons, C. S. A stochastic-convective transport representation of dispersion in one dimensional porous media systems. Water Res. Research 18 (1982), pp. 1193–1214.
76. Smirnov, S. U. Proishozdenie solenosti podzemnyh vod sedimentacionnyh bassejnov. Moskva: Nedra, 1971.
77. Smith, L., and F. W. Schwartz. Mass transport. 1. A stochastic analysis of macroscopic dispersion. Water Res. Research 16 (1980) 2, pp. 303–313.
78. Smith, L., and F. W. Schwartz. Mass transport. 2. Analysis of uncertainty in prediction. Water Res. Research 17 (1981) 2, pp. 351–369.
79. Smith, L., and F. W. Schwartz. Mass transport. 3. Role of hydraulic conductivity in prediction. Water Res. Research 17 (1981) 5, pp. 1463–1479.
80. Sommer, K. Wissensspeicher Chemie. Berlin: Verl. Volk und Wissen, 1977.
81. Strunz, H. Mineralogische Tabellen, 6. Aufl. Leipzig: Geest & Portig, 1976.
82. Stugren, B. Grundlagen der allgemeinen Ökologie, 3. Aufl. Jena: Fischer Verl., 1978.
83. Stumm, W., and J. J. Morgan. Aquatic chemistry. An introduction emphasizing chemical equilibria in natural waters. New York: Wiley & Sons, 1970.
84. Tjutjonowa, F. I. Physiko-chemische Prozesse in Grundwässern. Leipzig: Dt. Verl. f. Grundstoffindustrie, 1980.
85. Travis, C. C. Mathematical description of adsorption and transport of reactive solutes in soil: a review of selected literature. Oak Ridge, Tennessee: Oak Ridge National Laboratory, 1978.
86. Truesdell, A. H., and B. F. Jones. WATEQ, a computer program for calculating chemical equilibria on natural waters. U.S. Geological Survey Jour. Research (1973).
87. Truesdell, A. H., and B. F. Jones. WATEQ, a computer program for calculating chemical equilibria on natural waters. U.S. Geological Survey Jour. Research (1974), pp. 233–248.

88. Uhlmann, D. Hydrobiologie: Ein Grundriβ für Ingenieure und Naturwis-
senschaftler, 2. Aufl. Jena: Fischer-Verl., 1982.
89. De Wiest, R. J. M. Geohydrology. New York: Wiley & Sons, 1965.
90. Werner, A. Neuere Anschauungen auf dem Gebiete der anorganischen Chemie. Braunschweig: Vieweg & Sohn, 1923.

Part 2

Solution Methods

2.1

Analytical Solution Methods

The result of mathematical modeling is the description of the total migration process including the initial and boundary conditions as presented in Chapter 1.8. Schematizations, approximations, and transformations of this basic process description are usually required to formulate a model which is compatible with the planned solution procedure. Such a model is commonly called a *simulation model*.

The most important solution methods for migration models are analytical solution methods (Chapter 2.1) and numerical solution methods (Chapter 2.2). In this context the term "solution" means

- that in the direct or basic problem, discussed in Chapters 2.1 and 2.2, the dependent variables must be determined as functions of the independent variables, where the parameters used in the differential equation and the boundary and initial conditions are specified
- that in the inverse problem, discussed in Chapter 2.3, the dependent variables, and sometimes part of the parameters or boundary and initial conditions, are specified, and the unknown parameter and boundary or initial conditions are to be determined

In groundwater hydrology, the term "analytical solution" has become an expression synonymous with explicit closed solutions of mathematical simulation models. The application of analytical solutions to solute and heat migration problems in the soil- and groundwater zone requires simplifications of natural conditions, and may therefore cause serious errors of approximation. Nevertheless, analytical solutions offer substantial advantages over other solution methods for processing migration models. The most important advantages are the following:

177

1. Analytical solutions show the effect of parameters or boundary and initial conditions directly in the result.
2. Analytical solutions avoid numerical approximation errors.
3. Analytical solutions make data handling and data processing particularly easy.

2.1.1 REVIEW OF IMPORTANT SOLUTION METHODS

Migration processes are described by differential equations, as shown in Chapter 1.8. The most important methods for their analytical solution are

1. the integration of linear ordinary differential equations of first and second order
2. the solution of partial differential equations by substitution of variables as well as by Laplace and other integral transformations
3. the spatial and temporal superposition of basic solutions

As a premise for applying these solution methods, the basic migration model must be simplified to one linear ordinary or partial differential equation with one dependent variable.

2.1.1.1 Integration of Ordinary Differential Equations

Ordinary differential equations of first order

Two important elementary cases of the ordinary differential equation of first order are the ordinary differential equation with separated variables:

$$y' = dy/dx = f(x)/g(y)$$
$$\int g(y)dy = \int f(x)dx + C \tag{2.1a}$$

and the linear ordinary differential equation:

$$y' + f(x)y = g(x)$$
$$y(x) = \exp(-\int f(x)dx)(\int (g(x)\exp(\int f(x)dx))dx + C) \tag{2.1b}$$

where y is the dependent variable, x is the independent variable, and C is the integration constant.

Ordinary differential equations of second order

The linear ordinary differential equation of second order may be integrated and results in a closed form:

$$y'' + a(x)y' + b(x)y = c(x) \tag{2.2}$$

where $y'' = d^2y/dx^2$ and $y' = dy/dx$. Special cases of this equation are those possessing constant coefficients:

$$y'' + b_o y = 0$$

$$y = C_1 \exp(\sqrt{-b_o}x) + C_2 \exp(-\sqrt{-b_o}x) \tag{2.2a}$$

where C_1 and C_2 are the integration constants which fulfill the requirements of the two external boundary conditions of a second-order differential equation (see [2.13, p. 215, Figure 5.2]).

$$y'' + y'/r + b_o y = 0 \tag{2.2b}$$

This equation is Bessel's differential equation. Its solution yields with $b_o = -1/B^2$:

$$y(x) = C_1 I_o(r/B) + C_2 K_o(r/B)$$

where the constants C_1 and C_2 are given for the different types of boundary conditions and the modified Bessel functions of zero order, and of the first and second kind I_o and K_o, in Busch and Luckner [2.13, pp. 229 and 230, Tables 5.6 and 5.7].

$$y'' + a_o y' + b_o y = 0 \tag{2.2c}$$

The approach $y = \exp(rx)$ results in the conditional equation $r^2 + a_o r + b_o = 0$, including the solutions $r_{1,2} = -a_o/2 \pm \sqrt{(a_o/2)^2 - b_o}$. As long as $b_o < 0$, the two roots are real and provide two linearly independent solutions from which the general solution may be obtained as follows:

$$y = C_1 \exp(r_1 x) + C_2 \exp(r_2 x)$$

For the constants C_1 and C_2 see Busch and Luckner [2.13, p. 215, Table 5.2] and compare Equation 2.2a.

$$y'' + a_o y' + b_o y = f(x) \tag{2.2d}$$
$$y = C_1 \exp(r_1 x) + C_2 \exp(r_2 x) + y_s(x)$$

where $y_s(x)$ is a particular solution.

2.1.1.2 Solution of Partial Differential Equations

In both cases, substitution of variables or Laplace and other integral transformations, the partial differential equation is transformed into an ordinary differential equation, which is then integrated as explained previously in §2.1.1.1.

Substitution of variables

Consider the typical example of well hydraulics:

$$\partial^2 y/\partial r^2 + (1/r)\partial y/\partial r = a\partial y/\partial t \tag{2.3}$$

where $\partial y/\partial r(r \to 0,t) = -C/(2\pi r)$
$$y(\infty,t) = 0$$
$$y(r,0) = 0$$

The use of Boltzmann's transformation

$$y(r,t) = Cf(u) \text{ with } u = ar^2/t$$

converts this equation into the following ordinary differential equation (see, e.g., Busch and Luckner [2.13] and Luckner and Schestakow [2.63]):

$$f'' + f'(1/u + 1/4) = 0$$

where $f(\infty) = 0$ and $uf'(u \to 0) = -1/(4\pi)$. This leads finally to the well-known Theis solution [2.13, p. 232]:

$$y(r,t) = \frac{C}{4\pi} \int_{u/4}^{\infty} \exp(-u)\frac{du}{u} = \frac{C}{4\pi} W(\sigma)$$

where $W(\sigma)$ is the well function with $\sigma = u/4 = ar^2/(4t)$.

Laplace transformation

The Laplace transformation reduces difference equations, differential equations, and integral equations to less difficult mathematical problems. In using Laplace transformations, the conditions of the initial equation are called the original space, and the conditions of the transformed equation are called the image space. The following examples of L-transformations may be given:

- An ordinary differential equation in the original space yields after its transformation (symbolized by $\mathcal{L}$) an algebraic equation in the image space.
- A partial differential equation in two independent variables in the original space yields after its transformation one ordinary differential equation in the image space.
- A convolution equation in the original space

$$h(t) = f(t) * g(t) = \int_0^t f(\tau)g(t - \tau)d\tau$$

yields after its transformation a product equation in the image space:

$$H(s) = F(s)G(s)$$

The final solution in the original space must be determined by inverse Laplace transformation (symbolized by $\mathcal{L}^{-1}$) after the transformed mathematical model equation of the image space has been solved by an analytical or digital method (see Figure 2.0).

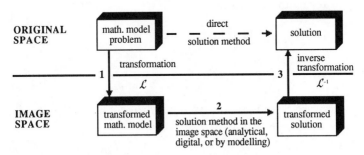

Figure 2.0. Methodological approach of Laplace transformation.

The Laplace transformation of a given function f(x) is defined:

$$\mathcal{L}\{f(x)\} = \overline{f}(s) = \int_0^\infty \exp(-sx)f(x)dx \qquad (2.4)$$

where x is the spatial or temporal independent variable to be reduced, i.e., transformed into a parameter. The Laplace transformation is linear and does not impact other independent variables (e.g., y) which are not transformed:

$$\mathcal{L}\left(a\frac{\partial^2 f(x)}{\partial y^2}\right) = a\frac{d^2\overline{f}(s)}{dy^2} \qquad \mathcal{L}\left(a\frac{\partial f(x)}{\partial y}\right) = a\frac{d\overline{f}(s)}{dy} \qquad (2.5)$$

$$\mathcal{L}\{af(x)\} = a\overline{f}(s) \text{ and } \mathcal{L}\{a\} = a/s$$

The function $\overline{f}(s)$ is defined in all points $s = \sigma + i\omega$ of the image space, where the Laplace integral converges. This holds only for

$$|f(x)| \leq M \exp(\rho x) \qquad \rho, M > 0, \text{ real} \qquad (2.6)$$

The functions
$$\exp(-\alpha x), \cos(\omega x), \sin(\omega x), x^n, \exp(-x^2), \ldots$$

for instance, fulfill this requirement, but not the functions

$$\exp(x^2), 1/x^2, \tan x, \ldots$$

The Laplace transformation of f(x), i.e., $\mathcal{L}\{f(x)\}$, is unique. The back- or inverse transformation $\mathcal{L}^{-1}\{\bar{f}(s)\}$ is only meaningful in the integration interval $[0, \infty]$. The Laplace transformation of the derivative f'(x) of the function f(x) requires that f'(x) fulfills the following conditions:

1. f'(x) must be continuous for $x > 0$
2. f'(x) must exist almost everywhere for $x > 0$
3. f'(x) must be transformable

This transformation results in

$$\mathcal{L}\{f'(x)\} = s\bar{f}(s) - f(0) \tag{2.7a}$$

where $\bar{f}(s) = \mathcal{L}\{f(x)\}$
$f(0) = \lim_{x \to 0^+} f(x)$

Higher derivatives may be transformed when comparable requirements are fulfilled. The transformation then results in

$$\mathcal{L}\{f''(x)\} = s^2\bar{f}(s) - sf(0) - f'(0) \tag{2.7b}$$
$$\mathcal{L}\{f'''(x)\} = s^3\bar{f}(s) - s^2f(0) - sf'(0) - f''(0) \tag{2.7c}$$

The quantification of f(0), f'(0), f''(0), ... often creates problems and results in restrictions to the application of the L-transformation. The second of the two following examples shows such a case:

$$\partial^2 P/\partial y^2 = a\partial P/\partial t \tag{2.7d}$$
$$P(0,t) = P_o \quad P(\infty,t) = 0 \quad P(y,0) = 0$$

The transformation of P(t), $\partial P(t)/\partial t$, $\partial^2 P(t)/\partial y^2$, and of the boundary conditions P(0,t) and P(∞,t) yields

$$d^2\bar{P}/dy^2 = a(s\bar{P} - 0)$$

where $\bar{P}(0,s) = P_o/s$, $\bar{P}(\infty,s) = 0$, and P(y,0) enters into $\partial P(t)/\partial t$. In this case a solution in the image space is given by the following equation:

$$\bar{P}(s) = (P_o/s)\exp(-y\sqrt{a}\,\sqrt{s}) \tag{2.8}$$

However, if in a second case P(y), $\partial^2 P(y)/\partial y^2$, ... are transformed, the following image equation results:

$$s^2\bar{P}(s) - sP_o - \partial P/\partial y|_{y=0} = a d\bar{P}/dt$$

which cannot be solved, because the term $\partial P/\partial y|_{y=0}$ cannot be specified by the remaining boundary condition $\bar{P}(\infty,t)$. The most common rules for calculating the Laplace transformation are listed in Table 2.1.

After the problem is solved in the image space, the inverse or back-

Table 2.1. Laplace Transformation: Theorems and Correspondence

Original Space P(t)	Image Space $\bar{P}(s)$
Theorems	
1. addition theorem $P_1(t) + P_2(t) \ldots$	$\bar{P}_1(s) + \bar{P}_2(s) + \ldots$
2. multiplication theorem $aP(t)$	$a\bar{P}(s)$
3. translation theorem $P(t - b)$ for $t > b > 0$ O for $t < b > 0$	$\exp(-bs)\,\bar{P}(s)$
4. similarity theorem $P(at)$ $(1/\alpha)\,P(t/\alpha)$	$(1/a)\,\bar{P}(s/a)$ $\bar{P}(\alpha s)$
5. convolution theorem $\displaystyle\int_0^1 P_1(t - \tau)\,P_2(\tau)\,d\tau$	$\bar{P}_1(s)\,\bar{P}_2(s)$
6. linear substitution theorem $\dfrac{1}{\alpha}\exp\left(-\dfrac{\beta t}{\alpha}\right) P\left(\dfrac{t}{\alpha}\right)$	$\bar{P}(\alpha s + \beta)$
7. frequency-shift theorem $\exp(-\alpha t)\,P(t)$	$\bar{P}(s + \alpha)$
Correspondence	
1. 1 and a, respectively	1/s and a/s, respectively
2. at	a/s^2
3. $a\exp(-\alpha t)$	$a/(\alpha + s)$
4. $\dfrac{a}{\alpha}(1 - \exp(-\alpha t))$	$\dfrac{a}{s(\alpha + s)}$
5. $\begin{cases} a \text{ for } t > b > 0 \\ 0 \text{ for } t < b > 0 \end{cases}$	$\dfrac{a}{s}\exp(-bs)$
6. $\begin{cases} a\exp(-\alpha(t - b))\ t > b > 0 \\ 0 \qquad\qquad t < b > 0 \end{cases}$	$\dfrac{a}{\alpha + s}\exp(-bs)$
7. $\begin{cases} a(1 - \exp(-\alpha(t - b)))\ t > b > 0 \\ 0 \qquad\qquad\qquad t < b > 0 \end{cases}$	$\dfrac{a}{s(\alpha + s)}\exp(-bs)$
8. Dirac's delta function $\delta(t) = \begin{cases} 0 \text{ for } t \neq 0 \\ \infty \text{ for } t = 0 \end{cases}$	1
9. $a\cdot\mathrm{erfc}\left(\dfrac{\alpha}{2\sqrt{t}}\right)$	$\dfrac{a}{s}\exp(-\alpha\sqrt{s})$
10. $\mathrm{erf}\left(\dfrac{\alpha}{2\sqrt{t}}\right)$	$\dfrac{1 - \exp(-\alpha\sqrt{s})}{s}$

Table 2.1 continued.

11. $\dfrac{\alpha}{2t} \dfrac{a}{\sqrt{\pi t}} \exp\left(-\dfrac{\alpha^2}{4t}\right)$	$a \exp(-\alpha\sqrt{s})$
12. $\dfrac{1}{2}\left[\exp(a^2t - a\alpha) \operatorname{erfc}\left(\dfrac{\alpha}{2\sqrt{t}} - a\sqrt{t}\right)\right.$	$\dfrac{\exp(-\alpha\sqrt{s})}{s - a^2}$
$\left. + \exp(a^2t + a\alpha) \operatorname{erfc}\left(\dfrac{\alpha}{2\,t} + a\sqrt{t}\right)\right]$	

transformation of the function $\bar{f}(s) = F(s)$ into the original space must be carried out. In most practical problems the inverse L-transformation is difficult to perform. This transformation is defined by

$$f(x) = \mathcal{L}^{-1}\{F(s)\} = \frac{1}{2\pi i}\int_\Omega \exp(sx)F(s)ds \qquad t > 0 \qquad (2.9)$$

For a large number of special functions $f(x)$ and $f(t)$, correspondence tables are available (see, e.g., Bronstein and Semendjajew [2.11], Doetsch [2.29; 2.28], and Table 2.1), which provide solutions of the integral equation (Equation 2.9). If, for instance, the correspondence no. 9 of Table 2.1 is applied to the example considered above, the following equation results:

$$P(y,t) = \mathcal{L}^{-1}\{(P_o/s)\exp(-y\sqrt{a}\ \sqrt{s})\} = P_o \operatorname{erfc}(y/\sqrt{4t/a})$$

If no correspondences are available, a solution can usually be obtained only by numerical integration of Equation 2.9. This transformation is commonly designated as the numerical L-back-transformation.

Further integral transformations

In most cases, the Laplace transformation and the substitution of variables do not provide solutions if the space of investigation is limited. In such cases, several forms of Fourier transformations may be applied. A compilation in Busch and Luckner [2.13, p. 224] gives a survey of solutions of the partial differential equation $\partial^2 y/\partial x^2 = a\partial y/\partial t$ for the four combinations of boundary conditions of first and second type.

2.1.1.3 Superposition of Basic Solutions

Problems of migration processes which may be solved analytically are characterized by simplest boundary conditions in the form of Dirac's delta function or a unit-step function, and only such solutions are considered in the following:

$$\delta(t) = 0 \text{ for } t \neq 0 \qquad \delta(t) = \infty \text{ for } t = 0 \qquad \int_{-\infty}^{\infty} \delta(t)dt = 1$$

$$p(t) = 0 \text{ for } t < 0 \qquad p(t) = 1 \text{ for } t \geq 0$$

Boundary conditions have the effect of inputs on the considered system. Inputs are the cause of the system's state changes, and affect the system's response. The corresponding system output to a delta function input and unit-step function input are the pulse-response function $h(t)$ and the step-response function $S(t) = \int h dt$, respectively.

Analytical solutions of migration processes obtained with such boundary conditions consequently represent pulse-and step-response functions. For linear migration models, the superposition of such analytical solutions may be performed provided that the parameters are independent of the dependent variables in the mathematical model. This is the case for simplified mathematical models of migration processes in the subsurface.

Temporal superposition

The convolution or Duhamel integral allows temporal superposition. Hence, the output $P(t_o)$ for an arbitrary temporal variable boundary condition (input) $R(t, x = 0)$ results in [2.64]

$$P(t_0) = \int_{\tau = -\infty}^{t_0} R(\tau)h(t_0 - \tau)d\tau = \int_{\tau = -\infty}^{t_0} \frac{dR}{d\tau}\bigg|_\tau S(t_0 - \tau)d\tau = \int_{\tau = -\infty}^{t_0} \frac{d^2R}{d\tau^2}\bigg|_\tau F(t_0 - \tau)d\tau \quad (2.10)$$

Spatial superposition

Superposition of solutions of linear differential equations leads to new solutions valid in the space where all superimposed single solutions are valid. Along the boundaries of the joint spatial domain, the total sum of boundary conditions of all partial solutions are effective. The effective initial condition of a superposed solution results in similar manner from the sum of the initial conditions of all single solutions.

All analytical solutions in groundwater hydraulics for multiwell systems are based on application of the principle of spatial superposition (see, e.g., Busch and Luckner [2.13, Figure 5.24]). Zero initial and boundary conditions, or boundary conditions of the second type common in well hydraulics, are usually the least difficult.

2.1.2 SOLUTION OF CONVECTIVE MODELS

In many migration processes in the subsurface, convective transport is dominant. This is the reason why transport due to heat conduction, solute diffusion, or hydrodynamic dispersion may sometimes be neglected (see also Chapter 1.3). Model approaches, reflecting convection as the only transport process, are called *convective model schemes.* Approximation errors of such schemes are relatively small, provided that no sudden temporal or spatial changes of the external and internal boundary conditions occur (see also §1.7.3).

Typical examples for such cases include contamination of groundwater caused by diffuse sources of pollution due to wet and dry atmospheric deposition, agricultural fertilization, or pesticide application. In the following, Example A investigates migration of NH_4^+-nitrogen in an aquifer caused by regional manure application, while in Example B the spreading of a tracer in a steady-state horizontal groundwater flow field is considered.

2.1.2.1 Example A

Flow model

The following one-dimensional steady-state flow model is considered:

$$\nabla(T\nabla h) = Td^2h/dx^2 = -\bar{w}_o \qquad (2.11a)$$

where   T = transmissivity (T = ∫kdz) in m²/s
  h = piezometric head in m
  q = flow rate; q = ∫vdz = -Tdh/dx in m²/s
  $\bar{w}_o$ = recharge in m/s

After integration this flow pattern yields

$$q = q_o + \bar{w}_o x \text{ and } dq/dx = \bar{w}_o \qquad (2.11b)$$

Water quality model

This model is derived following the abstraction succession presented in Figure 1.56. The following assumptions are made:

- equilibrium-exchange processes between the mixphases G' and o, o and st, and between st and G" (see also §1.5.1)
- convection in the mixphase o as the sole transport process (see Chapter 1.3)
- degradation of NH_4^+ (nitrification) in all mixphases as a first-order kinetic reaction (see Chapter 1.4)
- constant areal input of NH_4^+
- applicability of a linear storage process model (see §1.5.5)

The resulting model has the form:

$$\underbrace{- \frac{\partial}{\partial x}(qc) = -q\frac{\partial c}{\partial x} - c\frac{\partial q}{\partial x}}_{\text{transport}} = \underbrace{\epsilon\frac{-\partial c}{\partial t}}_{\text{storage}} + \underbrace{\bar{\lambda}^*c}_{\text{degradation}} - \underbrace{\bar{w}}_{\substack{\text{source term} \\ \text{(areal input)}}} \qquad (2.12)$$

and after transformation with $\bar{\lambda} = \bar{\lambda}^* + \partial q/\partial x = \bar{\lambda}^* + \bar{w}_o$:

$$-q\frac{\partial c}{\partial x} = \underbrace{\bar{\epsilon}\frac{-\partial c}{\partial t}}_{} + \underbrace{\bar{\lambda}c}_{} - \underbrace{\bar{w}}_{} \qquad (2.13)$$

$$\underset{\text{transport}}{} \quad \underset{\text{storage}}{} \quad \underset{\substack{\text{degradation} \\ \text{and} \\ \text{dilution}}}{} \quad \underset{\text{source}}{}$$

where c = concentration of the migrant NH_4^+-N in g per m^3 of water
q = flow rate in the aquifer in m^3 of water per second and m of width
$\bar{\epsilon}$ = storage coefficient for NH_4^+-N; $\bar{\epsilon} = \int(\theta_o + \theta_{st} + \alpha)dz$ in m where
$\quad \alpha = K_{dpdry}$
$\bar{\lambda}$ = velocity coefficient of NH_4^+-N conversion, i.e., of combined degradation $\bar{\lambda}^*$ and dilution $\bar{w}_o$ in m/s
$\bar{w}$ = NH_4^+-N areal source (input) in g per second and m^2 of manured land

The analytical solution of this migration model may be illustrated by the following examples:

One-dimensional nonstationary migration where w = 0 and $\bar{\lambda}$ = 0

$$\bar{\epsilon}\partial c/\partial t = -q\partial c/\partial x \qquad (2.14a)$$

$$c(0,t) = c_o \text{ and } c(x,0) = c_I$$

This partial differential equation is solved by the substitution of variables (see §2.1.1). The substitution of $qt - \bar{\epsilon}x$ by u in Equation 2.14a yields

$c = h(u)$ $c(0,t) = h(qt) = c_o$ and hence $h(u > 0) = c_o$
$\qquad\qquad c(x,0) = h(-\bar{\epsilon}x) = c_I$ and hence $h(u < 0) = c_I$
$\qquad\qquad \partial c/\partial t = qdh/du$ and $\partial c/\partial x = -\bar{\epsilon}dh/du$

and Equation 2.14a

$$\bar{\epsilon}qdh/du - \bar{\epsilon}qdh/du = 0$$

is fulfilled for all finite values of the derivative dh/du. Hence, the solution of Equation 2.14a is given by

$$c(x,t) = c_o \text{ for } qt - \bar{\epsilon}x > 0$$

and

$$c(x,t) = c_I \text{ for } qt - \bar{\epsilon}x < 0$$

and the velocity of the moving front whose position is determined by $qt = \bar{\epsilon}x$ follows as

$$u_f = x_f/t = q/\bar{\epsilon} \qquad (2.14b)$$

One-dimensional steady-state migration with w = constant

$$q\, dc/dx + \bar{\lambda}c = \bar{w} \qquad (2.15a)$$

This nonhomogeneous linear ordinary differential equation fits the type of Equation 2.1b and yields the following solution:

$$c = \exp\left(-\frac{\bar{\lambda}}{q}x\right)\left(\int \left(\frac{\bar{w}}{q}\exp\left(\frac{\bar{\lambda}}{q}x\right)\right)dx + C\right)$$

$$c = \bar{w}/\bar{\lambda} + C\, \exp(-(\bar{\lambda}/q)x)$$

where $c(x=0) = c_o$ yields $C = c_o - \bar{w}/\bar{\lambda}$ and hence

$$c = \frac{\bar{w}}{\bar{\lambda}} + \left(c_o - \frac{\bar{w}}{\bar{\lambda}}\right)\exp\left(-\frac{\bar{\lambda}}{q}x\right) \qquad (2.15b)$$

$$c(x = \frac{q}{\bar{\epsilon}}t) = \frac{\bar{w}}{\bar{\lambda}} + \left(c_o - \frac{\bar{w}}{\bar{\lambda}}\right)\exp\left(-\frac{\bar{\lambda}}{\bar{\epsilon}}t\right) \qquad (2.15c)$$

where again $c(x=0) = c_o$ and $c(x\to\infty) = \bar{w}/\bar{\lambda}$.

Nonstationary migration not influenced by transport processes

$$\bar{\epsilon}dc/dt + \bar{\lambda}c = \bar{w} \qquad (2.16a)$$

This ordinary differential equation has the same form as Equation 2.15a and hence the corresponding solutions:

$$c = \frac{\bar{w}}{\bar{\lambda}} + [c_I - \frac{\bar{w}}{\bar{\lambda}}]\exp[-\frac{\bar{\lambda}}{\bar{\epsilon}}t] \qquad (2.16b)$$

$$c(t = x\bar{\epsilon}/q) = \frac{\bar{w}}{\bar{\lambda}} + [c_I - \frac{\bar{w}}{\bar{\lambda}}]\exp[-\frac{\bar{\lambda}}{q}x] \qquad (2.16c)$$

where $c(x,0) = c_I$.

One-dimensional nonstationary migration

$$-q\frac{\partial c}{\partial x} = \bar{\epsilon}\frac{\partial c}{\partial t} + \bar{\lambda}c - \bar{w} \quad \text{where } c(0,t) = c_o \text{ and } c(x,0) = c_I \quad (2.17a)$$

$$\frac{\partial c}{\partial x} + \frac{\bar{\epsilon}}{q}\frac{\partial c}{\partial t} + \frac{\bar{\lambda}}{q}c = \frac{\bar{w}}{q}$$

After Laplace transformation by means of Equations 2.15 and 2.7, one obtains

$$\underbrace{\frac{\partial c}{\partial x} + [\frac{\bar{\lambda}}{q} + s\frac{\bar{\epsilon}}{q}]|\,\bar{c}}_{f} = \underbrace{\frac{\bar{\epsilon}}{q}c_I + \frac{\bar{w}}{sq}}_{g}$$

where $\bar{c}(0,s) = c_o/s$

This is a linear ordinary differential equation of the first order (see Equation 2.1b) which yields the following solution:

$$\bar{c} = g/f + C\,\exp(-fx)$$

With C resulting from $\bar{c}(0,s) = c_o/s$,

$$C = \frac{c_o}{s}\,\frac{c_I}{\bar{\lambda}/\bar{\epsilon}+s} - \frac{\bar{w}/\bar{\epsilon}}{s(\bar{\lambda}/\bar{\epsilon}+s)}$$

the solution in the image space becomes

$$\bar{c}(x,s) = \frac{c_I}{\bar{\lambda}/\bar{\epsilon}+s} + \frac{\bar{w}/\bar{\epsilon}}{s(\bar{\lambda}/\bar{\epsilon}+s)} \tag{2.17b}$$

$$+ \left[\frac{c_o}{s} - \frac{c_I}{\bar{\lambda}/\bar{\epsilon}+s} - \frac{\bar{w}/\bar{\epsilon}}{s(\bar{\lambda}/\bar{\epsilon}+s)}\right]\exp[-\frac{\bar{\lambda}}{q}x]\exp[-\frac{\bar{\epsilon}x}{q}s]$$

The inverse transformation of this equation into the original space by means of the correspondences no. 3 to 7 of Table 2.1 results in

$$c(x,t) = \frac{\bar{w}}{\bar{\lambda}} + \left(c_I - \frac{\bar{w}}{\bar{\lambda}}\right)\,\exp\left(\frac{\bar{\lambda}t}{\bar{\epsilon}}\right) \tag{2.17c}$$

$$+ \delta\left(c_o - \frac{\bar{w}}{\bar{\lambda}} - \left(c_I - \frac{\bar{w}}{\bar{\lambda}}\right)\exp\left(-\frac{\bar{\lambda}}{\bar{\epsilon}}\left(t - \frac{\bar{\epsilon}x}{q}\right)\right)\right)\exp\left(-\frac{\bar{\lambda}x}{q}\right)$$

where

$$\delta = \begin{cases} 1 \text{ for } t > \bar{\epsilon}\,x/q \\ \\ 0 \text{ for } t < \bar{\epsilon}\,x/q \end{cases} \quad \text{(see also Equation 2.14b)}$$

This concentration function (Equation 2.17c) is constrained by the following initial and boundary conditions:

$c(x,0) = c_I \qquad c(0,t) = c_o$ (see also Equation 2.17a)
$c(x\to\infty,t) = \bar{w}/\bar{\lambda} + (c_I - \bar{w}/\bar{\lambda})\exp(-\bar{\lambda}t/\bar{\epsilon})$ (see also Equation 2.16b)
$c(x,t\to\infty) = \bar{w}/\bar{\lambda} + (c_o - \bar{w}/\bar{\lambda})\exp(-\bar{\lambda}x/q)$ (see also Equation 2.15b)

Using the following sample data, Equation 2.17c may be quantified:

initial condition : $c(x,0) = c_I = 2$ g NH_4^+-N per m³ of water in g/m³

boundary condition : $c(0,t) = c_o = 10$g NH_4^+-N per m³ of water in g/m³

source/sink-term : $\bar{w} = 4 \cdot 10^{-6}$ g NH_4^+-N per m² area of land per s in g/(m²s)

parameters : $|q| = 4 \cdot 10^{-4}$ m³ per s and per m of width in m²/s; $\bar{\epsilon} = 4 \cdot 10$ m; $\bar{\lambda} = 8 \cdot 10^{-7}$ m/s

Figure 2.1 shows the resulting concentration plots. As can be seen, Equations 2.14b or 2.15c, and 2.16b or 2.16c, provide good approximations of the complete Equation 2.17c.

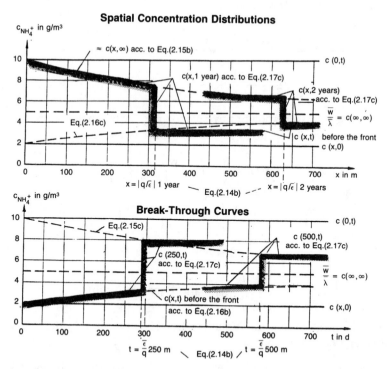

Figure 2.1.   Plots of special analytical solutions of Equation 2.17. The spatial concentration distributions for t = 1 year and t = 2 years are shown *above,* and the breakthrough curves at the positions x = 250 m and x = 500 m are plotted *below.*

2.1.2.2 Example B

Flow model

Considering two-dimensional steady-state groundwater flow in a homogeneous aquifer with negligible recharge, the differential equation governing the flow can be written as follows (compare Equation 2.11a):

$$\nabla(T\nabla h) = T\partial^2 h/\partial x^2 + T\partial^2 h/\partial y^2 = 0 \tag{2.18}$$

Three typical cases are discussed:

1. infinite parallel flow in which tracer-carrying water is injected by means of a well
2. semiinfinite flow with the boundary condition $h(x,0) = h_o$ and a pumping well at point $(0,-a)$
3. pumping and injection well in an infinite flow field

Water quality model

A water quality model reduced in comparison with Equation 2.13 is considered:

$$\bar{\epsilon}\partial c/\partial t = -\nabla(qc) = -\frac{\partial}{\partial x}(qc) - \frac{\partial}{\partial y}(qc) \tag{2.19}$$

If the orthogonal net lines ξ,η corresponding to the potential and flow lines ϕ,ψ are used instead of the orthogonal coordinate lines x,y, the following groundwater quality model results from the solute balance at the moving concentration front in a flow tube with width $B = B(\eta)$ and constant discharge $Q = \dot{V}_s = B(\eta)q(\eta)$:

$$\dot{V}_s c = B(\eta)\bar{\epsilon}d\eta_f/dt$$

and

$$q(\eta)c = \bar{\epsilon}d\eta_f/dt \tag{2.20a}$$

The local front position at time t_o and the time necessary for the front to arrive at point h_o are obtained by integration of the ordinary differential Equation 2.20a:

$$d\eta_f = \frac{q}{\epsilon}dt \rightarrow \begin{cases} \eta_f = (1/\bar{\epsilon}) \int_o^{t_o} q(\eta)dt \\ \\ t_f = \bar{\epsilon} \int_o^{\eta_o} (1/q(\eta))d\eta \end{cases} \tag{2.20b}$$

Taking into account that $q(\eta) = |T dh/d\eta|$ it follows:

$$\eta_f = \frac{T}{\epsilon} \int \frac{dh}{d\eta} \, dt \quad \text{and} \quad t_f = \frac{\bar{\epsilon}}{T} \int \frac{d\eta^2}{dh} \qquad (2.20c)$$

or with $\nabla h = dh/d\eta$:

$$t_f = \frac{\bar{\epsilon}}{T} \int \cdot \frac{dh}{|\nabla h|^2} \qquad (2.20d)$$

For the three selected examples, the following solutions are obtained.

Infinite parallel flow with injection well

The potential function ϕ results from superposition of the steady-state parallel and radial groundwater flow. Using exclusively the variable part of the expression $\ln r - \ln r_o$, the superposition yields

$$\phi = h_o - h = -\frac{q_o}{T} x - \frac{Q}{2\pi T} \ln \frac{r}{r_o} \rightarrow -\frac{q_o}{T} x - \frac{Q}{2\pi T} \ln \sqrt{x^2 + y^2} \qquad (2.21)$$

where $h = h_o$ is the well water level. Corresponding to its definition, the flow function ϕ^* can be expressed as

$$\psi^* = \int q(\eta) d\xi = \int q(x) dy = T \int \frac{\partial \phi}{\partial x} \, dy$$

and the related flow function $\psi = \psi^*/T$ follows by means of Equation 2.21 as

$$\psi = -\frac{q_o}{T} y - \frac{Q}{2\pi T} \arctan(y/x) \qquad (2.22)$$

The geometrical shape of the flow line $\phi = 0$ is consequently given by

$$y/x = \pm \tan (2\pi q_o y/Q) \begin{cases} + & \text{for } y > 0 \\ - & \text{for } y < 0 \end{cases} \qquad (2.23)$$

These flow lines approach the asymptotes $\pm y = Q/(2q_o)$ for $x \rightarrow \infty$. Therefore, the stagnation point $(x = x_s, y = 0)$ is obtained using the constraint $\partial h/\partial x = 0$ at $y = 0$ (see Equation 2.21):

$$x_s = -Q/(2\pi q_o) \qquad (2.24)$$

Figure 2.2a shows the hydraulic flow net with the flow lines $\psi = 0$ according to Equation 2.23. These two lines limit the zone in which the tracer can spread.

The determination of isochrones, or lines of equal travel time, in Figure 2.2b by means of Equation 2.20d requires the partial differentiation of the potential function ϕ. The expression $\nabla\phi = -\nabla h$

$$|\nabla h| = |\partial h/\partial \eta| = \sqrt{(\partial h/\partial x)^2 + (\partial h/\partial y)^2} \qquad (2.25)$$

is obtained as follows using Equation 2.21:

$$|\nabla h|^2 = \left[\frac{q_o}{T} + \frac{Q}{2\pi T}\frac{x}{x^2 + y^2}\right]^2 + \left[\frac{Q}{2\pi T}\frac{y}{x^2 + y^2}\right]^2 \qquad (2.26)$$

and Equation 2.20d yields the isochrones or front positions shown in Figure 2.2b [2.5; 2.67]:

$$\tau_f = x' + \ln\left(\frac{\sin\theta}{\sin(\theta + y')}\right) \qquad (2.27a)$$

where $\tau_f = \dfrac{2\pi q_o^2}{\bar{\epsilon} Q} t$

$\theta = \arctan\dfrac{y}{x} = \arctan\dfrac{y'}{x'}$

$x_s = \dfrac{Q}{2\pi q_o}$

$\bar{\epsilon} = \epsilon M$

$x' = \dfrac{2\pi q_o}{Q} x = x/x_s$

$y' = \dfrac{2\pi q_o}{Q} y = y/x_s$

Equation 2.27a has for $y = y' = 0$ the following limit (see Figure 2.2b):

$$\tau_f = x' - \ln(1 + x') \qquad (2.27b)$$

Optimal placement of a pumping well and an injection well in a parallel groundwater flow field is of special interest in the design of heat pump systems. In general, the cooled water must by law be reinfiltrated into the aquifer, and the two wells must be located on the often restricted property area. Determination of the minimal well distance $2a_{crit}$ which avoids return flow of cooled water to the pumping well is then of great impor-

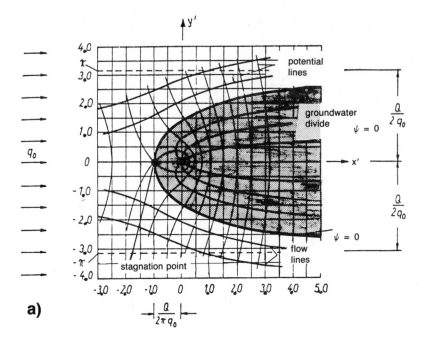

a)

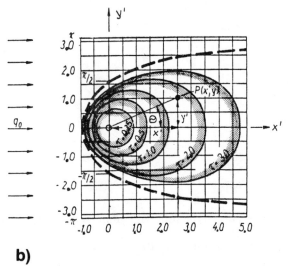

b)

Figure 2.2. Injection well in parallel groundwater flow: *(a)* flow net and spreading zone of traced infiltrate for t ∞; *(b)* isochrone lines inside the tracer spreading zone.

tance. If the connection line of the two wells and the groundwater flow direction form the angle ω, the potential and flow function according to Equations 2.21 and 2.22 may be written as

$$|\phi| = \frac{q_o}{T}(x \cos \omega + y \sin \omega) + \frac{Q}{2\pi T}\ln \left[\frac{(x-a)^2 + y^2}{(x+a)^2 - y^2} \right]^{1/2} \quad (2.28a)$$

$$|\psi| = \frac{q_o}{T}(y \cos \omega - x \sin \omega) + \frac{Q}{2\pi T} \left[\arctan \frac{y}{x-a} - \arctan \frac{y}{x+a} \right] \quad (2.28b)$$

A typical example of such a groundwater quality problem is shown in Figure 2.3 (see also Mehlhorn et al. [2.68] and Victor [2.95]).

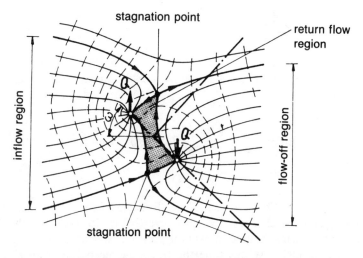

Figure 2.3. Example of a hydraulic flow net caused by operating an extraction and an injection well in a parallel groundwater flow field where the angle between flow direction and connection line of the wells is $\omega = 45°$ and the well distance 2a is less than the critical distance $2a_{crit}$ (see Victor [2.95]).

Pumping Well Near a Surface Water Body

The case of a pumping well near a fully penetrating stream is an example of the type of problem which allows a closed analytical solution first derived by Muskat [2.70]. Based on the explicit potential function h(x,y) and the flow function ϕ(x,y):

$$h_o - h = -\frac{Q}{2\pi T} \ln \sqrt{\frac{x^2 + (y - a)^2}{x^2 + (y + a)^2}}$$

$$\psi = \frac{-Q}{2\pi T} \arctan \frac{2ax}{x^2 + y^2 - a^2} \tag{2.29}$$

the derivative of second order $|\nabla h|^2$ is obtained [2.5]:

$$|\nabla h|^2 = \left(\frac{Q}{2\pi aT}\right)^2 (\cosh \alpha + \cos \beta)$$

where $\alpha = \dfrac{2\pi T(h_o - h)}{Q}$

$\beta = \dfrac{2\pi T\psi}{Q}$

The isochrones follow using Equation 2.20d and an aquifer depth of $M = T/k$:

$$t_f = \frac{2\pi a^2 \epsilon M}{Q \sin^2 \beta} \left[\frac{\sinh \alpha}{\cosh \alpha + \cos \beta} - 2 \cot (\beta) \arctan \left(\tanh \frac{\alpha}{2} \tan \frac{\beta}{2}\right)\right]$$

$$\tag{2.30}$$

Figure 2.4 shows flow lines satisfying Equation 2.29. The tracer front position for selected times $\tau_f = t_f \cdot Q/(2\pi a^2 \epsilon M)$ is also shown in this figure. The advancing front arrives at the pumping well at time $\tau_f = 1/3$ or $t_f = 2\pi a^2 \epsilon M/(3Q)$. At this time the tracer plume occupies an area of $2\pi a^2/3$.

Travel times of migrants between two isochrones are equal. Thus, it takes all water molecules starting from the isochrone $\tau_f = 0.1$ the travel time $\Delta t = 0.1 \cdot 2\pi a^2 \epsilon M/Q$ to pass the isochrone $\tau_f = 0.2$ (see Figure 2.4).

Pumping and Injection Well in an Infinite Flow Field

Muskat [2.70] solved this problem and obtained the following function for the isochrones:

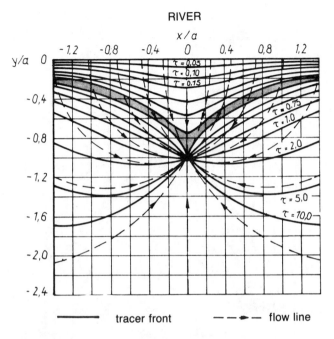

Figure 2.4. Advancing front of bank-filtered water toward a nearby pumping well. *Source:* Victor [2.95].

$$t_f = \frac{2\pi a^2 \epsilon M}{Q \sin^2 \beta} \left[\frac{\sinh \alpha}{\cosh \alpha + \cos \beta} - 2 \cot (\beta) \arctan \right.$$

$$\left(\tanh \frac{\alpha}{2} \tan \frac{\beta}{2} \right) - \frac{\sinh \alpha_o}{\cosh \alpha_o + \cos \beta} \qquad (2.31)$$

$$\left. + 2 \cot (\beta) \arctan \left(\tan \frac{\alpha_o}{2} \tan \frac{\beta}{2} \right) \right]$$

where $\alpha = \alpha_o$ refers to the water level in the infiltration well. Figure 2.5 shows isochrones and flow lines calculated for this example. The front arrives at the pumping well after a period of $t_f = 4\pi a^2 \epsilon M/(3Q)$.

2.1.3 SOLUTION OF DIFFUSIVE/DISPERSIVE MODELS

Migration models reflecting heat conduction, molecular solute diffusion, and hydrodynamic dispersion, but neglecting convective transport, are termed *diffusive/dispersive models* (see Chapter 1.3). Their use is appropriate when one of the following conditions is met:

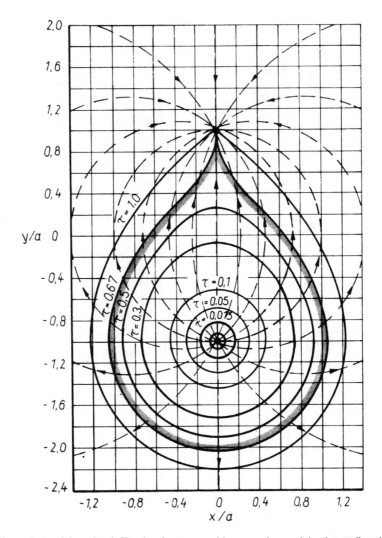

Figure 2.5. Advancing infiltration front caused by operating an injection well and an extraction well. *Source:* Muskat [2.70].

- Existing strata of low permeability (clay, for instance) allow no flow and hence no convective solute or heat transport.
- The convective transport term may be eliminated from the migration model by a suitable mathematical transformation.

One example for each case will be presented in the following.

2.1.3.1 Example A—Heat Conduction in a Clay Layer

Two permeable layers are separated by a clay layer of thickness M. The vertical flow v_z within the clay is assumed to be negligible. The temperature difference between the top and bottom of the clay layer amounts to ΔT_o. According to Equation 1.118, the heat migration model with its initial and boundary conditions may be expressed by

$$\frac{\partial^2(\Delta T)}{\partial z^2} = a \frac{\partial(\Delta T)}{\partial t} \tag{2.32a}$$

where
$$a = \frac{\phi(\rho c)_w + (1 - \phi)(\rho c)_s}{\lambda_o}$$

$\Delta T(x,0) = \Delta T_1 = 0$
$\Delta T(0,t) = \Delta T_o$
$\Delta T(M,t) = 0$

A compilation of solutions for this type of equation with boundary conditions of the first, second, and third kind can be found, for instance, in Busch and Luckner [2.13]. Thus, the heat flux resulting from Equation 2.32a with its initial and boundary conditions may be expressed as follows (see, e.g., Busch and Luckner [2.13, Table 5.4, p. 224]):

$$q_h(0,t) = \lambda_o \, \Delta T_o/M + \frac{\Delta T_o}{M} \sum_{n=1}^{\infty} (-1)^n \exp\left(- \frac{n^2\pi^2 t}{M^2 a}\right) \tag{2.32b}$$

2.1.3.2 Example B—NH_4^+-Transport in an Aquifer

Let us consider the model Equation 2.13 with dispersion:

$$\bar{D} \frac{\partial^2 c}{\partial x^2} - q \frac{\partial c}{\partial x} = \bar{\epsilon} \frac{\partial c}{\partial t} + \bar{\lambda} c - \bar{w}$$

which becomes for $\bar{\lambda} = 0$, $\bar{w} = 0$, $\bar{D}/\bar{\epsilon} = K$ (where $\bar{D} = \delta q = DM$), and $q/\bar{\epsilon} = U$

$$K \frac{\partial^2 c}{\partial x^2} - U \frac{\partial c}{\partial x} = \frac{\partial c}{\partial t} \tag{2.33a}$$

where
$c(x,0) = c_1 = 0$
$c(0,t) = c_o$
$c(\infty,t) = 0$

The first two conditions $c(x,0) = 0$ and $c(0,t) = c_o$ reflect a step function. The corresponding solution of Equation 2.33a is hence the step-response function $S(x,t)$. If, on the other hand, Equation 2.33a is solved using the following pulselike initial or boundary conditions:

$$c(x,0) = \frac{m\omega}{\epsilon} \delta(x) \quad \text{or} \quad c(o,t) = \frac{m/\omega}{\epsilon U} \delta(t) \quad c(\infty,t) = 0 \qquad (2.33b)$$

where $\delta(x) = \delta(t)/U$ is Dirac's delta function, $\delta(x)$ is in m^{-1} and $\delta(t)$ in s^{-1}
m = mass in kg or amount of substance in mol
$\epsilon = \theta_0 + \theta_{st} + K_d\rho_{dry}$ is the integral storage coefficient
ω = area normal to the x-axis; $\omega = MB$
U = front velocity (compare Equation 2.14b)

the pulse-response function h(x,t) which may be transformed into the step-response function $S(x,t) = \int h(x,t)dt$ by integration is obtained.

The substitution (compare Equation 2.14)

$$c(x,t) = \tilde{c}(x^*,t) \quad \text{where} \quad x^*(x,t) = x - Ut$$

$$\frac{\partial c}{\partial x} = \frac{\partial \tilde{c}}{\partial x^*} \frac{\partial x^*}{\partial x} = \frac{\partial \tilde{c}}{\partial x^*}$$

$$\frac{\partial^2 c}{\partial x^2} = \frac{\partial}{\partial x}[\frac{\partial c}{\partial x}] = \frac{\partial(\partial \tilde{c}/\partial x^*)}{\partial x^*} \frac{\partial x^*}{\partial x} = \frac{\partial^2 \tilde{c}}{\partial x^{*2}}$$

$$\frac{\partial c}{\partial t} = \frac{\partial \tilde{c}}{\partial x^*} \frac{\partial x^*}{\partial t} + \frac{\partial \tilde{c}}{\partial t} = - \frac{\partial \tilde{c}}{\partial x^*} U + \frac{\partial \tilde{c}}{\partial t}$$

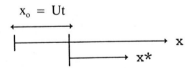

$$\partial^2 \tilde{c}/\partial x^{*2} = a\partial \tilde{c}/\partial t$$

transforms Eq. 2.33a into   (2.33c)

where a = 1/K. The boundary conditions are transformed into $\tilde{c}(x^* \to \infty,t) = 0$, and Dirac's δ-function into (compare also Figure 2.7)

$$\frac{m/\omega}{\epsilon U} \delta(t) = \tilde{c}(x^* \to 0, t \to 0)$$

where $\tilde{c}(x^* \neq 0,0) = c_I = 0$

After Laplace transformation of Equation 2.33c and its boundary conditions, the following equation results in the image space (compare Equation 2.2c):

$$d^2 \bar{c}/dx^{*2} - as\bar{c} = 0$$

where $\bar{c}(x^* \to \infty, s) = 0$ and $\bar{c}(x^* \to 0, s) = m/\omega\epsilon U$ (see Table 2.1, correspondence no. 8). This ordinary differential equation has the following solution in the image space:

$$\bar{c}(x^*, s) = C_1 \exp(+\sqrt{as}\ x^*) + C_2 \exp(-\sqrt{as}\ x^*)$$

$$C_1 = 0 \text{ because } \bar{c}(x^* \to \infty, s) = 0$$

$$C_2 = m/(\omega\epsilon U) \text{ because } \bar{c}(x^* \to 0, s) = m/(\omega\epsilon U) \qquad (2.33d)$$

$$\bar{c}(x^*, s) = \frac{m/\omega}{\epsilon U} \exp(-\sqrt{a}\ x^* \sqrt{s})$$

Inverse Laplace transformation back to the original space and substitution of $c(x,t)$ by $\tilde{c}(x^*,t)$ and $x^* = x - Ut$ yield the following functions:

$$\tilde{c}(x^*, t) = \frac{m/\omega}{2\epsilon\sqrt{\pi t K}} \exp\left(-\frac{x^{*2}}{4Kt}\right) \qquad (2.33e)$$

$$c(x,t) = h(x,t) = \frac{m/\omega}{2\epsilon\sqrt{\pi t K}} \exp\left(-\frac{(x - Ut)^2}{4Kt}\right)$$

With the original condition $dm/dt = \dot{m} = \omega\epsilon U c_o = Q c_o = \text{const.}$, the solution is finally obtained in the following form:

$$c(x,t) = \frac{U c_o}{2\sqrt{\pi K}} \int_{\tau=0}^{t} \exp\left[-\frac{(x - U\tau)^2}{4K\tau}\right] \frac{d\tau}{\sqrt{\tau}}$$

$$= \frac{c_o}{2}\left[\text{erfc}\ \frac{x - Ut}{2\sqrt{Kt}} - \exp\frac{Ux}{K}\ \text{erfc}\ \frac{x + Ut}{2\sqrt{Kt}}\right] \qquad (2.33f)$$

The function $\text{erfc}(\zeta)$ is plotted in Figure 2.6, and a typical graphical presentation of $\tilde{c}(x^*,t)$ according to Equation 2.33e is shown in Figure 2.7. If an approximation of Equation 2.33f is used neglecting the second term in the brackets

$$c(x,t) \approx \frac{c_o}{2}\ \text{erfc}\ \frac{x - Ut}{2\sqrt{Kt}}$$

and superimposed with the solutions of the convective model Equations 2.15 and 2.16, the following models result:

f	erf (f)	f	erf (f)	f	erf (f)	f	erf (f)	f	erf (f)
0,05	0,056	0,30	0,329	0,55	0,563	0,80	0,742	1,05	0,862
0,10	0,112	0,35	0,379	0,60	0,604	0,85	0,771	1,10	0,880
0,15	0,168	0,40	0,428	0,65	0,642	0,90	0,797	1,15	0,896
0,20	0,223	0,45	0,475	0,70	0,678	0,95	0,821	1,20	0,910
0,25	0,276	0,50	0,520	0,75	0,711	1,00	0,843	1,25	0,923

$$\text{erf}(-f) = -\text{erf}(f) \qquad\qquad \text{erf}(0) = 0$$

$$\text{erfc}(f) = 1{,}0 - \text{erf}(f) \qquad\qquad \text{erf}(f) = \frac{2}{\sqrt{\pi}} \int_{f}^{\infty} \exp(-\tau^2)\, d\tau$$

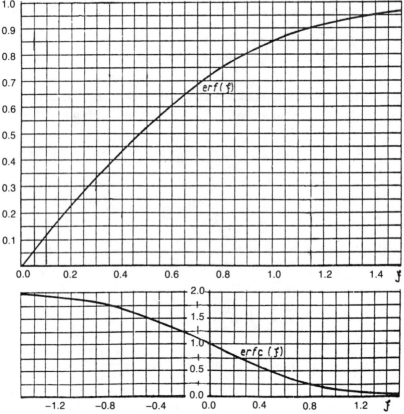

Figure 2.6. The functions erf(χ) and erfc(χ).

$$c(x,t_o) = c_1 + 0.5(c_2 - c_1)\, \text{erfc}\, \frac{x - Ut_o}{2\sqrt{Kt_o}} \qquad (2.33g)$$

where

$$c_1 = \frac{\overline{w}}{\lambda} + \left(c_I - \frac{\overline{w}}{\lambda} \right) \exp(-\overline{\lambda}\xi/q)$$

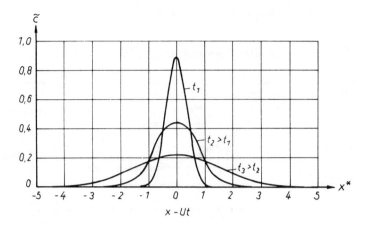

Figure 2.7. Graphical representation of Equation 2.33e.

$$c_2 = \frac{\overline{w}}{\lambda} + \left(c_I - \frac{\overline{w}}{\lambda}\right) \exp(-\overline{\lambda}x/q)$$

$$\xi = \begin{cases} x & \text{for } x \le Ut_o \\ Ut_o & \text{for } x > Ut_o \end{cases}$$

$$c(x_o,t) = c_1 + 0.5(c_2 - c_1)\,\text{erfc}\,\frac{x_o - Ut}{2\sqrt{Kt}} \qquad (2.33h)$$

where

$$c_1 = \frac{\overline{w}}{\lambda} + \left(c_I - \frac{\overline{w}}{\lambda}\right) \exp(-\overline{\lambda}t/\epsilon)$$

$$c_2 = \frac{\overline{w}}{\lambda} + \left(c_o - \frac{\overline{w}}{\lambda}\right) \exp(-\overline{\lambda}\tau/\epsilon)$$

$$\tau = \begin{cases} t & \text{for } t \le x_o/U \\ x_o/U & \text{for } t > x_o/U \end{cases}$$

Applied to the example shown in Figure 2.1 by using the dispersivities d according to Figure 1.33 as follows:

$K = \delta U = \delta q/\overline{\epsilon}$ (2.34)	$\overline{x}$	317.0	634.0	250.0	500.0 m
$U = q/\overline{\epsilon} = 1 \cdot 10^{-5}$ m/s (2.35)	δ	7.1	11.1	6.0	9.6 m

The calculated results are presented in Figure 2.8.

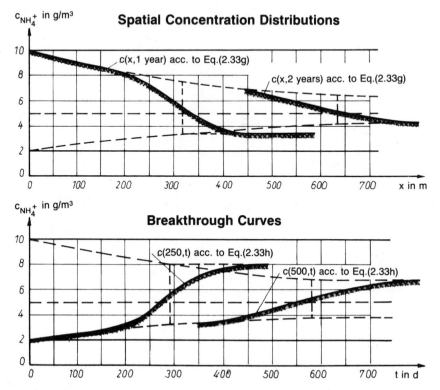

Figure 2.8.   Graphical representation of the approximate solutions of Equation 2.33 for the example shown in Figure 2.1.

2.1.4 SOLUTION OF DISPERSIVE-CONVECTIVE MODELS

Models reflecting diffusive/dispersive as well as convective transport (see Chapter 1.3) are called *dispersive-convective migration models*. The following examples represent the most significant analytical solutions of these models.

2.1.4.1 One-Dimensional Migration ($\bar{\epsilon} \neq 0$, $\bar{\lambda} = 0$, $\bar{w} = 0$)

The model investigated in §2.1.2.1 is reconsidered. For $\bar{\lambda} = 0$, $\bar{w} = 0$, and $D > 0$, Equation 2.13 takes the following form:

$$MD\frac{\partial^2 c}{\partial x^2} - q\frac{\partial c}{\partial x} = \bar{\epsilon}\,\frac{\partial c}{\partial t} \qquad \text{or} \qquad K\frac{\partial^2 c}{\partial x^2} - U\frac{\partial c}{\partial x} = \frac{\partial c}{\partial t} \quad (2.36a)$$

where   M = thickness of the aquifer in m

         D = dispersion coefficient with resp. to Darcy's velocity v in m^2/s
           ($D \approx \delta v$)

         K = dispersion coefficient with resp. to the velocity U; $K = D/(\theta_o + \theta_{st} + \alpha)$ in m^2/s ($K \approx \delta U$)

         U = migration velocity; $U = q/\bar{\epsilon} = v/(\theta_o + \theta_{st} + \alpha)$ in m/s

This partial differential equation with the initial and boundary conditions

$$c(x,0) = 0, \quad c(0,t) = c_o, \quad c(\infty,t) = 0, \text{ resp. } \partial c/\partial x|_{\infty,t} = 0$$

becomes by Laplace transformation an ordinary differential equation:

$$K \, d^2\bar{c}/dx^2 - U \, d\bar{c}/dx = s\bar{c} \tag{2.36b}$$

where $\bar{c}(0,s) = c_o/s$ and $\bar{c}(\infty,s) = 0$. This ordinary differential equation is of the same type as Equation 2.2c. Thus, the solution in the image space is obtained by means of the following expressions:

becomes by Laplace transformation an ordinary differential equation:

$$K \, d^2\bar{c}/dx^2 - U \, d\bar{c}/dx = s\bar{c} \tag{2.36b}$$

where $\bar{c}(0,s) = c_o/s$ and $\bar{c}(\infty,s) = 0$. This ordinary differential equation is of the same type as Equation 2.2c. Thus, the solution in the image space is obtained by means of the following expressions:

$$r_1 = \frac{U}{2K} + \sqrt{\frac{U^2}{4K^2} + \frac{s}{K}} \qquad r_2 = \frac{U}{2K} - \sqrt{\frac{U^2}{4K^2} + \frac{s}{K}}$$

$C_1 = 0$ resulting from the boundary condition $\bar{c}(\infty,s) = 0$

$C_2 = c_o/s$ resulting from the boundary condition $\bar{c}(0,s) = c_o/s$

in the following form:

$$\bar{c} = \bar{c}(x,s) = C_2 \exp(r_2 x) = \frac{c_o}{s} \exp\left[\frac{Ux}{2K} - x\sqrt{\frac{U^2 + 4sK}{4K^2}}\right]$$

$$= c_o \exp\left(\frac{Ux}{2K}\right) \frac{1}{s} \exp\left[\frac{-x}{\sqrt{K}}\left(\frac{U^2}{4K} + s\right)^{1/2}\right] \tag{2.36c}$$

$$= c_o \exp\left(\frac{Ux}{2K}\right) \frac{1}{\left(\frac{U^2}{4K} + s\right) - \frac{U^2}{4K}} \exp\left[-\frac{x}{\sqrt{K}}\left(\frac{U^2}{4K} + s\right)^{1/2}\right]$$

Using the frequency theorem according to Table 2.1, the solution in the original space results in the following equation:

$$c(x,t) = c_o \exp\left(\frac{Ux}{2K}\right) \exp\left(-\frac{U^2t}{4K}\right) \mathcal{L}^{-1}\left\{\frac{\exp[-(x/\sqrt{K})\sqrt{s}]}{s - U^2/(4K)}\right\}$$

and when correspondence no. 12 of Table 2.1 is applied, becomes finally:

$$c(x,t) = \frac{c_o}{2}\left[\text{erfc}\,\frac{x - Ut}{2\sqrt{Kt}} + \exp\left(\frac{Ux}{K}\right)\text{erfc}\,\frac{x + Ut}{2\sqrt{Kt}}\right] \quad (2.36d)$$

The same result is obtained when Equation 2.36c is back transformed using the convolution theorem and the frequency-shift theorem of Table 2.1:

$$c(x,t) = c_o \exp\left(\frac{Ux}{2K}\right)\mathcal{L}^{-1}\left\{\frac{1}{s}\right\} *\mathcal{L}^{-1}\left\{\exp\left[-\frac{x}{\sqrt{K}}\left(\frac{U^2}{4K} + s\right)^{1/2}\right]\right\} \quad (2.36d')$$

$$= c_o \exp\left(\frac{Ux}{2K}\right)\frac{x}{2\sqrt{\pi K}}\int_{\tau=0}^{t}\exp\left(-\frac{U^2\tau^2 + x^2}{4K\tau}\right)\frac{d\tau}{\tau^{3/2}}$$

The solution of this integral results in Equation 2.36d (see also Equation 2.46). Equation 2.36d simplifies when the second term in square brackets is negligible compared to the first term (compare Equation 2.50e):

$$c(x,t) = \frac{c_o}{2}\,[\text{erfc}(\xi) + \exp(\eta)\,\text{erfc}(\xi^*)] \approx \frac{c_o}{2}\,\text{erfc}(\xi) \quad (2.36e)$$

The error of this approximation may be estimated by $\gamma = 0.3/\sqrt{\eta}$ [2.87].

Other boundary conditions lead to the following solutions of Equation 2.36a [2.43; 2.77; 2.44; 2.17]:

- $c(x,0) = c_I$ $c(0,t) = c_o$ $\partial c/\partial x|_{\infty,t} = 0$

$$c(x,t) = c_I + \frac{c_o - c_I}{2}\left[\text{erfc}\,\frac{x - Ut}{2\sqrt{Kt}} + \exp\frac{Ux}{K}\text{erfc}\,\frac{x + Ut}{2\sqrt{Kt}}\right] \quad (2.36f)$$

- $c(x,0) = c_I$ $(-K\,\partial c/\partial x + Uc)|_{0,t} = \frac{\dot{m}}{MB\epsilon}$; $\partial c/\partial x|_{\infty,t} = 0$

$$c(x,t) = c_I + \frac{c_o - c_I}{2}\left[\text{erfc}\,\frac{x - Ut}{2\sqrt{Kt}} + 2\sqrt{\frac{U^2t}{\pi K}}\exp\left(-\frac{(x - Ut)^2}{4Kt}\right)\right.$$

$$\left. - \left(1 + \frac{Ux}{K} + \frac{U^2t}{K}\right)\exp\frac{Ux}{K}\,\text{erfc}\,\frac{x + Ut}{2\sqrt{Kt}}\right] \quad (2.36g)$$

- $c(x,0) = c_I$ $c(0,t) = c_o$ $\partial c/\partial x|_{L,t} = 0$

$$c(x,t) = c_I + (c_o - c_I)\left[1 - \sum_{n=1}^{\infty} \frac{2\beta_n \sin\left(\frac{\beta_n x}{L}\right) \exp\left(\frac{Ux}{2K} - \frac{U^2 t}{4K} - \frac{\beta_n K t}{L^2}\right)}{\beta_n^2 + \left(\frac{UL}{2K}\right)^2 + \frac{UL}{2K}}\right]$$

(2.36h)

where the eigenvalues β_n are the positive solutions of the equation

$$\beta_n \cot(\beta_n) + UL/(2K) = 0$$

Many more solutions are derived in van Genuchten and Alves [2.43].

Commonly it is not justified to set up the boundary condition $\partial c/\partial x|_{L,t} = 0$, which constrains the diffusion/dispersion term to zero. There is no physical reason to assume that the dispersive transport may be neglected at the end of a soil column. The use of the boundary condition $\partial c/\partial x|_{\infty,t} = 0$ is therefore advisable for analytical or digital simulations of migration processes in finite regions when these regions are artificially extended by 10 to 20% beyond the actual dimensions. Boundary conditions $\partial c/\partial x|_{\infty,t} = 0$ at the edges of such extended model domains will not notably impact the region of interest for naturally occurring Peclet numbers $Pe = UL/K \approx L\delta$ according to Figure 1.33, as shown in Figure 2.9.

The boundary conditions at $x = 0$, where the migrating substance is applied, may be chosen either (1) as a given concentration (or temperature in the case of heat migration), i.e., as a boundary condition of the first kind, e.g., as $c(0,t) = c_o$ or $c(0,t) = c_o + C_1 \exp(-\alpha t)$ or (2) as a given mass flux rate, i.e., as a boundary condition of the third kind, e.g., as $(-K \partial c/\partial x + Uc)|_{o,t} = \dot{m}/(MB\epsilon)$. Figure 2.9 illustrates the impact of these two alternative types of boundary conditions upon the concentration distribution in space and time. Deviations are particularly significant when the Peclet numbers $Pe = UL/K \approx L/\delta$ are small. Nevertheless, these deviations are often significant also for naturally occurring Peclet numbers (see Figure 1.33). Figure 2.9 shows also that Equation 2.36e

$$\bar{c}(\bar{x},\bar{t}) = 0.5 \; \text{erfc}[(\bar{x} - \bar{t})/(4\bar{t}/Pe)^{1/2}]$$

approximates Equation 2.36g remarkably well even for unnaturally small Pe numbers (see also Sauty [2.85]).

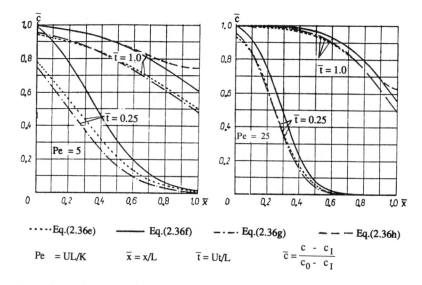

····Eq.(2.36e)   ——— Eq.(2.36f)   —·—· Eq.(2.36g)   — — — Eq.(2.36h)

$Pe = UL/K$ $\bar{x} = x/L$ $\bar{t} = Ut/L$ $\bar{c} = \dfrac{c - c_I}{c_0 - c_I}$

Figure 2.9. Spatial concentration distribution in a flow tube for Pe = 5 and Pe = 20 at time t = 1.0 L/U and t = 0.25 L/U according to Equations 2.36e to 2.36h.

2.1.4.2 One-Dimensional Migration ($\bar{\epsilon} \neq 0$, $\bar{\lambda} = 0$, $\bar{w} = 0$)

Let us consider the mathematical model:

$$DM\frac{\partial^2 c}{\partial x^2} - q\frac{\partial c}{\partial x} = \bar{\epsilon}\frac{\partial c}{\partial t} - \bar{w} \tag{2.37a}$$

With the boundary conditions $c(0,\infty) = c_o$ and $-\infty < \partial c/\partial x|_{\infty,\infty} < \infty$, this model provides the following steady-state solution (t → ∞):

$$c(x,\infty) = c_o + \bar{w}x/q \tag{2.37b}$$

With the boundary conditions $-\infty < (-DM \; dc/dx + qc)|_{o,\infty} = \dot{m}/(\epsilon B)$, $dc/dx|_{\infty,\infty} < \infty$, and $DM = q\delta$, the steady-state solution follows as

$$c(x,\infty) = c_o + \bar{w}(qx + DM)/q^2 = c_o + \bar{w}(x+\delta)/q \tag{2.37c}$$

Applying Laplace transformation to the transient case of Equation 2.37a, an ordinary linear differential equation of second order is obtained in the image space which is of the same type as Equation 2.2d. The solution of this equation must be inverse-transformed into the original space. The following equation finally results in the original space [2.43; 2.15; 2.42]:

- $c(x,0) = c_I$ $c(0,t) = c_o$ $\partial c/\partial x|_{\infty,t} = \text{finite}$

$$c(x,t) = c_I + \frac{c_o - c_I}{2} A(x,t) + B(x,t) \tag{2.37d}$$

$$A(x,t) = \text{erfc} \frac{\bar{\epsilon}x - qt}{\sqrt{4DM\bar{\epsilon}t}} + \exp \frac{qx}{DM} \text{erfc} \frac{\bar{\epsilon}x + qt}{\sqrt{4DM\bar{\epsilon}t}}$$

$$B(y,t) = \frac{\bar{w}}{\bar{\epsilon}} \left[t + \frac{\bar{\epsilon}x - qt}{2q} \text{erfc} \frac{\bar{\epsilon}x + qt}{\sqrt{4DM\bar{\epsilon}t}} \right. $$
$$\left. - \frac{\bar{\epsilon}x + qt}{2q} \exp \left(\frac{qx}{DM} \right) \text{erfc} \frac{\bar{\epsilon}x + qt}{\sqrt{4DM\bar{\epsilon}t}} \right]$$

or approximately

$$c(x,t) \approx c_I + \frac{c_o - c_I}{2} \tilde{A}(x,t) + \tilde{B}(x,t) \tag{2.37e}$$

$$\tilde{A}(x,t) = \text{erfc} \frac{\bar{\epsilon}x - qt}{\sqrt{4DM\bar{\epsilon}t}} \qquad \tilde{B}(x,t) = \frac{wt}{\bar{\epsilon}} + \bar{w} \frac{\bar{\epsilon}x - qt}{2q\bar{\epsilon}} \text{erfc} \frac{\bar{\epsilon}x + qt}{\sqrt{4DM\bar{\epsilon}t}}$$

- $c(x,0) = c_I$ $(-DM \, \partial c/\partial x + qc)|_{o,t} = \dot{m}/B$ $\partial c/\partial x|_{\infty,t} = \text{finite}$

$$c(x,t) = c_I + \frac{c_o - c_I}{2} A(x,t) + B(x,t) \tag{2.37f}$$

$$A(x,t) = \text{erfc} \frac{\bar{\epsilon}x - qt}{\sqrt{4DM\bar{\epsilon}t}} + \sqrt{\frac{4q^2t}{\pi DM\bar{\epsilon}}} \exp \left(- \frac{(\bar{\epsilon}x - qt)^2}{4DM\bar{\epsilon}t} \right)$$

$$- \left(1 + \frac{qx}{DM} + \frac{q^2t}{DM\bar{\epsilon}} \right) \exp \frac{qx}{DM} \text{erfc} \frac{\bar{\epsilon}x + qt}{\sqrt{4DM\bar{\epsilon}t}}$$

$$B(x,t) = \frac{\bar{w}}{\bar{\epsilon}} \left(t + \frac{1}{2q} \left(\bar{\epsilon}x - qt + \frac{DM\bar{\epsilon}}{q} \right) \text{erfc} \frac{\bar{\epsilon}x - qt}{\sqrt{4DM\bar{\epsilon}t}} \right.$$

$$- \sqrt{\frac{t}{4\pi DM\bar{\epsilon}}} \left(\bar{\epsilon}x + qt + \frac{2DM\bar{\epsilon}}{q} \right) \exp \left(- \frac{(\bar{\epsilon}x - qt)^2}{4DM\bar{\epsilon}t} \right]$$

$$+ \left(\frac{t}{2} - \frac{DM\bar{\epsilon}}{2q^2} + \frac{(\bar{\epsilon}x + qt)^2}{4DM\bar{\epsilon}} \right) \exp \frac{qx}{DM} \text{erfc} \frac{\bar{\epsilon}x + qt}{\sqrt{4DM\bar{\epsilon}t}} \right)$$

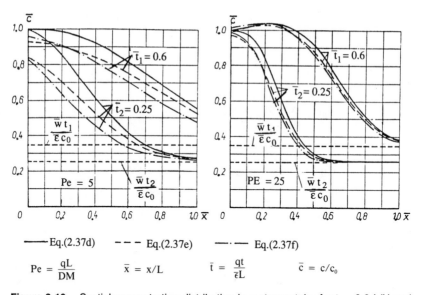

$$Pe = \frac{qL}{DM} \qquad \bar{x} = x/L \qquad \bar{t} = \frac{qt}{\bar{\epsilon}L} \qquad \bar{c} = c/c_0$$

Figure 2.10. Spatial concentration distribution in a stream tube for t = 0.6 L/U and t = 0.25 L/U, and Pe = 5 and Pe = 25 according to Equations 2.37d to 2.37f using the following input data: c_0 = 10 mg/L, c_i = 2 mg/L, q = $4 \cdot 10^{-4}$ m²/s, $\bar{\epsilon}$ = 40 m, $\bar{w}$ = $4 \cdot 10^{-3}$ mg/(ms), U = $q/\bar{\epsilon}$ = $1 \cdot 10^{-5}$ m/s, L = 250 m, δ = L/Pe, DM = qδ (see also Figure 2.1).

Figure 2.10 demonstrates again that Equation 2.37e approximates Equation 2.37f very well, and that all three solutions show close agreement for larger Pe numbers.

2.1.4.3 One-Dimensional Heat Migration

In designing heat pump systems based on bank filtration, it is important to estimate the temperature T of the extracted groundwater [2.60]. For this purpose, Equation 2.37a can be useful in the following special configuration:

$$K\frac{\partial^2 T}{\partial x^2} - U\frac{\partial T}{\partial x} = \frac{\partial T}{\partial t} - w^* \qquad (2.38a)$$

$$K = D/(\rho c)_\Sigma = \frac{\lambda_0 + \delta v_0 (\rho c)_w}{(\rho c)_\Sigma}$$

$$U = v_0 (\rho c)_w / (\rho c)_\Sigma$$

$$w^* = \frac{\lambda_0'}{M(\rho c)_\Sigma} \frac{\partial T'}{\partial z'}\bigg|_{z'=0} + \frac{\lambda_0''}{M(\rho c)_\Sigma} \frac{\partial T''}{\partial z''}\bigg|_{z''=0}$$

$$(\rho c)_\Sigma'' = (\rho c)_\Sigma' = (\rho c)_\Sigma \quad \text{and} \quad T = \Theta - \bar{\Theta}$$

where Θ = temperature in °C
         $\overline{\Theta}$ = mean annual temperature in °C

With the following boundary conditions and source–sink term:

$$T(x=0,t) = T^0 \sin(2\pi t/t_0)$$

$$T'(z' \to \infty, t) = T''(z'' \to \infty, t) = 0$$

$$t_0 = 365 \text{ days} = 1 \text{ year}$$

$$\lambda_0' \partial T'/\partial z'|_{z'=0} = \lambda_0'' \partial T''/\partial z''|_{z''=0} = \lambda_0^* \partial T^*/\partial z|_{z=0}$$

and, consequently,

$$w^* = -\omega T^0 \exp(-\alpha x)(\sin \varphi + \cos \varphi) \qquad (2.38b)$$

where $\omega = \dfrac{-2\lambda^*}{M(\rho c)_\Sigma} \sqrt{\dfrac{\pi(\rho c)_\Sigma}{\lambda^* t_0}}$

Equation 2.38a has the following asymptotic periodic solution

$$T(x,t) = T^0 \exp(-\alpha x) \sin\left(\frac{2\pi t}{t_0} - \beta x\right) \qquad (2.38c)$$

where $\beta x = 2\pi \Delta t/t$
         $\varphi = 2\pi t/t_0 - \beta x$
         Δt = phase shift
         α = decay coeff.

Equation 2.38c thus shows that the temperature $T(x,t)$ follows the boundary condition $T(0,t)$ in a delayed and decayed form.

2.1.4.4 One-Dimensional Migration ($\overline{\epsilon} \neq 0, \overline{\lambda} \neq 0, \overline{w} \neq 0$)

Let us consider the following mathematical migration model:

$$DM\frac{\partial^2 c}{\partial x^2} - q\frac{\partial c}{\partial x} = \overline{\epsilon}\,\frac{\partial c}{\partial t} + \overline{\lambda} c - \overline{w} \qquad (2.39a)$$

In the steady-state, i.e., for $t \to \infty$, Equation 2.39a is of the type 2.2d and has, therefore, the following solution:

$$c(x,\infty) = C_1 \exp(r_1, x) + C_2 \exp(r_2, x) + (c(\infty, \infty) \qquad (2.39b)$$

With the boundary condition $c(x \to \infty)$ = finite or $dc/dx|_{x \to \infty} = 0$, C_1 must be zero. The particular solution $c(\infty, \infty) = \overline{w}/\overline{\lambda}$ follows immediately from the condition $dc/dx|_{x \to \infty} = 0$, and C_2 is found from the condition $c(0, \infty) = c_o$ as $C_2 = c_o - \overline{w}/\overline{\lambda}$.

With the root $r_2 = (q - \sqrt{q^2 + 4\lambda DM})/(2DM)$, the steady-state solution finally results in

$$c(x,\infty) = \frac{\overline{w}}{\lambda} + \left(c_0 - \frac{w}{\lambda}\right) \exp\left(\frac{x}{2DM}(q - \sqrt{q^2 + 4\lambda DM})\right) \quad (2.39c)$$

Since the argument of the exp-function approaches for $D \to 0$ the term $\sigma - \lambda x/q$, which may be seen by transforming the square root into Taylor series, Equation 2.39c reduces to Equation 2.15b for $D \to 0$.

With a boundary condition of the third type at $x = 0$ ($-DM\, dc/dx + qc)|_{x=0} = qc_o$, instead of a boundary condition of the first type, the following solution is obtained:

$$c(x,\infty) = \frac{\overline{w}}{\lambda} + \left(c_0 - \frac{\overline{w}}{\lambda}\right) \frac{2q}{q + \sqrt{q^2 + 4\lambda DM}} \exp$$

$$\left[\frac{x}{2DM}\left(q - \sqrt{q^2 + 4\lambda DM}\right)\right] \quad (2.39d)$$

Transient solutions of Equation 2.39a are obtained via Laplace transformation [2.43; 2.42; 2.82]. The most notable solutions are the following:

- $c(x,0) = c_1 \qquad c(0,t) = c_o \qquad \partial c/\partial x|_{\infty,t} = 0$

$$c(x,t) = \frac{\overline{w}}{\lambda} + \left(c_1 - \frac{\overline{w}}{\lambda}\right) A(x,t) + \left(c_0 - \frac{\overline{w}}{\lambda}\right) B(x,t) \quad (2.39e)$$

$$A(x,t) = \exp\left(-\frac{\overline{\lambda t}}{\epsilon}\right)\left(1 - \frac{1}{2}\,\mathrm{erfc}\,\frac{\overline{\epsilon}x - qt}{\sqrt{4DM\overline{\epsilon}t}}\right.$$

$$\left. - \frac{1}{2}\exp\frac{qx}{DM}\,\mathrm{erfc}\,\frac{\overline{\epsilon}x + qt}{\sqrt{4DM\overline{\epsilon}t}}\right)$$

$$B(x,t) = \frac{1}{2}\exp\frac{(q - q^*)x}{2DM}\,\mathrm{erfc}\frac{\overline{\epsilon}x - q^*t}{\sqrt{4DM\overline{\epsilon}t}}$$

$$+ \frac{1}{2}\exp\frac{(q + q^*)x}{2DM}\,\mathrm{erfc}\frac{\overline{\epsilon}x + q^*t}{\sqrt{4DM\overline{\epsilon}t}}$$

where $q^* = \sqrt{q^2 + 4\lambda DM}$

which may be approximated by

$$c(x,t) \approx \frac{\overline{w}}{\lambda} + \left(c_I - \frac{\overline{w}}{\lambda}\right)\tilde{A}(x,t) + \left(c_o - \frac{\overline{w}}{\lambda}\right)\tilde{B}(x,t) \qquad (2.39f)$$

$$\tilde{A}(x,t) = \exp\left(-\frac{\overline{\lambda}t}{\epsilon}\right)\left(1 - \tfrac{1}{2}\,\mathrm{erfc}\,\frac{\overline{\epsilon}x - qt}{\sqrt{4DM\epsilon t}}\right)$$

$$\tilde{B}(x,t) = \tfrac{1}{2}\exp\frac{(q - q^*)x}{2DM}\,\mathrm{erfc}\,\frac{\overline{\epsilon}x - q^*t}{\sqrt{4DM\epsilon t}}$$

And also

- $c(x,0) = c_I \qquad (-DM\,\partial c/\partial x + qc)|_{o,t} = \dfrac{\dot{m}}{B} \qquad \partial c/\partial x|_{\infty,t} = 0$

$$c(x,t) = \frac{\overline{w}}{\lambda} + \left(c_I - \frac{\overline{w}}{\lambda}\right)A(x,t) + \left(c_o - \frac{\overline{w}}{\lambda}\right)B(x,t) \qquad (2.39g)$$

$$A(x,t) = \exp\left(-\frac{\overline{\lambda}t}{\epsilon}\right)\Bigg[1 - \tfrac{1}{2}\,\mathrm{erfc}\,\frac{\overline{\epsilon}x - qt}{\sqrt{4DM\epsilon t}}$$

$$- \sqrt{\frac{q^2 t}{\pi DM\epsilon}}\,\exp\left(-\frac{(\overline{\epsilon}x - qt)^2}{4DM\epsilon t}\right)\Bigg]$$

$$+ \tfrac{1}{2}\left(1 + \frac{qx}{DM} + \frac{q^2 t}{DM\epsilon}\right)\exp\frac{qx}{DM}\,\mathrm{erfc}\left(\frac{\overline{\epsilon}x + qt}{\sqrt{4DM\epsilon t}}\right)$$

$$B(x,t) = \frac{q}{q + q^*}\exp\frac{(q - q^*)x}{2DM}\,\mathrm{erfc}\,\frac{\overline{\epsilon}x - q^*t}{\sqrt{4DM\epsilon t}}$$

$$+ \frac{q}{q - q^*}\exp\frac{(q + q^*)x}{2DM}\,\mathrm{erfc}\,\frac{\overline{\epsilon}x + q^*t}{\sqrt{4DM\epsilon t}}$$

$$+ \frac{q^2}{2\overline{\lambda}DM}\exp\left(\frac{qx}{DM} - \frac{\overline{\lambda}t}{\epsilon}\right)\mathrm{erfc}\,\frac{\overline{\epsilon}x + qt}{\sqrt{4DM\epsilon t}}$$

where $\quad q^* = \sqrt{q^2 + 4\overline{\lambda}DM}$

Figure 2.11 illustrates that Equation 2.39f approximates Equation 2.37g remarkably well, and that all three solutions approach each other for large Pe numbers.

Figure 2.12 compares the solutions of Equations 2.34 and 2.35 shown in Figure 2.8 with the solution of Equation 2.39f for the example which was also used with the plotted curves in Figure 2.1. The good agreement

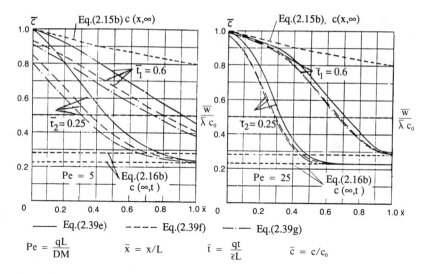

$$Pe = \frac{qL}{DM} \qquad \bar{x} = x/L \qquad \bar{t} = \frac{qt}{\epsilon L} \qquad \bar{c} = c/c_0$$

Figure 2.11. Spatial concentration distribution in a flow tube with the same parameters as in Figure 2.10, but with $\bar{\lambda} = 8 \cdot 10^{-7}$ m/s.

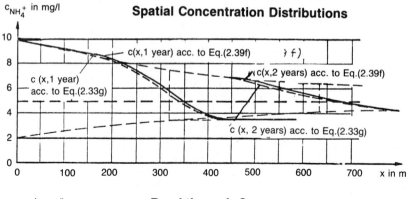

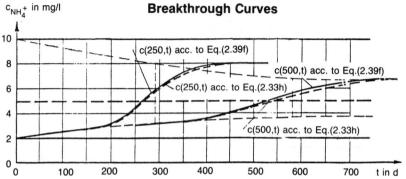

Figure 2.12. Comparison the solutions of Equations 2.34, 2.35 and 2.39f.

demonstrates that by appropriate problem analysis migration processes may be described by approximate and simple models with an accuracy sufficient for practical purposes.

2.1.4.5 One-Dimensional Migration with Kinetic Exchange

A migration model for a single migrant distributed in two phases, the mobile phase and the immobile phase, is considered in this section (see Equations 1.136c and 1.136d). According to Equation 2.39a, the corresponding mathematical model is given by the following two equations:

$$DM \frac{\partial^2 c}{\partial x^2} - q \frac{\partial c}{\partial x} = \bar{\epsilon}_1 \frac{\partial c}{\partial x} + \bar{\lambda}_1 c - \bar{\lambda} c_{st} \qquad (2.40a)$$

$$0 = \bar{\epsilon}_2 \frac{\partial c_{st}}{\partial t} - \bar{\lambda}_1 c + \bar{\lambda} c_{st} \qquad (2.40b)$$

where $qc - DM\ \partial c/\partial x|_{x=0} = \dot{m}/B$
$c(\infty, t) = 0$
$c(x, 0) = c_1$
$c_{st}(x, 0) = c_{st,1}$

Using Laplace transformation, the two Equations 2.40a and 2.40b yield a single ordinary linear differential equation. The inverse transformation of this differential equation is discussed in Luckner et al. [2.62]. The evaluation of the resulting analytical solution in the original space requires the aid of a computer (see FORTRAN program ALSUB 3 in Luckner et al. [2.62]). Comparable solutions of Equations 2.40a and 2.40b are given elsewhere [2.14; 2.76; 2.39; 2.4].

2.1.4.6 Transverse Migration in Parallel and Radial Flow

The following migration models in parallel and radial groundwater flow are considered:

Parallel Flow

Flow model

$$\vec{v} = v_x = -v_o = \text{constant} \qquad (2.41a)$$

Water quality model

$$K_\ell \frac{\partial^2 c}{\partial x^2} + K_{tr} \frac{\partial^2 c}{\partial y^2} - U_o \frac{\partial c}{\partial x} = \frac{\partial c}{\partial t} \qquad (2.41b)$$

where
$$c(0,y) = \begin{cases} c_o \text{ where } -\infty < y < +0 \\ 0 \text{ where } 0 < y < \infty \end{cases}$$

$$K = D/\epsilon$$
$$U_o = v_o/\epsilon$$
$$\epsilon = \theta_o + \theta_{st} + K_d\rho_{dry}$$
$$c(x,0) = c_o/2 \text{ (required due to symmetry; see Figure 2.13a)}$$

Under steady-state conditions, the terms $K_\ell \, \partial^2c/\partial x^2$ and $\partial c/\partial t$ become zero. Thus, Equation 2.41b takes the following form in the domain $y \geq 0$:

$$K_{tr} \frac{\partial^2c}{\partial y^2} = U_o \frac{\partial c}{\partial x} \quad \text{and} \quad \frac{\partial^2c}{\partial y^2} = a \frac{\partial c}{\partial x} \quad c = c(y,x) \qquad (2.42a)$$

where $c(x,0) = c_o/2$
$$c(x,\infty) = 0$$
$$c(0,y) = 0$$
$$a = U_o/K$$

Substituting $y \triangleq x$, $x \triangleq t$, $c \triangleq P$, and $c_o/2 \triangleq P_o$ in Equation 2.42a, it becomes evident that this equation with its boundary conditions is equivalent to Equation 2.7. Hence, the following solution may be derived:

$$c(x,y,\infty) = \frac{c_o}{2} \operatorname{erfc} \left(y \sqrt{\frac{U_o}{4K_{tr}x}} \right) \qquad (2.42b)$$

where $x > 0$
$$y > 0$$

The solution in the domain $y < 0$ is given by $c(-y) = c - c(+y)$. If the condition $c(x,y,0) = c_o$ for $-\infty < y < 0$ and the condition $c(x,y,0) = 0$ for $0 < y < +\infty$ are applied, the transport terms $K_\ell\partial^2c/\partial x^2$ and $U_o \, \partial c/\partial x$

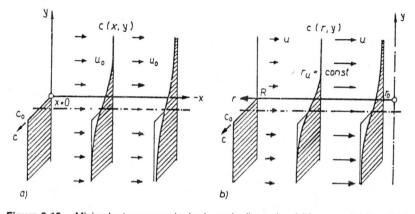

Figure 2.13.   Mixing by transverse hydrodynamic dispersion *(a)* in a parallel flow field (flow toward a trench) and *(b)* in a radial flow field (flow toward a well).

become zero since c does not depend on x, and Equation 2.42a takes the following form:

$$\partial^2 c/\partial y^2 = a\partial c/\partial t$$

where $a = 1/K_{tr}$. This equation provides the following solution:

$$c(y,t) = \frac{c_o}{2} \, \text{erfc} \, (y/\sqrt{4K_{tr}t}) \qquad (2.42c)$$

Radially Symmetric Flow

Flow Model

$$\vec{v} = -v_r \qquad 2\pi r v_r = \text{constant} \qquad (2.43a)$$

Water Quality Model (see Equation 2.38a)

$$2\pi r K_{tr} \frac{\partial^2 c}{\partial y^2} = -\frac{\partial}{\partial r}(2\pi r Uc) = \frac{\partial^2 c}{\partial y^2} = -a\frac{\partial c}{\partial r} \quad c = c(y,r) \, (2.43b)$$

where $2\pi r U = \text{const.}$
$\qquad\qquad a = U/K_{tr}$
$\qquad\qquad K_{tr} = U\delta_{tr}$
$\qquad\qquad c(0,r) = c_o/2$
$\qquad\qquad c(\infty,r) = 0$
$\qquad\qquad c(y,R) = 0$

Laplace transformation of this equation results in (see Equation 2.8)

$$d^2\bar{c}/dy^2 = as\bar{c} - ac(y,0)$$

with $\bar{c}(0,s) = c_o/(2s)$ and $\bar{c}(\infty,s) = 0$. Transforming the radial coordinate by $r' = R - r$ provides $ac(y,R) = ac(y,r'=0) = 0$, which leads again to Equation 2.42 and hence to the following solution:

$$c(y,r') = \frac{c_o}{2} \, \text{erfc} \left(y \sqrt{\frac{U}{4K_{tr}r'}}\right) = \frac{c_o}{2} \, \text{erfc} \, \frac{y}{\sqrt{4\delta_{tr}(R - r)}} \qquad (2.43c)$$

2.1.4.7 Migration in a Parallel Steady-State Groundwater Flow

The spreading of a tracer injected at a point into a parallel steady-state groundwater flow field results in the following migration model:

$$D_x \frac{\partial^2 c}{\partial x^2} + D_y \frac{\partial^2 c}{\partial y^2} + D_z \frac{\partial^2 c}{\partial z^2} - v_x \frac{\partial c}{\partial x} = \epsilon \frac{\partial c}{\partial t} \qquad (2.44)$$

With the pulselike tracer injection $m = c_o \int Q dt$ (in mol or kg) at the origin of the spatial coordinate system at $t = 0$, the following tracer concentrations result:

One-Dimensional Migration in X-Direction (See Equation 2.33e)

$$c(x,t) = \frac{m/(MB)}{2\sqrt{\pi D_x t \epsilon}} \exp\left[-\frac{(\epsilon x - vt)^2}{4\epsilon D_x t}\right] \tag{2.45a}$$

The time at which the maximum concentration c_{max} reaches x_o is given by

$$t_m = \frac{\epsilon}{v^2}\left(\sqrt{v^2 x_0^2 + D_x^2} - D_x\right) =$$

$$\frac{\epsilon}{v}\left(\sqrt{x_0^2 + \delta_x^2} - \delta_x\right) \approx \frac{\epsilon x_0}{v}\left(1 - \delta_x/x_0\right)!$$

Therefore, the plot $c(x_o,t)$ is not symmetrical with respect to $t_o = \epsilon x_o/v$.

Two-Dimensional Migration in the X,Y-Plane (See Equation 1.35)

$$c(x,y,t) = \frac{m/M}{4\pi t\sqrt{D_x D_y}} \exp\left[-\frac{(\epsilon x - vt)^2}{4\epsilon D_x t} - \frac{\epsilon y^2}{4 D_y t}\right] \tag{2.45b}$$

Figure 1.35 shows the tracer distribution $c(x, y-y_o, t)$ according to Equation 2.45b.

Three-Dimensional Migration in the X,Y,Z-Space

$$c(x,y,z,t) = \frac{m\sqrt{\epsilon}}{8\sqrt{\pi^3 t^3 D_x D_y D_z}} \exp\left[-\frac{(\epsilon x - vt)^2}{4\epsilon D_x t} - \frac{\epsilon y^2}{4 D_y t} - \frac{\epsilon z^2}{4 D_z t}\right] \tag{2.45c}$$

where M = extension of the flow field in z-direction
B = extension of the flow field in y-direction
ϵ = storage coefficient; $\epsilon = \theta_o + \theta_{st} + \alpha$ (compare Equation 1.78b)

In the case of continuous tracer injection $\dot{m}$ = constant (step function) at $t = 0$ and at the origin of the spatial coordinate system, the following corresponding solutions are obtained:

One-Dimensional Migration in X-Direction (See Equation 2.33f)

$$c(x,t) = \frac{\dot{m}/(MB)}{2\sqrt{\pi D_x \epsilon}} \int_o^t \exp\left[-\frac{(\epsilon x - v\tau)^2}{4\epsilon D_x \tau}\right] \frac{d\tau}{\sqrt{\tau}}$$

$$= \frac{\dot{m}}{2MBv}\left[\text{erfc}\,\frac{\epsilon x - vt}{\sqrt{4 D_x t \epsilon}} - \exp\frac{vx}{D_x}\,\text{erfc}\,\frac{\epsilon x + vt}{\sqrt{4 D_x t \epsilon}}\right] \tag{2.46a}$$

Two-Dimensional Migration in the X,Y-Plane [2.56]

$$c(x,y,t) = \frac{\dot{m}/M}{4\pi\sqrt{D_xD_y}} \int_0^t \exp\left[-\frac{(\epsilon x - v\tau)^2}{4\epsilon D_x\tau} - \frac{\epsilon y^2}{4D_y\tau}\right]\frac{d\tau}{\tau}$$

$$= \frac{\dot{m}/M}{4\pi\sqrt{D_xD_y}} \exp\frac{qx}{2D_xM} W(\sigma,b) \tag{2.46b}$$

where $W(\sigma,b)$ is the well function after Hantush (see, for instance, Busch and Luckner [2.13, Equation 5.50 and Figure 5.14]):

$$W(\sigma,b) = \int_u^\infty \exp\left(-\tau - \frac{b^2}{4\tau}\right)\frac{d\tau}{\tau}$$

$$b^2 = \left(\frac{xq}{2D_xM}\right)^2 + \left(\frac{yq}{2M\sqrt{D_xD_y}}\right)^2 \quad \sigma = \frac{b^2\epsilon D_xM^2}{q^2t}$$

Under steady-state conditions $(t \to \infty)$, Equation 2.46b approaches

$$c(x,y,\infty) = \frac{\dot{m}/M}{2\pi\sqrt{D_xD_y}} \exp\frac{qx}{2D_xM} K_o(b) \tag{2.46c}$$

where K_o is the modified Bessel function of the second kind and zero order, $W(0,b) = 2K_o(b)$ (see, e.g., Busch and Luckner [2.13, Table 5.7]).

Three-Dimensional Migration in the X,Y,Z-Space (See Equations 2.36d' and 2.36d)

$$c(x,y,z,t) = \frac{\dot{m}\sqrt{\epsilon}}{8\sqrt{\pi^3D_xD_yD_z}} \int_0^t \exp\left[-\frac{(\epsilon x - v\tau)^2}{4\epsilon D_x\tau} - \frac{\epsilon y^2}{4D_y\tau} - \frac{\epsilon z^2}{4D_z\tau}\right]\frac{d\tau}{\tau^{3/2}}$$

$$= \frac{c_oQ\sqrt{\epsilon}}{8\pi r\sqrt{D_yD_z}} \exp\frac{xv}{2D_x} \left[\exp\left(-\frac{rv}{2D_x}\right) \text{erfc}\frac{\epsilon r - vt}{\sqrt{4D_x\epsilon t}}\right.$$

$$\left. + \exp\frac{rv}{2D_x} \text{erfc}\frac{\epsilon r - vt}{4D_x\epsilon t}\right] \quad r = \sqrt{x^2 + y^2\frac{D_x}{D_y} + z^2\frac{D_x}{D_z}} \tag{2.46d}$$

This solution is of practical value for a tracer migration into an aquifer from a point source located at the surface. For this half-space problem, the factor 8 in Equation 2.46d must be replaced by 4 (see Figure 2.14).

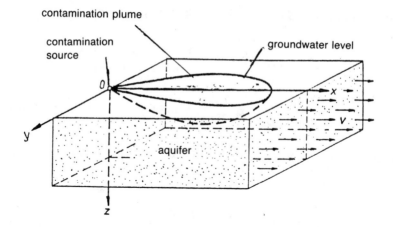

Figure 2.14. Migration of contaminants from a point source located at the ground surface into a homogeneous discharged aquifer (schematized after Technischer Bericht 82/5 [2.94]).

2.1.4.8 Axisymmetric Migration

Flow Model

This flow towards a well can be described by the following mathematical model (see, e.g., Bronstein and Semendjajew [2.11] and Leismann and Frind [2.55]):

$$\frac{\partial^2 s}{\partial r^2} + \frac{1}{r}\frac{\partial s}{\partial r} = \frac{S}{T}\frac{\partial s}{\partial t} \qquad (2.47a)$$

where

$$s(r,0) = 0$$
$$s(\infty,t) = 0$$
$$\lim_{r \to 0} \frac{2\pi T}{Q}\frac{\partial s}{\partial r} = 1$$
$$s = \text{drawdown}$$
$$S = \text{storativity}$$
$$T = \text{transmissivity}$$

Equation 2.47a with its initial and boundary conditions has the following solution:

$$s = \frac{Q}{4\pi T}W(\sigma) \qquad (2.47b)$$

where

$$\sigma = \frac{r^2 S}{4Tt}$$

The corresponding flow rate $q = q_r = \int v_r dz$ results in (see Beims [2.8])

$$s = \frac{Q}{4\pi} \frac{\partial W}{\partial r} = \frac{Q}{2\pi r} \exp\left(-\frac{r^2 S}{4Tt}\right) \qquad (2.47c)$$

Quality model

The axisymmetric quality model according to Equation 2.36 with DM $= \delta q$ is given by

$$\frac{\partial}{\partial r}\left(\delta q r \frac{\partial c}{\partial r}\right) - \frac{\partial}{\partial r}(qrc) = \bar{r}\bar{\epsilon}\frac{\partial c}{\partial t} \qquad \text{with } \bar{\epsilon} = \epsilon M \qquad (2.48a)$$

where δ is dispersivity measured in m (see Figure 1.33). Under steady-state flow conditions, i.e., when $q_o = Q/(2\pi r) = $ constant, Equation 2.48a takes the following form:

$$\frac{\partial^2 c}{\partial r^2} - \frac{1}{\delta}\frac{\partial c}{\partial r} = \frac{2\pi \bar{r}\bar{\epsilon}}{Q\delta}\frac{\partial c}{\partial t} \qquad (2.48b)$$

and in the transient flow case it takes the form:

$$\frac{\partial}{\partial r}\left(\left[\exp\left(-\frac{r^2 S}{4Tt}\right)\right]\frac{\partial c}{\partial r}\right) - \frac{1}{\delta}\frac{\partial}{\partial r}\left[c\exp\left(-\frac{r^2 S}{4Tt}\right)\right] = \frac{2\pi \bar{r}\bar{\epsilon}}{Q\delta}\frac{\partial c}{\partial t} \qquad (2.48c)$$

For the following initial and boundary conditions

$$c(r_o,0) = c_I, \qquad c(r_o,t) = c_o \qquad \text{and} \qquad c(\infty,t) = c_I$$

various approximate analytical solutions were derived. By comparison with numerical calculations, the following two approximate solutions showed suitable agreement [2.08; 2.54]:

- $$c(r,t) = c_I + \frac{c_o - c_I}{2} \text{ erfc } \frac{r^2 - r_F^2}{\sqrt{\frac{16}{3}\delta r_F^3}} \qquad (2.49a)$$

where $r_F^2 = Qt/(\pi\bar{\epsilon})$; Raimondi used $\sqrt{(16/3)\delta r^2 r_F}$ instead of $\sqrt{(16/3)\delta r_F^3}$ (see Sauty [2.85]).

- $$c(r,t) = c_I + (c_o - c_I)[1 - F_1(\nu,\sigma)] \qquad (2.49b)$$

where
$$F_1 = \frac{1}{\Gamma(\nu)}\int_o^\sigma \exp(-\tau)\tau^{\nu-1} d\tau$$
$$\nu = Q/(4\pi MD)$$
$$\sigma = r^2\epsilon/(4Dt)$$
$$D = \delta v = \frac{\delta Q}{2\pi r M}$$
$\Gamma = $ gamma function as derived e.g. in Botschewer and Oradovskaja [2.10]

Frequently Equation 2.49a is preferred because of its convenience and higher accuracy (see Fig. 2.15).

For dispersion on the macroscopic scale, where $D = \delta^* v^2$ (with δ^* in seconds), the following solution may be derived (see Schestakow [2.87]):

$$c(r,t) = c_1 + \frac{c_o - c_1}{2} \, \text{erfc} \, \frac{r^2 \pi \epsilon - Qt}{2Q\sqrt{\delta^* \epsilon t / M'}} \quad \text{with } \delta^* = (2\pi rM)^2 D/Q^2$$

$$(2.49c)$$

This solution agrees well with Equation 2.49a (see Figure 2.15).

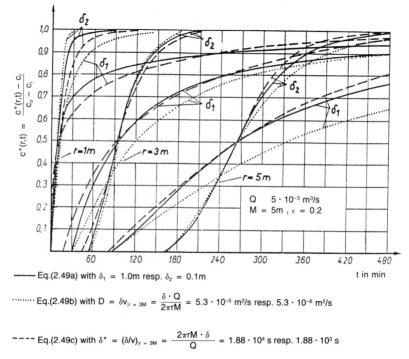

—— Eq.(2.49a) with $\delta_1 = 1.0$m resp. $\delta_2 = 0.1$m

········ Eq.(2.49b) with $D = \delta v_{|r = 3M} = \dfrac{\delta \cdot Q}{2\pi r M} = 5.3 \cdot 10^{-5}$ m²/s resp. $5.3 \cdot 10^{-6}$ m²/s

---- Eq.(2.49c) with $\delta^* = (\delta/v)_{|r = 3M} = \dfrac{2\pi r M \cdot \delta}{Q} = 1.88 \cdot 10^4$ s resp. $1.88 \cdot 10^3$ s

Figure 2.15. Comparison of solutions according to Equations 2.49a to 2.49c for two characteristic examples.

2.1.5 COMPUTATION OF ANALYTICAL SOLUTIONS

As shown in §2.1.2 to 2.1.4, analytical solutions of migration problems are so complex that the use of hand-held calculators or hand-held computers provides significant advantages. For practical applications several numerical formulations and algorithms of the previously discussed analytical solutions were developed and tested. The following paragraphs and the supplementary manual [2.59] present a selection of these formulations and algorithms. Based on the fact that today's programmable pocket computers are commonly operated with BASIC interpreters and all PCs and laptops are equipped with these interpreters, the notation of the following program examples was carried out in BASIC, and the programs were tested on a SHARP pocket computer PC-1401.

2.1.5.1 Calculation of Exp(α)Erfc(β)

The effective determination of the product function

$$EXC(\alpha,\beta) = \exp(\alpha)\mathrm{erfc}(\beta) \qquad (2.50a)$$

where $EXC(0,\beta) = \mathrm{erfc}(\beta)$
$EXC(\alpha,0) = \exp(\alpha)$

is essential for the calculation of nearly all analytical solutions of migration problems by means of pocket computers, and also for PCs and laptops. The following numerical approximations are recommended (see, e.g., Abramowitz [2.1, Equation 7.1.14 and 7.1.26]).

For $0 < \beta \le 3.0$

$$EXC\,(\alpha,\beta) \approx \exp(\alpha - \beta^2)(a_1\tau + a_2\tau^2 + a_3\tau^3 + a_4\tau^4 + a_5\tau^5) \quad (2.50b)$$

where $\tau = 1/(1.0 + 0.3275911\beta)$
$a_1 = 0.254829592$
$a_2 = -0.284496736$
$a_3 = 1.421413741$
$a_4 = -1.453152027$
$a_5 = 1.061405429$

For $\beta > 3$

$$EXC(\alpha,\beta) \approx \frac{1}{\sqrt{\pi}} \exp\,(\alpha - \beta^2)/(\beta + 0.5/(\beta + 1.0/(\beta + 1.5/$$
$$(\beta + 2.0/(\beta + 2.5/(\beta + 1.0)))))) \qquad (2.50c)$$

For $\beta < 0$

$$EXC(\alpha,\beta) = 2 \exp\,(\alpha) - EXC\,(\alpha,-\beta) \qquad (2.50d)$$

For $|\alpha| > 220$ and $\beta < 0$ or for $|\alpha\pi - \beta^2| > 220$ and $\beta > 0$

$$EXC(\alpha,\beta) \approx 0.0 \qquad (2.50e)$$

Table 2.2a shows the BASIC code for evaluating the function $EXC(\alpha,\beta)$ according to Equations 2.50a to 2.50e.

Table 2.2a. BASIC Program EXC(α, β)

1000	:	"EXC"
1010	:	EC = 0
1020	:	IF ABS AL > 220 AND BE < = 0 RETURN
1030	:	IF BE < > o GOTO 1060
1040	:	EC + EXP AL
1050	:	RETURN
1060	:	C = AL–BE * BE
1070	:	IF ABS C > 220 AND BE > 0 RETURN
1080	:	IF C < –220 GOTO 1180
1090	:	AB = ABS BE
1100	:	IF AB > 3 GOTO 1160
1110	:	TA = 1/(1 + .3275911 * AB)
1120	:	EC = 1.453152027–1.061405429 * TA
1130	:	EC = TA * (.284496736–TA * (1.421413741–TA * EC))
1140	:	EC = TA * (.254829592–EC)
1150	:	GOTO 1180
1160	:	EC = AB + 2/(AB + 2.5/(AB + 1))
1170	:	EC = .5641896/(AB + .5/(AB + 1/(AB + 1.5/EC)))
1180	:	EC = EC * EXP C
1190	:	IF BE < 0 LET EC = 2 * EXP (AL–EC)
1200	:	RETURN

2.1.5.2 A BASIC Program for Evaluating Equation 2.39e

The program evaluates the analytical solution of the following migration model:

$$DM \frac{\partial^2 c}{\partial x^2} - q \frac{\partial c}{\partial x} = \bar{\epsilon} \frac{\partial c}{\partial t} + \bar{\lambda} c - \bar{w} \qquad (2.51)$$

with the initial and boundary conditions

$$c(x,0) = c_I, \ \partial c/\partial x|_{\infty,t} = 0, \text{ and } c(0,t) = c_o.$$

Inputs of this program are the migration parameters:

$$D, M, q, \bar{\epsilon}, \bar{\lambda}, \bar{w}, c_I, c_o, \text{ and the coordinates } (x,t)$$

at which the concentration $c(x,t)$ is to be calculated. Table 2.2b shows the listing of the program with an example.

All the other presented analytical solutions can be programmed in the same manner. Luckner et al. [2.59] provide a comprehensive example, and may assist in both training and practical application.

Table 2.2b. BASIC Program Eq. 2.39e

```
10   :   "2/39 E":CLEAR
20   :   USING "###.####"
30   :   INPUT "CA = ";C1
40   :   INPUT "C0 = ";C0
50   :   INPUT "W = ";W
60   :   INPUT "LAM = ";L
70   :   IF L = 0 LET L = 1E–50
80   :   INPUT "EPS = ";E
90   :   INPUT "D = ";D
100  :   INPUT "M = ";M
110  :   D = D * M
120  :   INPUT "Q0 = ";Q0
130  :   Q = √(SQU Q0 + 4 * L * D)
140  :   INPUT "X = ";X
150  :   INPUT "TP = ";T
160  :   BE = (E * X + Q * T)/√(4 * D * E * T)
170  :   AL = (Q0 + Q) * X * .5/D
180  :   GOSUB "EXC"
190  :   B = .5 * EC
200  :   BE = (E * X–Q * T)/√(4 * D * E * T)
210  :   AL = (Q0–Q) * X * .5/D
220  :   GOSUB "EXC"
230  :   B = .5 * EC  + B
240  :   BE = (E * X + Q0 * T)/√(4 * D * E * T)
250  :   AL = Q0 * X/D
260  :   GOSUB "EXC"
270  :   A = –.5 * EC
280  :   BE = (E * X–Q0 * T)/√(4 * D * E * T)
290  :   AL = 0
300  :   GOSUB "EXC"
310  :   A = EXP(–L * T/E) * (1–.5 * EC + A)
320  :   C = W/L + (C1–W/L) * A + (CO–W/L) * B
330  :   PRINT "C = ";C
340  :   GOTO 140
```

2.1.5.3 Numerical Laplace Transformation

Numerical L-transformation is required if the considered function in the original space $f(x)$ cannot be transformed into the image space by an analytical solution of Equation 2.4 or by means of correspondence tables such as Table 2.1. This situation may occur when special boundary conditions must be fulfilled, or when solution functions in the original space were recorded by measuring devices and are to be used for parameter estimation in the image space. Parameter estimation in the image space in many cases is less difficult than in the original space. Such an approach requires the transformation of the mathematical model and its solution recorded by measuring devices only from the original to the image space, but it requires no inverse L-transformation.

The most obvious numerical approximation of Equation 2.4 with $\Delta x = $ constant and $k = n + 1/2$ is given by the following equation:

$$\mathcal{L}\{f(x)\} = \bar{f}(s) = \int_o^\infty \exp(-sx)\, f(x)\, dx \approx \sum_{n=o}^\infty \exp(-sk\Delta x)\, f(k\Delta x)\ (2.52a)$$

The estimation of convergence of this integral or of the convergence of this series is required (see Equation 2.6).

Test example: $P(t) = t$

Analytical solution: $\mathcal{L}\{P(t)\} = \bar{P}(s) = 1/s^2$

Numerical solution: $\mathcal{L}_N\{P(t)\} = P_N(s) = \sum_{n=0}^\infty \exp(-sk\Delta t)f(k\Delta t)$

with $k = n + 1/2$

For $\Delta t = 1$ it follows:

$$\bar{P}_N(1) = 0.30 + 0.33 + 0.21 + 0.11 + 0.05 + 0.02 + 0.01$$
$$= 1.03 \text{ but the exact value would be } \bar{P}(1) = 1.00$$
$$\bar{P}_N(2) = 0.18 + 0.07 + 0.02 = 0.27 \text{ instead of } \bar{P}(2) =$$
$$0.25$$
$$\bar{P}_N(4) = 0.07 \text{ instead of } \bar{P}(4) = 0.06 \text{ being the exact value}$$

As this simple example shows, the approximation accuracy of Equation 2.52a leaves much to be desired. More accurate solutions are required. Let us hence consider the following polynomial approach used in groundwater flow analysis (see Luckner and Schestakow [2.63, Equation 3.35]):

$$\mathcal{L}_N\{P(t)\} = \bar{P}_N(s) = \frac{1}{s}\sum_{n=0}^4 A_n(t_n)P(t_n) \qquad t_n = \bar{t}_n/s \qquad (2.52b)$$

where

$n = 0$	1	2	3	4
$t_n = 0$	0.617	2.11	4.61	8.40
$A_n = 0.167$	0.564	0.238	0.0306	$1.04 \cdot 10^{-3}$

For $s = 1$ follows, e.g., $t_n = \bar{t}_n$ and for $s = 2$, $t_n = \bar{t}_n/2$ and therefore:

$$\bar{P}_N(1) = 1(0.167 \cdot 0 + 0.564 \cdot 0.617 + 0.238 \cdot 2.11 + ...)$$
$$= 0.35 + 0.50 + 0.14 + 0.01 = 1.00 \text{ (sufficiently accurate)}$$
$$\bar{P}_N(2) = 0.5(0.564 \cdot 0.308 + 0.238 \cdot 1.055 + ...) = 0.25 \text{ (sufficiently accurate)}$$

2.1.5.4 Numerical Inverse Laplace Transformation

Numerical inverse L-transformation is necessary when a solution of the mathematical model in the original space is desired, the solution of this model in the image space is known, and a correspondence between both solutions allowing normal back-transformation does not exist or is unknown (compare §2.1.1, Figure 2.1a and Table 2.1). The most obvious numerical solution of this transformation problem is the formulation of a system of linear equations based on Equation 2.52a

$$\sum_{n=0}^{\infty} \exp(-sk\Delta x)\, f(k\Delta x) = \bar{f}(s)$$

and the calculation of the desired values $f(k\Delta x)$ from this system.

Test Example 1: $P(s) = 1/s^2$

For the data set $s = 1.000,\ 1.200,\ 1.728,\ 2.074$, the following system yields

$$
\begin{pmatrix}
0.3545 & 0.0446 & 0.0056 & 0.0007 & 0.0001 \\
0.4215 & 0.0749 & 0.0133 & 0.0024 & 0.0004 \\
0.4868 & 0.1153 & 0.0273 & 0.0065 & 0.0015 \\
0.5488 & 0.1653 & 0.0498 & 0.0150 & 0.0045 \\
0.6065 & 0.2231 & 0.0821 & 0.0302 & 0.0111
\end{pmatrix}
\cdot
\begin{pmatrix}
P_{0.5} \\ P_{1.5} \\ P_{2.5} \\ P_{3.5} \\ P_{4.5}
\end{pmatrix}
=
\begin{pmatrix}
0.233 \\ 0.335 \\ 0.482 \\ 0.694 \\ 1.000
\end{pmatrix}
$$

The following comparison between the numerical values resulting from this equation system and the exact values $P(t) = t$

$$P(k\Delta t) = \begin{cases} \end{cases}$$

t	0.5	1.5	2.5	3.5	4.5
numerical	0.415	1.167	7.711	−18.802	38.079
exact	0.500	1.500	2.500	3.500	4.500

refers to the problems arising with numerical inverse L-transformations.

More reliable results are obtained with polynomial approaches (see e.g. Luckner and Schestakow [2.63, Equation 3.38a]):

$$P(t) = \sum_{n=1}^{4} B_n(\bar{t})\, s_n \bar{P}(s_n) \tag{2.53}$$

where $\bar{t} = t/s_1$
$s_1 = s_{max}$
$s_2 = s_1/2$
$s_3 = s_1/4$
$s_4 = s_1/8$

and the coefficients $B_n(\bar{t}_n)$ are given as follows (see Luckner and Schestakow [2.63, p. 85]):

$\bar{t} =$	1.0	2.0	3.0	4.0	5.0	6.0	7.0	8.0
B1	−1.48	−2.23	−1.80	−1.07	−0.304	0.425	1.08	1.63
B2	4.79	4.84	2.98	0.825	−1.21	−3.00	−4.51	−5.71
B3	−2.83	−1.87	−0.091	1.63	3.09	4.25	5.12	5.71
B4	0.525	0.253	−0.100	−0.385	−0.578	−0.679	−0.695	−0.633

Equation 2.53 yields for $s_n \bar{P}(s_n) = 1/s_n$ and $s_1 = 1.0$ the following solutions:

$P(1) = -1.48 + 9.58 - 11.32 + 4.20 = 0.98$	instead of 1.00
$P(2) = -2.23 + 9.68 - 7.48 + 2.02 = 1.99$	instead of 2.00
$P(4) = -1.07 + 1.66 + 6.52 - 3.08 = 4.03$	instead of 4.00

Test Example 2

$$\mathcal{L}^{-1}\{(10/s) \exp(-6\sqrt{s})\} = 10 \text{ erfc } (3/\sqrt{x})$$
(compare Equation 2.42 and Figure 2.16)

Substituting t by x in Equation 2.53 yields with $s_1 = 0.2$, $s_2 = 0.1$, $s_3 = 0.05$, and $s_4 = 0.025$:

$$c(5) = -1.01 + 7.18 - 7.40 + 2.03 = 0.80 \text{ instead of } 0.58$$

and with $s_1 = 0.1$, $s_2 = 0.05$, $s_3 = 0.025$, and $s_4 = 0.0125$:

$$c(10) = -2.22 + 12.52 - 10.96 + 2.68 = 2.02 \text{ instead of } 1.79$$

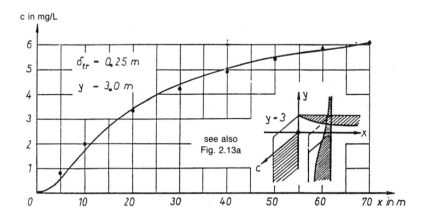

●   solution according to Eq. 2.54
———— analytical solution (exact reference solution)

Figure 2.16. Plot resulting from the numerical inverse Laplace transformation of the function $\bar{c}(s) = (10/s)\exp(-6\sqrt{s})$.

$c(20) = -3.34 + 12.65 - 7.25 + 1.29 = 3.36$ instead of 3.43
$c(30) = -2.70 + 7.79 - 0.35 - 0.51 = 4.23$ instead of 3.37
$c(40) = -1.60 + 2.16 + 6.31 - 1.97 = 4.90$ instead of 5.06
$c(50) = -0.46 - 3.16 + 11.97 - 2.95 = 5.40$ instead of 5.53
$c(60) = 0.69 - 7.84 + 16.46 - 3.47 = 5.84$ instead of 5.81
$c(70) = 1.62 - 11.79 + 19.83 - 3.55 = 6.11$ correct solution

Further methods of numerical inverse Laplace transformation have been developed to provide significantly improved accuracy (see, e.g., Piessens [2.80]). Also, state-of-the-art numerical Fourier transformations and Fourier cosine transformations have proved to be useful tools in solving migration problems [2.31; 2.32; 2.18]. Talbot's inverse L-transformation is particularly recommended [2.93]. This highly accurate numerical transformation method is based on the residual theorem given with Equation 2.9, and results in the following approximation when the trapezoidal rule is used in solving the integral of Equation 2.9:

$$f(x) = \mathcal{L}^{-1}\{F(s)\} \approx \frac{2}{N} \sum_{k=0}^{N-1} ReQ(z_k) \qquad z_k = \frac{2k\pi i}{N} \qquad (2.54)$$

where $\quad S \rightarrow \lambda s + \sigma \quad S(z) = z(1 - \exp(z))$
$$Q(z) = \lambda \exp(\lambda s + \sigma) \, tF(\lambda s + \sigma) \, S'(z) \qquad (2.54a)$$

The accuracy of $f(x)$ may be increased up to a half of the computer precision by proper selection of N. Our tests have shown that $N = 8$ or $N = 16$ should be used. Figure 2.17 illustrates the program structure. The user need only provide the function $F(s)$ in the image space in the form of a subroutine, which can also be given as a table with an interpolation procedure.

Talbot's program routine given in Table 2.3a as well as the subroutine calculating the function $F(s)$ must use complex numbers because s is defined as $s = \sigma + iw$ in the image space. Hence, an additional subrou-

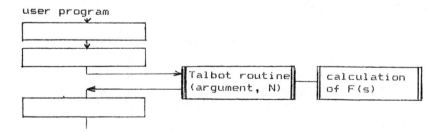

Figure 2.17. Flowchart showing the utilization of numerical inverse Laplace transformation by means of the Talbot routine according to Equation 2.54.

tine is provided with Table 2.3b for multiplication, division, and other calculation procedures of complex numbers, which may be used when such routines are not available with the user's computer.

Let us now demonstrate the numerical inverse L-transformation by means of Talbot's routine listed in Table 2.3, and consider an example characterized by the following correspondence:

$$\frac{e^{-\sqrt{s}}}{s} \quad \bullet\!-\!\!-\!\!-\!\!\circ \quad \text{erfc}\frac{1}{2\sqrt{t}}$$

The subroutine calculating the function $F(s) = (c^{-\sqrt{s}})/s$ may be written as follows:

```
SUBROUTINE FUNC(SR,SI,ER,EI)
DOUBLE PRECISION SR,SI,ER,EI,HR,HI,PR,PI,A,B
CALL COMP(SR,SI,A,B,HR,HI,4)
HR = -HR
HI = -HI
CALL COMP(HR,HI,A,B,PR,PI,5)
CALL COMP(PR,PI,SR,SI,ER,EI,3)
RETURN
END
```

This subroutine FUNC, together with the subroutines TALBOT and COMP shown in Tables 2.3a and 2.3b, provides the following numerical test results which illustrate the very high accuracy:

t	$\mathcal{L}^{-1}\{e^{-\sqrt{s}}/s\}$	erfc $(1/(2\sqrt{t}))$	relative error
0.25	.1572991881	.1572992071	.12 D–06
1.00	.4795002962	.4795001222	.36 D–06
4.00	.7236658771	.7236736098	.10 D–04

Let us now consider a second example. Talbot's algorithm provides for the image function

$$\bar{P}(s) = \frac{1}{s} \exp(-bs) \qquad \text{(see Table 2.1 correspondence no. 5)}$$

the correct result 1 only for $t > b > 0$, but not the correct result 0 for $t < b > 0$. This type of function requires the application of the translation theorem as a first step (see Table 2.1). For the example considered this results in

$$\mathcal{L}^{-1}\{\exp(-bs)\,\bar{P}(s)\} = \begin{cases} P(t - b) & \text{for} \quad t > b > 0 \\ 0 & \text{for} \quad t < b > 0 \end{cases}$$

Table 2.3. Talbot Routine for the Numerical Inverse Laplace Transformation with the Subroutines TALBOT and COMP

```
        SUBROUTINE TALBOT(FT,T,N)
        DOUBLE PRECISION FT, T, NU, LAMDA,
    *           SR(100),SI(100),DSR(100),DSI(100),
    *           ZZR,ZZI,SUMR,SUMI,BR,BI,B1R,B1I,B2R,B2I,
    *           Z,PI,TAU,PIBYN,THETA,ALPHA,PSI,CP,SP,
    *           VH,FR,FI,PR,PPI,V2,QR,QI
        Z = .0D0
        PI = 3.1415926536
        NU = 1.D0
        TAU = 6.D0
        PIBYN = PI/N
        ZZR = .0D0
        ZZI = .0D0
        LAMDA = TAU/T
        NM1 = N−1
        DO 1 K = 1,NM1
        THETA  = K * PIBYN
        ALPHA = THETA * DCOS(THETA)/DSIN(THETA)
        SR(K) = ALPHA
        SI(K) = THETA
        DSR(K) = NU * .5D0
        DSI(K) = (THETA + ALPHA * (ALPHA−1.D0)/THETA) * .5D0
    1   CONTINUE
        PSI = TAU * PIBYN
        CP = 2.D0 * DCOS(PSI)
        SP = DSIN(PSI)
        BR = ZZR
        BI = ZZI
        B1R = BR
        B1I = BI
        DO 2 KA = 1,NM1
        K = N−KA
        VH = TAU * SR(K)
        IF(DABS(VH).GT.75.D0)VH = DSIGN(75.D0,VH)
        V2 = DEXP(VH)
        B2R = B1R
        B2I = B1I
        B1R = BR
        B1I = BI
        CALL FUNC(LAMDA * SR(K),LAMDA * SI(K),FR,FI)
        CALL COMP(DSR(K),DSI(K),FR,FI,PR,PPI,2)
        BR = CP * B1R−B2R + V2 * PR
        BI = CP * B1I−B2I + V2 * PPI
    2   CONTINUE
        CALL COMP(BR,BI,Z,SP,QR,QI,2)
        CALL FUNC(LAMDA + ZZR,ZZI,FR,FI)
        SUMR = .5D0 * DEXP(TAU) * FR + CP * BR−2.D0 * (B1R−QR)
        SUMI = .5D0 * DEXP(TAU) * FI + CP * BI−2.D0 * (B1I−QI)
        FT = LAMDA * SUMR/N
        RETURN
        END


        SUBROUTINE COMP(A,B,C,D,X,Y,F)
        DOUBLE PRECISION A,B,C,D,X,Y,NENN,DFAK,R,PHI,H1,H2,U,V,W,
    *                    WF,EPS,PI,PI2
```

Table 2.3. continued.

```
         INTEGER F
C COMP   ENABLES COMPLEX CALCULATION (X,Y) = F((A,B),(C,D))
C F      :CONTROL PARAMETER
C          1 ADDITION   2 MULTIPLICATION   3 DIVISION
C          4 SQUARE ROOT   5 EXPONENTIAL FUNCTION
C          6 LOGARITHMIC FUNC.   7 ABSOLUTE VALUE
         EPS-.1D-8
         PI = 3.1415926536
         PI2 = 6.2831853072
         GOTO (10,20,30,40,50,40,40),F
   10    X = A + C
         Y = B + D
         RETURN
   20    X = A * C-B * D
         Y = A * D + B * C
         RETURN
   30    NENN = C * C + D * D
         IF (NENN.NE.0.D0) GOTO 31
         WRITE(3,32)
   32    FORMAT(13HDENOMINATOR = 0//)
         X = 1.D6
         Y = 1.D6
         RETURN
   31    X = (A * C + B * D)/NENN
         Y = (C * B-A * D)/NENN
         RETURN
   40    R = DSQRT(A * A + B * B)
         IF(DABS(A).LT..1D-10) GOTO 401
         PHI = DATAN(B/A)
         IF(A.LT.0.D0.) PHI = PHI + PI
         IF(A.GT.0.D0.AND.B.LT.0.D0) PHI = PHI + PI2
         GOTO 402
   401   PHI = 1.5707963268
         IF(B.LT.0.D0) PHI = PHI + PI
   402   CONTINUE
         GOTO(90,90,90,42,50,60,90),F
   42    PHI = .5D0 * PHI
         R = DSQRT(R)
         X = R * DCOS(PHI)
         Y = R * DSIN(PHI)
         RETURN
   50    IF(DABS(A).LE.75.D0) GOTO 52
         H1 = 1.0D30
         IF(A.LT.0.D0)H1 = 1.0D-30
         GOTO 54
   52    H1 = DEXP(A)
   54    X = H1 * DCOS(B)
         Y = H1 * DSIN(B)
         RETURN
   60    X = DLOG(R)
         Y = PHI
         RETURN
   90    X = R
         Y = PHI
         RETURN
         END
```

and

$$\bar{f}(s) = \begin{cases} \bar{P}(s) = 1/s & \text{for} \quad t > b > 0 \\ 0 & \text{for} \quad t < b > 0 \end{cases}$$

For the more complex example of Equation 2.17b, this approach leads to the following solution:

$$\bar{c}(x,s) = \begin{cases} \dfrac{\bar{w} + c_1\bar{\epsilon}s}{(\bar{\lambda} + \bar{\epsilon}s)s} & \text{for } x < qt/\bar{\epsilon} \\[3mm] \dfrac{\bar{w} + c_1\bar{\epsilon}s}{(\bar{\lambda} + \bar{\epsilon}s)s} + \dfrac{c_0\bar{\lambda} - \bar{w} + (c_0 - c_1)\bar{\epsilon}s}{(\bar{\lambda} + \bar{\epsilon}s)s} \exp\left(-\dfrac{\bar{\lambda} + \bar{\epsilon}s}{q}x\right) & \text{for } t > qt/\bar{\epsilon} \end{cases}$$

2.1.5.5 Convolution

The convolution integral may be expressed in different ways, as shown with Equation 2.10. The introduction of integration limits of the convolution integral with its unlimited upper boundary has been proven effective in order to minimize the expense of the numerical solution:

$$P(t_0) = \int_{\tau=t_0-t_E}^{t_0-t_L} R(\tau)\, h(t_0 - \tau)d\tau = \int_{\tau=t_0-t_E}^{t_0-t_L} (dR/d\tau)|_\tau S(t_0 - \tau)d\tau$$

$$= \int_{\tau=t_0-t_E}^{t_0-t_L} (d^2R/d\tau^2)|_\tau F(t_0 - \tau)d\tau$$

where $P(t_0)$ = output (effect or change) at time t
$R(\tau)$ = input (change of the change) at time τ
$dR/d\tau|_\tau$ = change of the input function at time τ
$d^2R/d\tau^2|_\tau$ = change of the input function slope (gradient) at τ
$h(t_0 - \tau)$ = pulse-response function at time $t_0-\tau$
$S(t)$ = step-response function $S = \int h\, dt$
$F(t)$ = step-response function $F = \int S\, dt$
a = slope of $R = R(t)$; $a(t) = dR/dt|_t$
t_E = effect time, time passing from the application of an input pulse with negligible length $\Delta t \to 0$ until the end of a significant effect on the considered system.
t_L = lag time; time passing from the application of an input pulse with negligible time length until the beginning of a significant effect ont he considered system.

Example

The pulse input of a tracer mass m into an aquifer is considered. The effect (output) caused by this input was determined by measurements.

The following recorded output data are then the pulse-response function h(t):

t	0.5	1.5	2.5	3.5	4.5	5.5	6.5	$t_L = 1.0$	$t_E = 7.0$
h(t)	0	0.2	1.2	2.2	1.4	0.8	0.2	$\Delta t = 1.0$	

The step-response function S(t) and the slope-response function F(t) may be calculated from these data by numerical integration. Using the trapezoidal rule, the values and the plots shown in Figure 2.18 result.

The selection of the formulation of the convolution integral according

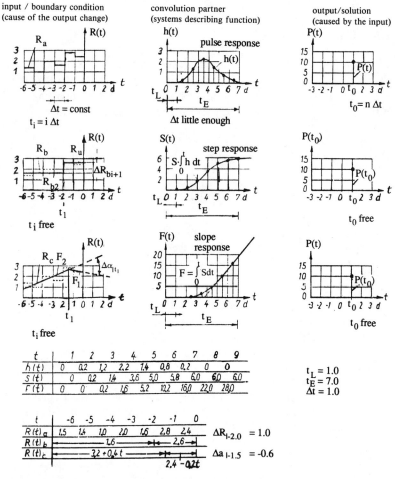

Figure 2.18. Example illustrating a consistent formulation of functions coupled in the convolution integral 2.55a to 2.55c or their numerical equivalents according to Equations 2.56a to 2.56c.

to Equations 2.55a, 2.55b, or 2.55c which should be used for the output calculations depends primarily on the form of the input function $R(\tau)$ and on the available form of the convolution partners h(t), S(t), and F(t).

As shown in Figure 2.18, Equation 2.55a or its numerical equivalent Equation 2.56a is preferred when R(t) can be easily approximated by a step function with $\Delta t = \Delta\tau = $ constant, and when the target time points t_0 at which the output are to be calculated can be chosen as $t_0 = n\Delta t (n = 1, 2, \ldots)$. This is typical when measurements are recorded on a daily or monthly basis and the forecasts are required at the same fixed time pattern.

Equation 2.55b or its numerical equivalent 2.56b is preferred when the input function R(t) can be approximated best by a step function with irregular time structure (e.g., often used in pump test analysis), and the target times t_0 satisfy no fixed time pattern. Finally, the use of Equation 2.55c or 2.56c is preferable when R(t) can be approximated best by a polygon line and target times of calculations of t_0 are not in a regular time pattern.

Using the numerically approximated terms of the convolution integral, the example shown in Figure 2.18 provides the following results (outputs $P(t_0)$) at time $t_0 = 1$ day:

$$P_K = \sum_{i=K-M}^{K-(L+1)} R_{i+1/2}H_{K-(i+1/2)} \qquad (2.56a)$$

with $\quad H_j = h\Delta t \quad \tau = i\Delta t \quad \Delta t = 1.0 \quad K = t_0/\Delta t \quad L = t_L/\Delta t = 1$
$\qquad M = t_E/\Delta t = 7$
$\qquad P_1 = R_{-5.5}H_{6.5} + R_{-4.5}H_{5.5} + \ldots + R_{-0.5}H_{1.5}$
$\qquad P_1 = 1.4 \cdot 0.2 + 1.0 \cdot 0.8 + 2.0 \cdot 1.4 + 1.6 \cdot 2.2$
$\qquad \quad + 2.8 \cdot 1.2 + 2.4 \cdot 0.2 = 11.24$

$$P(t_0) = \sum_{j=1}^{n}(R_j - R_{j+1}) S_\tau \qquad (2.56b)$$

$$\text{for} \quad t_0 - t_j \begin{cases} \geq t_E \text{ yields } \tau = t_E, R_1 = 0 \\ < t_E \text{ yields } \tau = t_0 - t_j, R_1 \neq 0 \\ < t_L \text{ yields } S_\tau = 0 \end{cases}$$

with $\quad t_1 = -2.0 \quad R_1 = 2.6 \quad R_2 = 1.6 \quad t_T = 1.0 \quad t_w = 7.0$
$\qquad P(1) = (R_1 - R_2) S_{t_0-t_1} + (R_2 - 0) S_{t_E}$
$\qquad P(1) = 1.0S(3.0) + 1.6S(7.0) = 1.0 \cdot 1.4 + 1.6 \cdot 6.0 = 11.00$

$$P(t_0) = R_{t_0-t_E} \left.\frac{dF}{dt}\right|_{t>t_E} = \sum_{j=1}^{n}(a_j - a_{j+1})F\tau \qquad (2.56c)$$

$$\text{for } t_0 - t_j \begin{cases} \geq t_E \text{ yields } \tau = t_E & a_1 = 0 \\ < t_E \text{ yields } \tau = t_0 - t_j & a_1 \neq 0 \\ < t_L \text{ yields } F_\tau = 0 \end{cases}$$

with $t_1 = -1.5$ $a_1 = -0.2$ $a_2 = 0.4$ $t_L = 1.0$ $t_E = 7.0$

$P(1) = (a_1 - a_2) F_{t_0 - t_1} + (a_2 - 0) F_{t_E} + R_{t_0 - t_E} dF/dt|_{t > t_E}$

$P(1) = 0.6F(2.5) + 0.4F(6.5) + 0.8 \cdot 6.0$

$\qquad = -0.6 \cdot 0.2 + 0.4 \cdot 16.0 + 4.8 = 11.08$

Practical Example

Model 2.56a is used to predict temperatures of pumped water from a bank filtration gallery. The available data set includes measured temperatures of the river water as well as the gallery water in °C taken at two-week intervals:

Θ-river	14.2	16.0	17.7	19.4	17.2	16.0	17.6	18.6	14.8	12.0	13.7
Θ-gall.	8.5	10.0	11.4	14.0	14.1	14.7	15.4	15.8	15.6	14.9	14.1
Θ-river	11.5	8.6	4.0	2.4	2.1	2.6	3.4	1.5	2.7	6.0	8.2
Θ-gall.	13.4	12.2	11.3	10.5	8.8	8.0	6.8	6.3	4.5	4.8	5.6
Θ-river	9.3	9.6	11.8	16.0	16.4	16.6	16.2	19.5			
Θ-gall.	5.0	7.0	9.0	10.3	11.0	11.9	12.8	13.5			

With $\Delta t = 0.5$ month ≈ 15 days (see Figure 2.19)

The following estimates are used:

- effective distance between bank and gallery as 80 m
- average transport velocity $u = v_0(\rho c)_w/(\rho c)_\Sigma$ as 1.4 m per day
- thermal dispersion coefficient $K = [\lambda_0 + \delta v_0(\rho c)_w/(\rho c)_\Sigma]$ as 2 m² per day, where $v_0 = \Delta Q/\Delta A$ taken from the stream tube considered

If all temperatures are scaled by using the yearly mean water temperature $\theta_a = 10.4$°C resulting from the monthly means (estimated from the discrete measured data), the new variable $\overline{T}$ is as follows:

$$\overline{T} = \frac{\Theta - \Theta_a}{|\Theta_{max}| - \Theta_a} \tag{2.57}$$

Assuming now that the graphically averaged data of the extreme river and gallery water temperatures represent approximately a steady-state migration regime, the degradation factor λ may be estimated by Equation 2.39c ($w = 0$, $c_o = 1.0$):

$$\bar{T}(80 \text{ m}, t \to \infty) \approx \left\{ \begin{array}{l} (15.0 - 10.4)/(17.8 - 10.4) = 0.62 \\ \\ (5.4 - 10.4)/(2.7 - 10.4) = 0.65 \end{array} \right\} \approx 0.64$$

$$\bar{T}(x, t \to \infty) = \exp\left[x\left(\frac{u}{2K} - \sqrt{\frac{u_o^2 + 4K\lambda}{4K^2}} \right) \right] \quad \text{(See Equation 2.39c)}$$

$$\lambda = \frac{K(\ln \bar{T}_\infty)^2}{x^2} - \frac{u \ln \bar{T}_\infty}{x} \approx 8 \cdot 10^{-3} \text{ per day}$$

The step-response function S(t) can be expressed by Equation 2.39f as follows ($w = 0$, $c_I = 0$, $c_o = 1.0$):

$$S(t) = \bar{T}(80, t) \approx 0.5 \exp\frac{(q - q^*)x}{2DM} \text{ erfc } \frac{\bar{\epsilon}x - q^*t}{\sqrt{4DM\bar{\epsilon}t}}$$

$$= \frac{1}{2} \exp\frac{(u - u^*)x}{2K} \text{ erfc } \frac{x - u^*t}{2\sqrt{Kt}}$$

where $u^* = \sqrt{u^2 + 4\lambda K}$

For the example considered, this step-response function and its numerical derivative, the pulse-response function, provide the following results:

t in d	22.5	30.0	37.5	45.0	52.5	60.0	67.5	75.0	82.5	90.5
S(t)			0.010	0.075	0.229	0.408	0.536	0.603	0.630	0.639
h(t) in d^{-1}			0.001	0.009	0.021	0.017	0.024	0.010	0.003	0.001
h = hΔt		0.010	0.065	0.154	0.179	0.128	0.067	0.027	0.009	

These functions are shown in Figure 2.19. From the plots the lag time t_L may be estimated as 30 days. This will allow forecasting the gallery water temperature up to $30 + 1/2\Delta t = 37.5$ days in advance without influence of the future input values. The effect time t_E may be read from the plots to be approximately 90 days.

If now the gallery water temperatures are calculated 37.5 days in advance by Equation 2.56a using the first $K - (L + 1) - (K - M) + 1 = 4$ measured data for the calculation of the first forecasted temperature at $t_0 = 37.5$, followed by using the data 2 to 5 for calculating the forecast at $t_0 = 37.5 + \Delta t$, and then by the data 3 to 6 to calculate T at $t_0 = 37.5 + 2\Delta t$, and so on, the predicted values shown in Figure 2.19 result. This plot makes clear that the predicted values are in good agreement with the measured data. Hence, the forecasting model Equation

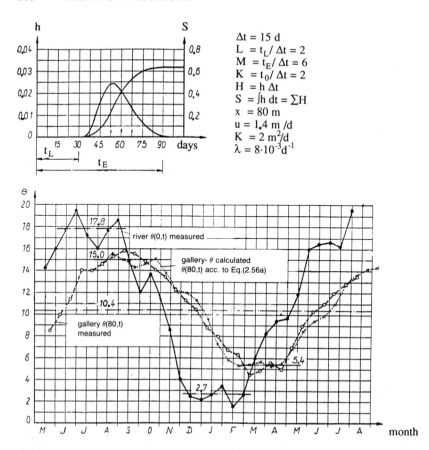

$\Delta t = 15\ d$
$L = t_L / \Delta t = 2$
$M = t_E / \Delta t = 6$
$K = t_0 / \Delta t = 2$
$H = h\ \Delta t$
$S = \int h\ dt = \Sigma H$
$x = 80\ m$
$u = 1.4\ m\ /d$
$K = 2\ m^2/d$
$\lambda = 8 \cdot 10^{-3} d^{-1}$

Figure 2.19. Examination of Equation 2.56a in predicting the pumped water temperature of a bank filtration gallery with a forecast horizon of 37.5 days.

2.56a is well suited to do the desired job as long as the prediction horizon does not exceed about 5 to 6 weeks. Such models, for instance, are used in the operation of heat pumping systems [2.60].

2.2

Numerical Solution Methods

The solution of boundary and initial value problems describing physically based models by use of numerical techniques and computers is known as *digital simulation* [2.53]. The study of migration processes relies upon digital simulation even more heavily than the study of flow processes because electric analog models can rarely be counted as useful tools in solving migration problems. Likewise, analog-digital or hybrid techniques are of little importance in this area [2.45]. Therefore, digital simulation and the analytical techniques introduced in Chapter 2.1 are the only practical methods for solving migration problems.

Digital simulation of migration processes has proven rather more complicated than the simulation of flow processes in the subsurface. Aspects of numerical stability, consistency, and convergence are much more important, and the effort to guarantee optimal accuracy often leads to a large computational expenditure in terms of time and memory demand. To keep costs at a reasonable level, the selection of the model and the desired simulation accuracy should therefore be compatible with the accuracy of the available data quantifying the model.

Because of the inherently approximate nature of numerical solutions, validation and assessment of accuracy, generally accomplished by comparison with appropriate analytical solutions, are important. In the following sections, several techniques that differ significantly with respect to expenditure and accuracy are discussed.

2.2.1 FUNDAMENTALS OF BUILDING DISCRETE MIGRATION MODELS

Migration in the subsurface is a continuous process in space and time. The infinitesimal mathematical migration models with their differential operators, derived in part 1, reflect this basic continuity. In order to apply digital models, the spatial and temporal continuum must first be discretized. The discretization of the independent variables, such as velocities, dispersivity, reaction rates, or distribution coefficients, means that the dependent variable, such as concentration or temperature, will be approximated at only a limited number of points in space and time. These points are commonly called *grid points* or *nodes.*

The solution region is usually divided into subregions with constant parameters. A discrete continuum problem is therefore characterized by

- a set of discrete points (nodes) at which the solution is approximated,
- the grid connecting these points, and
- constant-parameter subregions

Figure 2.20 shows some examples of typical two-dimensional spatial grids. Although for time-variant problems, spatial grids could vary in time. Time-invariant grids are usually used for computational reasons. Some exceptions are discussed in O'Neill [2.78], Jensen and Finlayson [2.52], and Lynch and O'Neill [2.65].

Once the grid is defined, the governing equation is written in algebraic form at each of the grid points. The grid points corresponding to the nth equation in a simple space-time (x-t) system are shown in Figure 2.21.

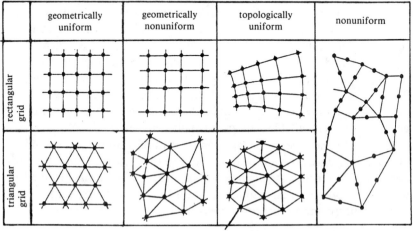

	geometrically uniform	geometrically nonuniform	topologically uniform	nonuniform
rectangular grid				
triangular grid				

each grid point has 6 neighbors

Figure 2.20.   Typical spatial grids discretizing two-dimensional regions.

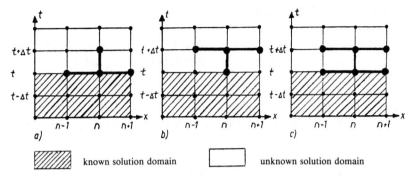

Figure 2.21. Examples of typical grid point patterns reflecting the nth algebraic equation of a one-dimensional migration problem at time t + Δt: *(a)* explicit solution scheme; *(b)* implicit solution scheme; *(c)* Crank-Nicolson solution scheme.

The following requirements are commonly made of discrete migration models:

1. The basic continuum model is to be approximated by a space and time discrete model having the best possible accuracy compatible with the given conditions. The spatial and temporal step lengths, h and Δt, impact the accuracy via the order of approximation $O(h^n, \Delta t^m)$.
2. The discretization and partition of the investigated region into subregions must allow a clear and meaningful quantification of the physical and chemical parameters and the boundary conditions.
3. The derived space and time discrete model should be stable, consistent, convergent, and economical in terms of computing costs.

A practical model, which will often be based on a compromise between these sometimes contradictory requirements, should be sufficiently accurate to provide the necessary insight into the migration process, and should at the same time operate with reasonable economy.

2.2.2 FORMULATION OF SPACE-DISCRETE MATHEMATICAL MODELS

Space-discrete mathematical models can be formulated by applying mathematical methods to the infinitesimal basis model (§2.2.2.1 and §2.2.2.2), or by applying the basic physicochemical laws and balance equations directly to discrete model elements (§2.2.2.3). The most frequently used mathematical methods for migration problems are the finite difference method (FDM) and the finite element method (FEM). Multiple cell models (MCM), discussed in §2.2.2.3, are also sometimes used.

2.2.2.1 Finite Difference Method

The finite difference method is based on approximating the differential operators in the infinitesimal continuum model by difference operators. The difference operators are determined by means of a Taylor series expansion, which also provides the approximation order through the leading truncation term. Thus, for instance, the expansion of the concentration term at point (x_o, y_o, t_o) in the x-direction yields

$$c(x_o + h, y_o, t_o) = c(x_o, y_o, t_o) + \frac{h}{1!} \left. \frac{\partial c}{\partial x} \right|_{x_o, y_o, t_o}$$

$$+ \frac{h^2}{2!} \left. \frac{\partial^2 c}{\partial x^2} \right|_{x_o, y_o, t_o} + \ldots + \frac{h^n}{n!} \left. \frac{\partial^n c}{\partial x^n} \right|_{x_o, y_o, t_o} + O(h^n) \quad (2.58)$$

An approximation $c(x_o, y_o + h, t_o)$ or $c(x_o, y_o, t_o + \Delta t)$ can be written in an analogous way.

Upon writing Equation 2.58 for the increments $+h$ and $-h$ up to $n = 2$ and subtracting the latter equation from the first, the differential operator $Fc = \partial c / \partial x$ becomes the difference operator $F_D c$:

$$F_D C = \frac{c(x_o + h) - c(x_o - h)}{2h} \quad (2.59a)$$

with $\partial c / \partial x = F_D c + O(h^2)$. This second-order form is known as a *centered approximation*.

On the other hand, the forward and backward difference operators obtained with a Taylor series expansion up to $n = 1$ are only first-order accurate. These operators are

$$\frac{\partial c}{\partial x} = \frac{c(x_o + h) - c(x_o)}{h} + O(h) \quad (2.59b)$$

$$\frac{\partial c}{\partial x} = \frac{c(x_o) - c(x_o - h)}{h} + O(h) \quad (2.59c)$$

Expanding Equation 2.58 up to $n = 2$ for the increments $+h$ and $-h$, and adding both equations, results in

$$\frac{\partial^2 c}{\partial x^2} = \frac{c(x_o + h) - 2c(x_o) + c(x_o - h)}{h^2} + O(h^2) \quad (2.59d)$$

These expressions show that the accuracy will decrease by one order if the spatial discretization h is nonuniform.

A higher approximation order is attainable, at increased computational costs, by including additional grid points. Table 2.4 gives an overview of difference operators relevant for digital simulation of migration processes, with ξ symbolizing one of the independent variables.

Finally, higher approximation orders may also be attained by multiple collocation (multiple FDM) [2.19]. Using this method, one considers the central mesh point and its neighbors, and forms difference equations for each neighboring point. These equations are multiplied by weighting factors and added, so that a difference equation for the central point results. The weighting factors are defined by Taylor fitting in such a manner that minimal discretization errors arise.

The subdomain collocation method, which can be considered a transition method between the finite difference method and the finite element method, is also used occasionally [2.38; 2.71; 2.96].

Table 2.4. Difference Operators with an $0(h^n)$ Order of Accuracy

$$\frac{\partial c}{\partial \xi}: \quad \frac{c(\xi+h) - c(\xi)}{h} + O(h); \quad \frac{c(\xi) - c(\xi-h)}{h} + O(h); \quad \frac{c(\xi+h) - c(\xi-h)}{2h} + O(h^2)$$

$$\frac{c(\xi+h) - c(\xi-h')}{h+h'} + O(h^2); \quad \frac{2c(\xi+h) + 3c(\xi) - 6c(\xi-h) + c(\xi-2h)}{6h} + O(h^3)$$

$$\frac{\partial^2 c}{\partial \xi^2}: \quad \frac{c(\xi+h) - 2c(\xi) + c(\xi-h)}{h^2} + O(h^2); \quad \frac{\dfrac{(\xi+h) - c(\xi)}{h} - \dfrac{c(\xi) - c(\xi-h)}{h'}}{(h+h')/2} + O(h)$$

$$a\frac{\partial c}{\partial \xi}: \quad \frac{c(\xi+h) - c(\xi)}{\int\limits_{\xi}^{\xi+h} (1/a)\, d\xi} + O(h); \quad \frac{c(\xi+h) - c(\xi-h)}{\int\limits_{\xi-h}^{\xi+h} (1/a)\, d\xi} + O(h^2); \quad \int (1/a)\, d\xi \text{ comp. } [1.53]$$

$$(1/a)\, d\xi: \quad \text{comp. } [1.53,\, p.\, 89] \qquad \begin{array}{c}\text{arith. mean}\\(\text{trapez. rule})\end{array}: \quad \int\limits_{\xi}^{\xi+h}(1/a)\, d\xi = \frac{2h}{a(\xi) + a(\xi+h)}$$

$$\frac{\partial c}{\partial \xi}\left(a\frac{\partial c}{\partial \xi}\right): \quad \frac{\left(1/\int\limits_{\xi}^{\xi+h}\frac{1}{a}\, d\xi\right)[c(\xi+h) - c(\xi)] - \left(1/\int\limits_{\xi}^{\xi+h'}\frac{1}{a}\, d\xi\right)[c(\xi) - c(\xi - h')]}{(h+h')/2} + O(h)$$

Note: Parameter a is an integral quantity in the range of ξ to $\xi + h$; a may be also nonlinear, i.e., $a = f(c)$.

$\vec{\nabla}\nabla c = v_x \partial c/\partial x + v_y \partial c/\partial y$ for a two-dimensional field with the spatial coordinates x, y where v_x and v_y are treated as scalar quantities a.

2.2.2.2 Finite Element Method

Step 1—Finitization

This method starts with subdivision of the investigated region into many small interconnected subregions, the finite elements. This step is known as *finitization*. Triangular, quadrilateral, and rectangular elements with areas A^m and boundaries Γ^m are preferentially used to represent the solution domain A with the boundary Γ of arbitrary geometry. The sides of these elements may also be curvilinear according to the number of grid points arranged on them. Figure 2.22 shows typical examples. A solution domain may contain different elements—for example, triangles and quadrilaterals—depending on the geometry of the domain (see right-hand side of Figure 2.20). The physical and chemical parameters of the migration problem are usually provided as element-wise constant, but can also be given in functional form [2.81].

Step 2—Equation System of Nodal Variables

At the nodes of the element grid lines, the solution function c or T is represented by the nodal variables c_i or T_i, where $i = 1, 2, \ldots, N$, with N being the number of nodes in the solution domain. Within every element the solution is approximated by an interpolation function using these nodal variables, which are obtained from a more general mathematical problem description based on either a variational or a weighted residual approach as described below. This description is transformed into a set of algebraic equations, which are solved simultaneously by an algorithm, taking into account the banded structure of the coefficient matrix. The discretization error inherent in this approach is closely related to the order of the interpolation function used.

Finite elements are called *isoparametric* if their geometric shape function and their interpolation function are of the same order, for instance, linear or quadratic (see Figure 2.22). Isoparametric elements are frequently used in the simulation of migration processes.

The FEM approach provides continuous solution functions within the elements and along the edges of the neighboring elements. But the continuity along the edges does not hold a priori also for the derivatives of the solution function, as for instance, for Darcy's velocity v if the hydraulic head h or the pressure p are used as nodal variables. However, discontinuities of velocities across element boundaries are usually acceptable when the elemental velocities are subsequently integrated over the elements, as is done in the sequential finite element solution of flow and migration—see Frind [2.34].

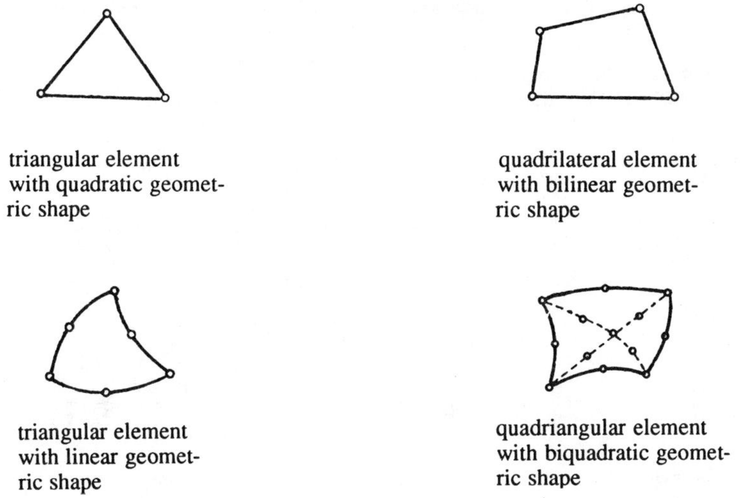

triangular element
with quadratic geomet-
ric shape

quadrilateral element
with bilinear geomet-
ric shape

triangular element
with linear geomet-
ric shape

quadriangular element
with biquadratic geomet-
ric shape

Figure 2.22. Examples of isoparametric elements for two-dimensional migration problems.

On the other hand, when velocities are used directly, as in the method of characteristics (MOC) or in the method of random walk (MORW), the overall accuracy of the solutions of c or T is dependent upon the continuity of the velocities. In such a case the preferred approach is the simultaneous use of h or p, v_x, v_y, and c or T as solution functions. Alternatively, it is often possible to achieve continuity of both the primary variable and their first derivative by using Hermite functions [2.90; 2.21; 2.20]. With both methods, the computing expense increases substantially. For example, in comparison to a (h,c)-scheme, the number of unknowns doubles for a two-dimensional migration problem using a (h,v_x,v_y,c)–scheme.

Interpolation Functions

The solution function within a finite element is preferably represented by polynomials of various degrees. The solution (interpolation) function for temperature T in the mth finite element, for instance, has the following form (using summation notation for simplicity):

$$T^m = \sum_{k=1}^{n} N_k T_k^m = N_k T_k^m \quad k = 1,2 \ldots w \quad (2.60)$$

where n is the number of nodes on the element.

The element basis functions N_k can be easily derived from the polyno-

Table 2.5. Selection of Isoparametric Two-Dimensional Elements with Basis Functions

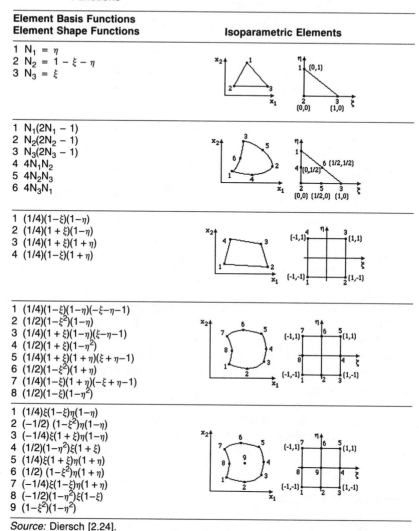

Element Basis Functions Element Shape Functions	Isoparametric Elements
1 $N_1 = \eta$ 2 $N_2 = 1 - \xi - \eta$ 3 $N_3 = \xi$	
1 $N_1(2N_1 - 1)$ 2 $N_2(2N_2 - 1)$ 3 $N_3(2N_3 - 1)$ 4 $4N_1N_2$ 5 $4N_2N_3$ 6 $4N_3N_1$	
1 $(1/4)(1-\xi)(1-\eta)$ 2 $(1/4)(1 + \xi)(1-\eta)$ 3 $(1/4)(1 + \xi)(1 + \eta)$ 4 $(1/4)(1-\xi)(1 + \eta)$	
1 $(1/4)(1-\xi)(1-\eta)(-\xi-\eta-1)$ 2 $(1/2)(1-\xi^2)(1-\eta)$ 3 $(1/4)(1 + \xi)(1-\eta)(\xi-\eta-1)$ 4 $(1/2)(1 + \xi)(1-\eta^2)$ 5 $(1/4)(1 + \xi)(1 + \eta)(\xi + \eta-1)$ 6 $(1/2)(1-\xi^2)(1 + \eta)$ 7 $(1/4)(1-\xi)(1 + \eta)(-\xi + \eta-1)$ 8 $(1/2)(1-\xi)(1-\eta^2)$	
1 $(1/4)\xi(1-\xi)\eta(1-\eta)$ 2 $(-1/2)\,(1-\xi^2)\eta(1-\eta)$ 3 $(-1/4)\xi(1 + \xi)\eta(1-\eta)$ 4 $(1/2)(1-\eta^2)\xi(1 + \xi)$ 5 $(1/4)\xi(1 + \xi)\eta(1 + \eta)$ 6 $(1/2)\,(1-\xi^2)\eta(1 + \eta)$ 7 $(-1/4)\xi(1-\xi)\eta(1 + \eta)$ 8 $(-1/2)(1-\eta^2)\xi(1-\xi)$ 9 $(1-\xi^2)(1-\eta^2)$	

Source: Diersch [2.24].

mial used. Commonly, dimensionless local (natural) coordinates are applied to formulate the basis functions. These coordinates are chosen in such a way that the real finite elements are transformed into simplified standard geometrical shapes (see Table 2.5). This is done in order to facilitate the evaluation of integrals over such elements, for instance, by using Gaussian quadrature [2.98].

At the six-node triangular element of standard type with quadratic basis and shape functions (see also geometric transformation functions), three local areal coordinates are used:

$$\xi_i = A_i^m / A^m \quad i = 1, 2, 3 \tag{2.61}$$

where A_i^m is the area of the subtriangle in the element m, and A^m is the total elemental area. Figure 2.23a shows a point P (ξ_1, ξ_2, ξ_3) in such an element and the local coordinates of the six element nodes. The six element basis functions N_k must be linearly independent of each other, which is guaranteed if Lagrange's interpolation matrix [2.98] is used:

$$
\begin{array}{lll}
N_1 = \xi_1(2\xi_1 - 1) & N_2 = \xi_2(2\xi_2 - 1) & N_3 = \xi_3(2\xi_3 - 1) \\
N_4 = 4\xi_1\xi_2 & N_5 = 4\xi_2\xi_3 & N_6 = 4\xi_1\xi_3
\end{array}
\tag{2.62}
$$

Normalizing a one-dimensional two-noded finite element within the boundaries $[-1, +1]$ results in the two linear basis functions, usually called *chapeau functions* N_k, shown on the left-hand side of Figure 2.23b:

$$N_1 = 0.5(1 - \xi) \quad N_2 = 0.5(1 + \xi) \tag{2.63a}$$

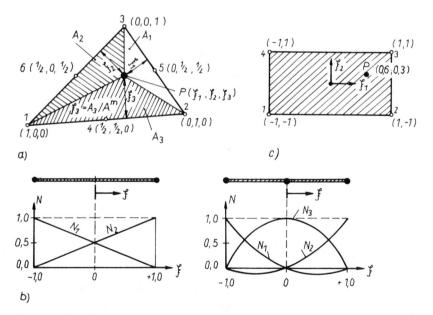

Figure 2.23. Examples of element basis functions in natural local coordinates: *(a)* 6-node triangle element of standard type with quadratic basis functions; *(b)* finite element for one-dimensional problems with linear and quadratic basis functions; *(c)* 4-node quadrilateral element with bilinear basis functions.

Similarly, a three-noded one-dimensional element corresponds to the three quadratic basis functions N_k (see right-hand side of Figure 2.23b):

$$N_1 = 0.5\xi(\xi - 1) \qquad N^2 = 1 - \xi^2 \qquad N_3 = 0.5\xi(\xi + 1) \quad (2.63b)$$

The bilinear basis functions of a quadrilateral element result from Lagrange's polynomial of first order (see Figure 2.23c):

$$N_1(\xi,\eta) = (\xi - 1)(\eta - 1)/4 \qquad N_3(\xi,\eta) = (1 + \xi)(1 + \eta)$$
$$N_2(\xi,\eta) = (1 + \xi)(1 - \eta)/4 \qquad N_4(\xi,\eta) = (1 - \xi)(1 + \eta) \qquad (2.63c)$$

These basic functions can be used together with the interpolation function of Equation 2.60 to determine the value of the solution function at any field point. For instance, inserting Equation 2.63c into Equation 2.60 yields at the point $\xi_1 = 0.6$ and $\xi_2 = 0.3$ the following temperature:

$$T^m(0.6, 0.3) = 0.07T_1 + 0.28T_2 + 0.52T_3 + 0.13T_4$$

Table 2.5 gives an overview of commonly used basis functions of linear and quadratic orders for isoparametric two-dimensional finite elements.

Geometric Transformation Functions

The element basis functions can also be used as transformation or shape functions in order to transform global coordinates into local coordinates and vice versa. This is known as the *isoparametric concept*. By analogy with Equation 2.60, shape functions are given in local coordinates ξ and η as follows:

One-dimensional elements

$$x = \sum_{k=1}^{n} N_k(\xi)x_k = N_k(\xi)x_k \qquad (2.64a)$$

Two-dimensional elements

$$x = \sum_{k=1}^{n} N_k(\xi,\eta)x_k \quad \text{and} \quad y = \sum_{k=1}^{n} N_k(\xi,\eta)y_k \qquad (2.64b)$$

where n is the number of nodes on the element. Thus, the coordinates for the linear one-dimensional element, for instance, are expressed as

$$x = N_1(\xi)x_1 + N_2(\xi)x_2 = (1 - \xi)x_1/2 + (1 + \xi)x_2/2 \quad (2.64c)$$

The global derivatives of the element shape functions, expressed in terms of local coordinates ξ,η, may be written, using the chain rule, as follows:

$$\frac{\partial N_k}{\partial y} = \frac{\partial N_k}{\partial x}\frac{\partial x}{\partial y} + \frac{\partial N_k}{\partial y}\frac{\partial y}{\partial \xi} \quad \text{and} \quad \frac{\partial N_k}{\partial \eta} = \frac{\partial N_k}{\partial x}\frac{\partial x}{\partial \eta} + \frac{\partial N_k}{\partial y}\frac{\partial y}{\partial \eta} \quad (2.64d)$$

or, respectively,

$$\frac{\partial N_k}{\partial x} = \frac{\partial N_k}{\partial y}\frac{\partial y}{\partial x} + \frac{\partial N_k}{\partial \eta}\frac{\partial \eta}{\partial x} \quad \text{and} \quad \frac{\partial N_k}{\partial y} = \frac{\partial N_k}{\partial \xi}\frac{\partial \xi}{\partial y} + \frac{\partial N_k}{\partial \eta}\frac{\partial \eta}{\partial y} \quad (2.64e)$$

or in matrix form using the Jacobian matrix [J]

$$\begin{pmatrix} \partial N_k/\partial \xi \\ \partial N_k/\partial \eta \end{pmatrix} = [J]\begin{pmatrix} \partial N_k/\partial x \\ \partial N_k/\partial y \end{pmatrix} \quad \text{or} \quad \begin{pmatrix} \partial N_k/\partial x \\ \partial N_k/\partial y \end{pmatrix} = [J]^{-1}\begin{pmatrix} \partial N_k/\partial \xi \\ \partial N_k/\partial \eta \end{pmatrix} \quad (2.64f)$$

where

$$[J] = \begin{pmatrix} \dfrac{\partial x}{\partial \xi} & \dfrac{\partial y}{\partial \xi} \\ \dfrac{\partial x}{\partial \eta} & \dfrac{\partial y}{\partial \eta} \end{pmatrix} \qquad [J]^{-1} = \begin{pmatrix} \dfrac{\partial \xi}{\partial x} & \dfrac{\partial \eta}{\partial x} \\ \dfrac{\partial \xi}{\partial y} & \dfrac{\partial \eta}{\partial y} \end{pmatrix} \qquad (2.64g)$$

In addition to transforming the derivatives with respect to global coordinates (x,y) to derivatives with respect to local coordinates (ξ,η), the differential area dA must also be transformed as

$$dA = \det[J]\, d\xi\, d\eta \qquad (2.64h)$$

where det [J] is the determinant of the Jacobian matrix. An integral of a function F over an elemental area A is therefore transformed by

$$\iint_{A(x,y)} F(x,y)dxdy = \iint_{A(\xi,\eta)} F(\xi,\eta)\det[J]d\xi d\eta \qquad (2.64i)$$

Equation System of Nodal Variables

The FEM requires the governing equation of migration processes to be expressed in an integral form. This form can be obtained using either a variational formulation (Rayleigh-Ritz formulation) or a weighted residual formulation (e.g., Galerkin formulation). Both options are considered briefly in the following.

Variational formulation

For most boundary value problems, equivalent variational forms can be found. Unfortunately, the formulation for migration processes entails some difficulties since the governing equation is not self-adjoint (symmetric), unlike that of flow processes, due to the directed term of convective transport. The necessary symmetry must therefore be forced by transformation. Using the transformation function $\psi = c \exp(\beta)$ with $\beta = -v_i x_i / D_{ii}$, it can be proven that

$$\frac{\partial}{\partial x_i}\left(D_{ij}\frac{\partial c}{\partial x_j}\right) - v_i\frac{\partial c}{\partial x_i} = \epsilon\frac{\partial c}{\partial t} + \lambda c - w \qquad i = 1,2 \text{ and } j = 1,2 \tag{2.65}$$

where w is $\overline{w}/M$, and D_{ij} is the second-order dispersion tensor, is equivalent to the following variational problem [2.22]:

$$J(c) = \iint\limits_A \left(\frac{D_{ii}}{2}\frac{\partial^2 c}{\partial x_i^2} + \left(\epsilon\frac{\partial c}{\partial t} - w + \frac{\lambda c}{2}\right)c\right)\exp(\beta)dA$$

$$- \oint\limits_\Gamma D_{ii}\frac{\partial c}{\partial x_i}\ell_i c\,\exp(\beta)ds \overset{!}{=} \min \tag{2.66}$$

where $\beta = v_i\,x_i/D_{ii}$

Variational formulations (Rayleigh-Ritz formulations) for solving two-dimensional solute transport processes with linear basis functions for triangular elements have been described, for instance, in Guymon et al. [2.46] and Guymon [2.47]. Comparisons of the variational formulation with the Galerkin-FEM formulation (see below) are given in Prakash [2.83] and Smith et al. [2.91]. Since the latter has important advantages in solving migration problems, the focus in the following will be on the Galerkin-FEM approach.

Galerkin formulation

The unknown solution function in a migration equation, for instance the function c in Equation 2.65, can be replaced by the interpolation function according to Equation 2.60. This replacement will result in an approximation error, or residual ϕ. If we represent the migration equation by $L(c)$, the corresponding residual $L(c^m)$ for the element m can be expressed by

$$L(c) \approx L(c^m) = L\left(\sum_{k=1}^{n} N_k c_k^m\right) = \phi \neq 0 \tag{2.67}$$

The key step in Galerkin's approach is to minimize the residual ϕ. This is achieved by choosing n linearly independent weighting functions w_k (k = 1, 2, . . ., n) in each element, integrating over the domain A, and setting the result to zero.

$$\iint_A w_k \phi \, dA = \sum_m \iint_{A^m} w_k L(c^m) \, dA = 0 \qquad (2.68)$$

Equation 2.68 is not easily justified on a physical basis and will therefore be used here as a given axiom. The result of the operation is that the total residual will be a minimum globally over all nodes. Locally at individual nodes, the residual will in general be nonzero.

The possible choice of weighting functions w_k leads to several alternative approaches. In the Galerkin method, the weighting functions are identical with the basis functions ($w_k = N_k$). In addition to the symmetrical basis functions N_k (Petrov-Galerkin method) mentioned above, asymmetrical weighting functions are sometimes used in solving migration problems, especially in the case of dominant convection ("Upwind"-FEM schemes) — see, e.g., Diersch [2.23]. However, such functions often cause a reduction of the approximation order.

Equation 2.65 is again used as an example to derive the *Galerkin migration model.* Thus, the basic equation is given as follows:

$$L(c) = \epsilon \frac{\partial c}{\partial t} + v_i \frac{\partial c}{\partial x_i} - \frac{\partial}{\partial x_i} \left(D_{ij} \frac{\partial c}{\partial x_j} \right) + \lambda c - w = 0 \qquad (2.69)$$

where
$$i = 1,2$$
$$j = 1,2$$
$$x_1 = x$$
$$x_2 = y$$
$$w = \bar{w}/M$$

Upon substitution of Equation 2.69 into Equation 2.68, and application of Green's theorem (integration by parts — see, e.g., Diersch [2.22; 2.24]), the algebraic equations for determining the discrete values c_k at the nodes k are obtained. Now, the complete system of equations can be written in matrix form as follows:

$$[S]\{c(t)\} + [O]\{dc/dt|_t\} + \{F(t)\} = 0 \qquad (2.70)$$

where [ ] and { } designate matrices and vectors, respectively. According to Diersch [2.22; 2.24], the individual coefficients in Equation 2.70 become

$$S_{kl} = \sum_{m=1}^{M} \iint_{A^m} \left[D_{ij} N_{k,i} N_{l,j} + N_k \sum_{\nu=1}^{n} (N_\nu v_i^m) N_{l,i} + \lambda N_k N_l \right] dA \qquad (2.70a)$$

$$O_{kl} = \sum_{m=1}^{M} \iint_{A^m} \epsilon N_k N_l \, dA \tag{2.70b}$$

$$F_k(t) = -\sum_{m=1}^{M} \left(\int_{\Gamma^m} \left(N_k D_{ij} \sum_{j=1}^{n} (N_{\nu,j} c_\nu^m(t)) l_i \right) ds + \iint_{A^m} w N_k \, dA \right) \tag{2.70c}$$

where $N_{k,i} = (\partial N_k / \partial \xi_\nu)/(\partial \xi_\nu / \partial x_i)$ and k, 1 = 1, 2, . . .
ν = node number in the finite element m
l_i = direction cosines
Γ^m = boundary of the mth element
$v_i = (N_\nu v_{\nu i}^m)$ interpolation term for v_i in the element m
M = number of the finite elements in the investigated region

The matrices [S] and [O] are of size ((N−N*), (N−N*)), where N is the number of all discrete nodes in the simulated region and N* is the number of nodes at which the solution function c is a priori provided through boundary conditions of the first kind. The double integrals according to Equation 2.70 are commonly evaluated numerically by Simpson's rule or Gaussian quadrature. Frequently a 3×3 Gaussian quadrature scheme is used [2.16], but a 2×2 scheme is generally found adequate for linear elements. For special cases where D = constant and w = 0, Appendix 4 shows the appropriate difference terms for the model Equations 2.69 and 2.70.

Effort Assessment and Recommendations

Although the Galerkin-FEM approach allows the use of many different element types and interpolation functions, today linear elements with identical basis (or interpolation) functions, identical shape (or geometrical transformation) functions, and identical weighting functions are almost exclusively used in finite element migration modeling.

This development has several reasons. One is that the ability to present a domain with a small number of large elements, thought important in the late 1970s, is no longer considered to be a relevant factor. During the past two decades of finite element groundwater simulation, any real advantages of higher-order elements have never been convincingly demonstrated. Although one can use fewer elements of higher-order type to represent a given domain, the accuracy and cost of an equivalent linear element grid having the same number of nodes is usually about the same. From the practical point of view, natural aquifer systems, in contrast to surface water systems or engineered systems, are generally highly variable and heterogeneous and therefore require a fairly fine discretization, which is, in any case, best accomplished using simple linear elements. A further reason is the recognition that the order of accuracy for higher-order elements is not the same at all nodes [2.82]. Finally, programs based on linear elements are undeniably simpler.

State-of-the-art finite element programs for migration modeling therefore tend to utilize the simplest possible numerical methods consistent with the required accuracy, which is usually obtained with triangular, rectangular, or linear quadrilateral elements. The side lengths of these elements are determined on the basis of the cell Peclet criterion Pe = $\Delta l/\delta \leq 2$ (see Equation 2.77a) where Δl is the length in the flow direction, and δ is the dispersivity. The time step Δt is based on the cell Courant criterion Cr = $|v|\Delta t/(\epsilon \Delta x) \leq 1$ (see Equation 2.77b). A discretization based upon these criteria can generally be relied upon to deliver accurate results.

The large numbers of algebraic equations that are generated with fine discretizations can be solved by means of highly efficient matrix solution techniques such as the principal direction/alternating direction technique [2.12], which has been applied to grids of up to a million nodal points — see Frind and Sudicky [2.37], or the preconditioned conjugate gradient method complemented by a symmetric matrix transformation — see Frind et. al. [2.35]. With the numerical aspects under control, modelers are then free to focus their attention on the complex physical, chemical, and biological aspects of the migration process itself.

2.2.2.3 Multiple Cell Method (MCM)

The MCM is similar to the FEM in that the solution region is subdivided into subregions or elements, here known as *planning elements* or *cells*. An MCM model can comprise hundreds of cells, or even just one single cell (see, e.g., Mercado [2.69]). Analogous to the formulation of mathematical models for an REV (representative elementary volume), the governing equations for cells are formulated directly based on fundamental physical, chemical, and biological laws.

For example, the complete migration process is described by the flow model:

$$\frac{\partial}{\partial x_i} T \frac{\partial h}{\partial x_i} = S \frac{\partial h}{\partial t} - \bar{w}_o \qquad (2.71)$$

where i = 1, 2 and x_1 = x, x_2 = y, together with the quality model of Equation 2.65. The procedure is illustrated in Figure 2.24, where it is assumed that the same cells are used for solving both the flow and the quality problem. As for the FEM, all parameters of cell m must be given as quantities averaged over the area A^m (e.g., as S←$\bar{S}^m$, ϵ←$\bar{\epsilon}^m$; see Figure 2.24a). The digital simulation of the flow problem then provides the source term $(\bar{S}\partial h/\partial t - \bar{w}_o)^m$ lumped at the cell center, and the flux compo-

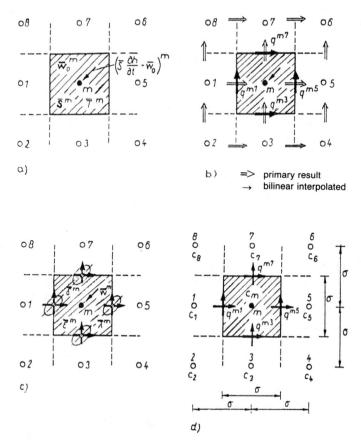

Figure 2.24. Cell scheme of the MCM with quadratic cells: *(a)* flow parameters averaged within the cell; *(b)* result of the flow simulation; *(c)* migration parameters averaged within the cell; *(d)* result of the migration simulation.

nents normal to the cell boundaries lumped at the center of each cell side (see Figure 2.24b).

The flux components parallel to the cell edges are calculated by bilinear interpolation using Equations 2.60 and 2.63c with the four neighboring flux values (see Figure 2.24b). The dispersivity values (see §1.3.3) also required on the cell boundaries may be linearly interpolated from δ-values of the neighboring cells by means of the same Equations 2.60 and 2.63c.

In the next step, the solute or heat balance equation must be formulated for each cell. The required c for quantifying the convective solute inflows and outflows at the edges of the cells is evaluated by linear interpolation between the two neighboring cell centers. This approach corresponds to the use of central differences in the FDM, and the use of c values of cell centers from which the flux originates corresponds to the upstream (backward) difference scheme of the FD approach.

The partial first derivatives of c normal and parallel to the cell edges are required to evaluate the solute inflows and outflows due to hydrodynamic dispersion. These values may again be determined from the four neighboring values by bilinear interpolation. The following balance equation results for cell m when the eight neighboring cells are numbered counterclockwise as shown in Figure 2.24:

$$[q^{m1}c_1 + q^{m3}c_3 - q^{m5}c_m - q^{m7}c_m] \, \sigma$$

convective transport (as backward difference)*

$$+ \left[D_{xy}^{m1} \frac{c_1 - c_m}{\sigma} + D_{xy}^{m1} \frac{c_2 + c_3 - c_8 - c_7}{4} - D_{xx}^{m5} \frac{c_m - c_5}{\sigma} - D_{xy}^{m5} \frac{c_3 + c_4 - c_7 - c_6}{4\sigma} \right.$$

dispersive transport

$$\left. + D_{yy}^{m3} \frac{c_3 - c_m}{\sigma} + D_{yx}^{m3} \frac{c_1 + c_2 - c_4 - c_5}{4\sigma} - D_{yy}^{m7} \frac{c_m - c_7}{\sigma} + D_{yx}^{m7} \frac{c_1 + c_8 - c_5 - c_6}{4\sigma} \right] \sigma =$$

dispersive transport

$$\sigma^2 \bar{\epsilon}^m \frac{\partial c_m}{\partial t} + \sigma^2 \bar{\lambda}^m c_m \qquad\qquad - \sigma^2 \bar{\bar{w}}^m \qquad\qquad\qquad (2.72)$$

storage   decay & dilution   source/sink term

*If the sign of q results from the calculation $q^{mk}c_k$ must be replaced by $0.5(q^{mk} + |q^{mk}|)c_k + (q^{mk} - |q^{mk}|)c_m$.

A linear system with $N - N^*$ equations results from this approach, where $N - N^*$ is the number of cells for which c_k is to be determined. The banded asymmetric coefficient matrix of this system contains, besides the main diagonal, two upper and two lower occupied codiagonals. The digital migration models formed in this manner are nearly equivalent to models formulated with normal FD schemes, or with FE schemes using linear basis functions, if the terms $S\partial h/\partial t$, $\partial c/\partial t$, and λc are averaged over the cell m by using Simpson's rule. However, the most important advantage of the MCM approach is the simplicity and clarity of its derivation.

2.2.3 FORMULATION OF TIME-DISCRETE MATHEMATICAL MODELS

2.2.3.1 Time Discretization

The time derivatives $\partial c/\partial t$ or $\partial T/\partial t$ can be approximated numerically by using either finite elements or finite differences (see, e.g., Diersch [2.22]). Since here the FEM offers no significant advantages over the simpler FDM, only the latter will be considered in this book.

Commonly a first-order difference quotient is applied ($n = 1$ in Equation 2.58), and the migration model is weighted between the two time levels t and $t + \Delta t$. Under these conditions, Equation 2.65 takes the following form:

$$(1 - \gamma)\left[\frac{\partial}{\partial x_i}\left(D_{ij}\frac{\partial c}{\partial x_j}\right) - v_i\frac{\partial c}{\partial x_i} - \lambda c + w\right]^t \qquad (2.73)$$

$$+ \gamma\left[\frac{\partial}{\partial x_i}\left(D_{ij}\frac{\partial c}{\partial x_j}\right) - v_i\frac{\partial c}{\partial x_i} - \lambda c + w\right]^{t+\Delta t} =$$

$$[(1 - \gamma)\epsilon^t + \gamma\epsilon^{t+\Delta t}](c^{t+\Delta t} - c^t)/\Delta t$$

where γ is a weight factor ($0 \leq \gamma \leq 1$). Depending on the choice of γ, discrete schemes are designated as

- $\gamma = 0$ explicit scheme (see Figure 2.21a)
- $\gamma = 1/2$ Crank-Nicolson scheme (see Figure 2.21c)
- $\gamma = 2/3$ Galerkin scheme (see Figure 2.21c)
- $\gamma = 1$ implicit scheme (see Figure 2.21b)

The approximation order can be shown to be of $O(\Delta t)$ for $\gamma = 0, 2/3$, or 1, and $O(\Delta t^2)$ for $\gamma = 1/2$.

From Equations 2.58 and 2.59, it is evident that the approximation of the convective transport term $v\partial c/\partial x$ by forward or backward differences causes a numerical error known as *numerical dispersion* of the following form:

$$O(h) = \frac{v}{h}\frac{h^2}{2!}\frac{\partial^2 c}{\partial t^2} = \underbrace{\left(\frac{v\Delta x}{2}\right)}_{D_{num}}\frac{\partial^2 c}{\partial x^2} \qquad (2.74)$$

In the same manner, the $O(\Delta t)$ approximation of the storage term $\epsilon\partial c/\partial t$ with $u = \partial x/\partial t = v/\epsilon$ according to Equations 1.24 and 1.78a causes a numerical dispersion of

$$O(h) = \frac{\epsilon}{\Delta t} \frac{\Delta t^2}{2!} \frac{\partial^2 c}{\partial t^2} = \frac{\epsilon}{2} \Delta t \frac{\partial x^2}{\partial t^2} \frac{\partial^2 c}{\partial x^2} = \underbrace{\left(\frac{\Delta t v^2}{2\epsilon}\right)}_{D_{num}} \frac{\partial^2 c}{\partial x^2} \qquad (2.75)$$

so that the total numerical dispersion caused by the first-order difference operators for $v\partial c/\partial x$ and $\epsilon\partial c/\partial t$, the truncation error, becomes

$$D_{num.} = D_{real} + |v|\Delta x/2 + \Delta t v^2/(2\epsilon) \qquad (2.76a)$$

Therefore, in order to guarantee the following conditions

$$\frac{|v|\Delta x}{2D} \approx \frac{\Delta x}{2\delta} <\, <1 \quad \text{and} \quad \frac{\Delta t v^2}{2\epsilon D} \approx \frac{\Delta t |v|}{2\epsilon\delta} <\, <1 \quad (2.76b)$$

a fine discretization may be required. The numerical error can be compensated when applying a correction term:

$$D_{corr.} = D_{real} - [|v|\Delta x/2 + \Delta t v^2/(2\epsilon)] > 0 \qquad (2.76c)$$

Based upon these considerations of truncation errors, the dimensionless cell Peclet number and Courant number may be derived, which are used to define the discretization of migration models where the derivatives $\partial c/\partial x_i$ and $\partial c/\partial t$ are approximated by higher-order expressions.

Cell Peclet number:

$$Pe_{\Delta x} = |v|\Delta x/D \approx \Delta x/\delta \qquad (2.77a)$$

Courant number:

$$Cr = \Delta t |v|/(\epsilon \Delta x) \qquad (2.77b)$$

Equation 2.75 suggests that a fine spatial discretization is only justified if simultaneously the error inherent in the time discretization can be brought to the same approximation order. Hence, schemes with at least $O(\Delta t^2)$ are of special interest. This is the reason why the Crank-Nicolson scheme with $O(\Delta t^2)$ is often used in migration modeling. The same holds also for the Douglas-Jones scheme, a two-step approach based on the Crank-Nicolson scheme, avoiding iteration when the migration parameters are nonlinear, i.e., functions of c or T. These two steps of the Douglas-Jones scheme may be characterized as follows:

1. Predictor step
   - implicit solution scheme with $\gamma = 1$
   - a half time step $\Delta t/2$
   - a parameter set at time t
   - results at time $t + \Delta t/2$

2. Corrector step
- Crank-Nicolson scheme with $\gamma = 1/2$
- a full time step Δt,
- parameters, source-sink terms, and convection term at time $t + \Delta t/2$ using the results of the predictor step
- final results at time $t + \Delta t$

Thus, the corrector step for Equation 2.65 may be expressed as follows (see also Figure 2.25a):

$$\frac{1}{2}\left[\frac{\partial}{\partial x_i}\left(D_{ij}\frac{\partial c}{\partial x_j}\right)\right]^t + \frac{1}{2}\left[\frac{\partial}{\partial x_i}\left(D_{ij}\frac{\partial c}{\partial x_j}\right)\right]^{t+\Delta t}$$
$$+ \left(v_i\frac{\partial c}{\partial x_i} - \lambda c + w\right)^{t+\Delta t/2} = \epsilon^{t+\Delta t/2}\frac{c^{t+\Delta t} - c^t}{\Delta t} \tag{2.78}$$

The DJ approach was used, for instance, in solving Equation 2.40 in Luckner and Nitsche [2.61] and Nitsche [2.75].

When the difference operator according to Equation 2.59a or 2.59d is derived using $n = 3$, it can be seen that these operators may also be considered as approximates of third-order $O(\Delta h^3)$. Some modelers regard the use of a third-order approximation for the time derivatives $O(\Delta t^3)$ as desirable. The simplest approach is given by the following equation (see also Figure 2.25b):

$$\left.\frac{\partial c}{\partial t}\right|_t = \frac{2c^{t+\Delta t} + 3c^t - 6c^{t-\Delta t} + c^{t-2\Delta t}}{6\Delta t} \tag{2.79a}$$

This approach requires no significant additional computational expense compared with the Crank-Nicolson approach, except an increased memory demand for storing the solution $c^{t-2\Delta t}$.

A more sophisticated third-order approximation $O(\Delta t^3)$ developed for the Crank-Nicolson scheme is given by (see Equation 2.73, and also Genuchten [2.40] and Genuchten and Gray [2.41])

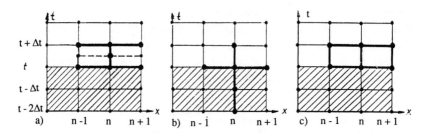

Figure 2.25. Time discrete schemes (see also Figure 2.21): *(a)* Douglas-Jones scheme; *(b)* third-order time approximation; *(c)* Stoyan scheme.

$$\frac{c^{t+\Delta t} - c^t}{\Delta t} \approx \left(a_1 \frac{\partial c}{\partial t} + a_2 \frac{\partial^2 c}{\partial t^2} \right)^{t+\Delta t} + \left(a_1 \frac{\partial c}{\partial t} + a_3 \frac{\partial^2 c}{\partial t^2} \right)^t \quad (2.79b)$$

Substituting in this equation $c = t^3$ according to the desired approximation order, one obtains

$$((t + \Delta t)^3 - t^3)/\Delta t = 3a_1(t + \Delta t)^2 + 6a_2(t + \Delta t)$$
$$+ 3a_1 t^2 + 6a_3 t \quad (2.79c)$$

which yields the following expressions for the coefficients of Equation 2.79b:

$$6a_1 t^2 = 3t^2 \quad 6a_1 t\Delta t + 6a_2 t + 6a_3 t = 3t\Delta t \quad 3a_1 \Delta t^2 + 6a_2 \Delta t = \Delta t^2$$
$$a_1 = 1/2 \quad a_2 = -\Delta t/12 \quad a_3 = +\Delta t/12$$

Thus Equation 2.79b becomes

$$\frac{c^{t+\Delta t} - c^t}{\Delta t} \approx \frac{1}{2} \left(\frac{\partial c}{\partial t} - \frac{\Delta t}{6} \frac{\partial^2 c}{\partial t^2} \right)^{t+\Delta t} + \frac{1}{2} \left(\frac{\partial c}{\partial t} + \frac{\Delta t}{6} \frac{\partial^2 c}{\partial t^2} \right)^t \quad (2.79d)$$

where $\partial c/\partial t$ must be replaced by the spatial terms of the corresponding one-, two-, or three-dimensional migration equation, and $\partial^2 c/\partial t^2$ by the derivatives of these terms [2.40; 2.41] (see also §2.2.4.1).

Finally, a further improvement in approximation accuracy may be obtained by FD approaches through weighting the time derivatives not only at the spatial node n, but also at the neighboring spatial nodes (see Figure 2.25c). More generally, this applies also to the source-sink terms λc and w. In the simplest form, numerical integration and averaging of $\partial c/\partial t$, λc, and w over the area lumped at node n by means of Simpson's rule yields the required weighting factors. For a one-dimensional process this approach results in the following approximation:

$$\partial c/\partial t = (1/6)[(\partial c/\partial t)_{n-1} + 4(\partial c/\partial t)_n + (\partial c/\partial t)_{n+1}] \quad (2.79e)$$

Corresponding approximations for two-dimensional migration problems are given in Appendix 4.

A similar FD scheme for one-dimensional migration processes according to Figure 2.25c was proposed by Stoyan [2.92]. This scheme subjects all spatial and temporal derivatives to controlled weighting. For convection-dominated processes, this scheme converges to the corresponding analytical solution.

2.2.3.2 Error Analysis and Cost Assessment

A valid numerical scheme must be convergent; in other words, it must approach the exact solution with decreasing step length. This required basic property of a numerical scheme is not discussed further in this text.

The theoretical fundamentals may be found in textbooks on numerical solution methods of differential equations. In practical applications, empirical tests with decreasing spatial and temporal step length are recommended in order to validate a computer program.

Discretization errors appear primarily in the form of numerical dispersion and numerical oscillation. Numerical dispersion is mainly caused by discretization of the convection and storage terms, that is, by approximation of the first-order derivatives $\partial c/\partial t$ and $\partial c/\partial x_i$ with terms of first-order accuracy. Such approximations can produce considerable errors, particularly when the dispersion coefficients D are small (see Equation 2.76a). For migration models, these errors depend on the characteristic cell numbers given by Equation 2.77. These errors may be reduced by introducing corrected dispersion coefficients according to Equation 2.76b.

Numerical oscillation arises only conditionally. The danger of oscillation arises mostly with convection-dominated migration (i.e., with small δ-values). The Crank-Nicolson scheme, central difference schemes applied to convection and storage terms, and subdomain collocation schemes are susceptible to oscillation, while implicit solution schemes, backward difference schemes applied to convection and storage terms, and the FEM-Ritz formulation are largely oscillation-free.

Besides discretization errors, stability errors should not be underestimated. A numerical migration model is termed *unconditionally stable* if an error is damped from time step to time step and from spatial step to spatial step, and it is termed *conditionally stable* if damping occurs only under certain conditions. These conditions may be given by analytical expressions for special cases (see, for instance, Diersch [2.22]). Using such expressions or empirical test data the stability conditions may be expressed in terms of critical time step lengths dependent upon the spatial discretization and the migration parameters. The implicit solution scheme with $\gamma = 1$ is absolutely stable, but with decreasing γ the tendency to instability increases.

With the recommended use of linear isoparametric elements in the FEM-Galerkin formulation (see §2.2.2.2), the optimal solution can be expected with a time-centered scheme when the spatial and temporal discretization is selected such that the Peclet criterion $Pe \leq 2$ and the Courant criterion $Cr \leq 1$ (see Equations 2.77a and 2.77b) are satisfied. In that case, numerical dispersion and oscillation are practically always eliminated. The same recommendation may also be given for the practical application of schemes of higher than second-order approximation because changing step lengths, changes of parameters, or nonlinear parameters decrease the approximation order. The important advantage of having predictable accuracy based upon linear finite elements (or corresponding finite difference terms), centered-time schemes, a spatial discretization satisfying Pe $= \Delta x/\delta \leq 2$, and a temporal discretization satisfying Cr $= \Delta t|v|/(\epsilon \Delta x) \leq 1$ makes this approach one of today's most reliable and popular numerical procedures.

Finally, the solution of the system of simultaneous equations, the matrix equations, may also cause significant numerical errors. Poorly conditioned equation systems, for which the main diagonal elements of the coefficient matrix are not sufficiently dominant compared to those of the codiagonals, are the main source of these errors. Also the approximation of the simultaneous solution of all matrix equations by partial step methods may cause numerical errors. Numerical errors can propagate over the solution domain, particularly in the case of conditionally stable schemes. Methods that may be susceptible to such errors are ADI methods, and especially methods where the band width of the coefficient matrix is reduced or made symmetric by transfer of selected coefficient elements to the right-hand side of the equation.

Computer round-off errors may cause problems when systems with very large grids, with extreme grid aspect ratios, or with insufficient numerical significance in specified boundary conditions are digitally simulated — see Frind and Matanga [2.36]. Problems of this type are, in most cases, eliminated by increasing the precision level of the computations.

The cost of numerical solutions depends primarily on the number of nodes in the spatial grid scheme, multiplied by the number of dependent variables (e.g., h or p; possibly v_x, v_y, c, or T). This product shows the number of equations in the system to be solved. The execution time required for one solution of this system may be a linear or nonlinear function of the number of equations. This time depends, of course, on the computer, on the matrix solution method, and on the software used. By multiplying this execution time expenditure by the number of time or iteration steps necessary for the full simulation, and adding 5–20% overhead for the generation of the coefficient matrix and the right-hand side of the equation system, the complete cost may be estimated with sufficient accuracy.

The memory demand on the computer for simulation of two- and three-dimensional migration processes also depends primarily on the memory demand of the subroutine solving the equation system.

2.2.4 ALGORITHMS FOR ONE- AND QUASI ONE-DIMENSIONAL MIGRATION PROCESSES

From among a large number of well-proven algorithms, the following three were selected as typical examples for a more detailed discussion:

- algorithm A, based on a third-order FD time approximation $O(\Delta t^3)$, a second-order FD space approximation $O(\Delta x^2)$, and integral lumped storage and source-sink terms using Simpson's rule (§2.2.4.1)
- algorithm B, based on Stoyan's method [2.92] (§2.2.4.2)
- algorithm C, based on an FE approximation of spatial terms with linear basis functions and a Galerkin weighting of time levels (§2.2.4.3)

These algorithms are given in this text for parameters and boundary conditions being constant in time and space as well as for equidistant spatial and temporal step lengths to ensure sufficient clarity. Their generalization and extension to quasi one-dimensional algorithms are also discussed in this chapter (§2.2.4.4).

2.2.4.1 Algorithm A

The migration model to be digitally simulated is described by Equation 2.39a, where $\bar{\bar{D}} = DM$:

$$\frac{\partial c}{\partial t} = \frac{\bar{\bar{D}}}{\bar{\epsilon}} \frac{\partial^2 c}{\partial x^2} - \frac{q}{\bar{\epsilon}} \frac{\partial c}{\partial x} - \frac{\bar{\lambda}}{\bar{\epsilon}} c + \frac{\bar{w}}{\bar{\epsilon}} \qquad (2.80)$$

Equation 2.79e is first applied to express the time discretization. Substituting in this equation for the term $\partial c/\partial t$ the right-hand side of Equation 2.80, and for term $\partial^2 c/\partial t^2$ the term $(q/\bar{\epsilon})^2 \partial^2 c/\partial x^2$ (see Equation 2.75), the time-discretized model takes the form:

$$\frac{\Delta t}{2} \left[\left(\bar{\bar{D}} - \frac{q^2 \Delta t}{6\bar{\epsilon}} \right) \frac{\partial^2 c}{\partial x^2} - q \frac{\partial c}{\partial x} - \bar{\lambda}c + \bar{w} \right]^{t+\Delta t} - \bar{\epsilon}c^{t+\Delta t} \qquad (2.81a)$$

$$+ \frac{\Delta t}{2} \left[\left(\bar{\bar{D}} + \frac{q^2 \Delta t}{6\bar{\epsilon}} \right) \frac{\partial^2 c}{\partial x^2} - q \frac{\partial c}{\partial x} - \bar{\lambda}c + \bar{w} \right]^{t} + \bar{\epsilon}c^{t} = 0$$

A uniform discretization of the spatial derivatives with simple FD approximations according to Equations 2.59a and 2.59d leads to

$$\partial c/\partial x|_n \approx (c_{n+1} - c_{n-1})/2\Delta x$$

$$\partial^2 c/\partial x^2|_n \approx (c_{n+1} - 2c_n + c_{n-1})/\Delta x^2$$

For $\bar{w} \neq f(c)$, the added source-sink and storage term $s = \bar{\lambda}c \pm \bar{\epsilon}c/\Delta t$ is given, according to Equation 2.79f, as $\bar{s} = (s_{n-1} + 4s_n + s_{n+1})/6$. The fully discretized form of Equation 2.80 for the nth node is therefore obtained as

$$\frac{1}{2}\left(\frac{\bar{D}\Delta t}{\Delta x^2} - \frac{q^2\Delta t^2}{6\bar{\epsilon}\Delta x^2}\right)(c_{n+1} - 2c_n + c_{n-1})^{t+\Delta t} - \frac{q\Delta t}{4\Delta x}(c_{n+1} - c_{n-1})^{t+\Delta t}$$

$$-\frac{\bar{\epsilon} + \bar{\lambda}\Delta t/2}{6}(c_{n+1} + 4c_n + c_{n-1})^{t+\Delta t}$$

$$+\frac{1}{2}\left(\frac{\bar{D}\Delta t}{\Delta x^2} + \frac{q^2\Delta t^2}{6\bar{\epsilon}\Delta x^2}\right)(c_{n+1} - 2c_n + c_{n-1})^{t} \qquad (2.81b)$$

$$-\frac{q\Delta t}{4\Delta x}(c_{n+1} - c_{n-1})^{t} + \frac{\bar{\epsilon} - \bar{\lambda}\Delta t/2}{6}(c_{n+1} + 4c_n + c_{n-1})^{t} + \bar{w}\Delta t = 0$$

For simplicity the following terms are introduced (compare von Genuchten and Alves [2.43]):

$$V = q\Delta t/\Delta x \qquad D = \bar{D}\Delta t/\Delta x^2 \qquad RN = (\bar{\epsilon} + \bar{\lambda}\Delta t/2)/6 \quad DZ = \bar{w}\Delta t$$
$$R0 = (\bar{\epsilon} - \bar{\lambda}\Delta t/2)/6 \quad DN = D - V^2/(6\bar{\epsilon}) \quad D0 = D + V^2/(6\bar{\epsilon})$$

which leads to

$$-(1/2)DN(c_{n+1}-2c_n+c_{n-1})^{t+\Delta t} + (V/4)(c_{n+1}-c_{n-1})^{t+\Delta t}$$
$$+ RN(c_{n+1}-4c_n+c_{n-1})^{t+\Delta t} = \qquad (2.81c)$$
$$R0(c_{n+1}+4c_n+c_{n-1})^{t} + (1/2)D0(c_{n+1}-2c_n+c_{n-1})^{t} - (V/4)(c_{n+1}-c_{n-1})^{t} + DZ$$

Equation 2.81c is the nth equation of a tridiagonal equation system with the unknowns $X_i = c_i^{t+\Delta t}$ and the coefficients DN2 of the coefficient matrix on the main diagonal and BN and EN on the codiagonals:

$$BN_nX_{n-1} + DN2_nX_n + EN_nX_{n+1} = F_n \qquad (2.81d)$$

where $F_n = B0_nc_{n-1}^{t} + D2_nc_n^{t} + E0_nc_{n+1}^{t} + DZ_n$. For the special case considered here with constant parameters, the coefficients take on the following values:

$$BN = -0.5DN - 0.25V + RN; \quad EN = -0.5DN + 0.25V + RN;$$
$$DN2 = + DN + 4RN; \qquad B0 = 0.5D0 + 0.25V + R0;$$
$$EO = 0.5D0 - 0.25V + R0; \qquad D2 = -D0 + 4R0$$

2.2.4.2 Algorithm B

Equation 2.80 is again considered, and equidistant temporal and spatial discretizations are applied. The discretization scheme is reflected by Figure 2.25c. With this scheme the spatial derivatives of Equation 2.82 are approximated as follows [2.92]:

$$\bar{D}\frac{\partial^2 c}{\partial x^2} - q\frac{\partial c}{\partial x} \approx [\bar{D} - q\Delta x(0.5 + \alpha)]\,(c_{n+1}^{(\gamma)} - c_n^{(\gamma)})/\Delta x^2$$

$$- [\bar{D} + q\Delta x(0.5 - \alpha)]\,(c_n^{(\gamma)} - c_{n-1}^{(\gamma)})/\Delta x^2 \qquad (2.82)$$

where

$$U = \begin{cases} \dfrac{-q\Delta x}{2D} & \text{for } \bar{D} + |q| > 0 \\[2mm] 0 & \text{for } \bar{D} = q = 0 \end{cases}$$

$$\chi(U) = U \coth U$$
$$\approx (3 + 3|U| + 3U^2 + 2|U|^3)/(3 + 3|U| + 2U^2)$$
$$\alpha = [\chi(U) - 1]/2U$$
$$c^{(\gamma)} = \gamma c^{t+\Delta t} + (1 - \gamma)c^t$$

The time derivative is approximated, according to van Genuchten and Gray [2.41], by

$$\bar{\epsilon}\frac{\partial c}{\partial t} \approx 2[\gamma\bar{\epsilon}\Delta x \max(o,\alpha)]_{n+1,n}\,(c_{n+1}^{t+\Delta t} - c_{n+1}^t)/(\Delta t\Delta x) \qquad (2.83a)$$

$$+ [\{\bar{\epsilon}\Delta x[0.5 + \alpha - 2\gamma \max(0,\alpha)]\}_{n+1,n}$$

$$+ \{\bar{\epsilon}\Delta x[0.5 - \alpha - 2\gamma \max(0,-\alpha)]\}_{n,n-1}](c_n^{t+\Delta t} - c_n^t)/(\Delta t\Delta x)$$

$$+ 2[\gamma\bar{\epsilon}\Delta x \max(0,-\alpha)]_{n,n-1}\,(c_{n-1}^{t+\Delta t} - c_{n-1}^t)/(\Delta t\Delta x)$$

where

$$\max(0,\alpha) = \begin{cases} 0 \text{ for } \alpha < 0 \\ \alpha \text{ for } \alpha > 0 \end{cases} \quad \text{and} \quad \max(0,-\alpha) = \begin{cases} 0 \text{ for } \alpha > 0 \\ \alpha \text{ for } \alpha < 0 \end{cases}$$

The source-sink terms are discretized as follows:

$$\bar{\lambda}c \approx 2[\omega\bar{\lambda}\Delta x \max(0,\alpha)]_{n+1,n}\,c_{n+1}^{\{\mu\}}/\Delta x$$
$$+ \{\bar{\lambda}\Delta x[0.5 + \alpha - 2\omega \max(0,\alpha)]\}_{n+1,n}c_n^{(\nu)}/\Delta x$$
$$+ \{\bar{\lambda}\Delta x[0.5 - \alpha - 2\omega \max(0,-\alpha)]\}_{n,n-1,n}c_n^{(\nu)}/\Delta x$$
$$+ 2[\omega\bar{\lambda}\Delta x \max(0,-\alpha)]_{n,n-1}c_{n-1}^{\{\mu\}}/\Delta x] \qquad (2.83b)$$

$$\bar{w} \approx \{[\Delta x\bar{\omega}^{(\gamma)}(0.5 + \alpha)]_{n+1,n} + [\Delta x\bar{w}^{(\gamma)}(0.5 - \alpha)]_{n,n-1}\}/\Delta x \qquad (2.83c)$$

The weighting factors γ, ω, μ, and ν may be determined for any spatial step length and for any parameter set (also "degenerated" parameters $\bar{D}$, q, $\bar{\epsilon}$, and $\bar{\lambda}$) as described in von Genuchten and Gray [2.41]. For "normal" parameters, i.e., particularly for equal velocity directions in all elements, the application of a uniform factor is recommended:

$$\gamma = \omega = \mu = \nu = 0.5 \qquad (2.83d)$$

Therefore, one obtains again for the spatial grid point n the characteristic governing equation of a tridiagonal equation system according to Equation 2.81d with the unknowns $X_i = c_i^{t+\Delta t}$:

$$B_n X_{n-1} + A_n X_n + C_n X_{n+1} = D_n \qquad (2.83e)$$

where FX1 $= [\bar{D} - q\Delta x(0.5 + \alpha)]/\Delta x$
  FX2 $= [\bar{D} + q\Delta x(0.5 - \alpha)]/\Delta x$
  FT1 $= 2\gamma\bar{\epsilon}\Delta x \max (0,\alpha)/\Delta t$
  FT2 $= \bar{\epsilon}\Delta x[0.5 + \alpha - 2\gamma \max (0,\alpha)]/\Delta t$
  FT3 $= \bar{\epsilon}\Delta x [0.5 - \alpha - 2\gamma \max (0,-\alpha)]/\Delta t$
  FT4 $= 2\gamma\bar{\epsilon}\Delta x \max (0,-\alpha)/\Delta t$
  FS1 $= 2\omega\lambda\Delta x \max (0,\alpha)$
  FS2 $= \lambda\Delta x [0.5 + \alpha - 2\omega \max (0,\alpha)]$
  FS3 $= \lambda\Delta x [0.5 - \alpha - 2\omega \max (0,-\alpha)]$
  FS4 $= 2\omega\lambda\Delta x \max (0,-\alpha)$
  FW1 $= \Delta x (0.5 + \alpha)\bar{w}$
  FW2 $= \Delta x (0.5 - \alpha)\bar{w}$

The coefficients of Equation 2.83e may now be written as follows:

$B_n = FX2\gamma - FT4 - FS4\mu$
$C_n = FX1\gamma - FT1 - FS1\mu$
$A_n = FX1\gamma - FX2\gamma - FT2 - FT3 - FS2\nu - FS3\nu$
$D_n = [-FX2(1 - \gamma) - FT4 + FS4(1-\mu)]c_{n-1}^t + [+FX1(1-\gamma)$
$\qquad + FX2(1-\gamma) - FT2 - FT3 + FS2(1-\nu) + FS3(1-\nu)]c_n^t$
$\qquad + [FX1(1-\gamma) - FT1 + FS1(1-\mu)]c_{n+1}^t$
$\qquad -FW1\gamma - FW1(1-\gamma) - FW2\gamma - FW2(1-\gamma)$

For constant parameters, again $B_n = B$, $C_n = C$, and $A_n = A$. This algorithm yields reliable solutions of Equation 2.39a also for the critical cases of $q = 0$ or $DM = \bar{D} = 0$ and nearly any temporal and spatial step lengths. However, this algorithm is difficult to extend to the two-or three-dimensional case and hence is of restricted practical value.

2.2.4.3 Algorithm C

For the one-dimensional Equation 2.80, the matrix or vector elements according to Equations 2.70a to 2.70c are specified as follows:

$$S_{kl} = \sum_{m=1}^{M} \int_x \left(\bar{D} \frac{dN_k}{dx} \frac{dN_l}{dx} + qN_k \frac{dN_l}{dx} + \bar{\lambda}N_k N_l \right)dx \qquad (2.84a)$$

$$O_{kl} = \sum_{m=1}^{M} \int_x \bar{\epsilon}N_k N_l dx \qquad (2.84b)$$

$$F_k = \sum_{m=1}^{M} \left(N_k q_c \big|_{\mp} + \int_x N_k \bar{w} dx \right) \quad \text{with } q_c = -\bar{D} \text{ grad } \tilde{c} \quad (2.84c)$$

where $q_c \big|_{\mp}$ represents the specified solute flux directed positively outward over the boundary ($q_c \big|_{-}$ denotes the left-hand side and $q_c \big|_{+}$ the right-hand side boundary value).

If one-dimensional finite elements are selected with a linear interpolation function over the element m (see Figure 2.23b), Equation 2.60 yields

$$c^m(x,t) = \sum_k N_k[x\xi)]c_1^m(t) \quad (2.85)$$

The indices k and l in Equations 2.84 and 2.85 take the values 1, 2. The two basis functions are described by Equation 2.63a in natural (or local) coordinates in the domain $[-1, +1]$. Equation 2.64a provides the relationship between these natural coordinates and the global coordinate x, according to the isoparametric concept:

$$x^m = \sum_{k=1}^{2} N_k(\xi)x_k^m \quad (2.86)$$

where x symbolizes the coordinates of node k in the element m.

By means of Equation 2.64d the global derivatives in Equation 2.84 can be expressed by local derivatives:

$$\frac{dN_k}{dx} = \frac{dN_k}{d\xi} \frac{d\xi}{dx} \quad k = 1,2 \quad (2.87)$$

From Equation 2.64c it follows that

$$dx^m/d\xi = -x_1/2 + x_2/2 = h/2 \quad \text{with} \quad h = x_2 - x_1$$

Using these results one obtains:

$$\frac{dN_1}{dx} = \frac{dN_1}{d\xi} \frac{2}{h} \quad 1 = 1,2 \quad \text{and} \quad dx = \frac{h}{2} d\xi$$

Hence, Equation 2.84 yields

$$S_{kl} = \sum_{m=1}^{M} \int_{-1}^{+1} \left(\bar{D} \frac{dN_k}{d\xi} \frac{dN_1}{d\xi} \frac{2}{h} + qN_k \frac{dN_1}{d\xi} + \bar{\lambda} N_k N_1 \frac{h}{2} \right) d\xi \quad (2.88a)$$

$$O_{kl} = \sum_{m=1}^{M} \int_{-1}^{+1} \bar{\epsilon} N_k N_1 \frac{h}{2} d\xi \quad (2.88b)$$

$$F_k = \sum_{m=1}^{M} \left(N_k q c \Big|_{+}^{-} + \int_{-1}^{+1} N_k \bar{w} \frac{h}{2} d\xi \right) \qquad k,1 = 1,2 \qquad (2.88c)$$

Equations 2.88 are integrated using Equation 2.63a and $dN_1/d\xi = -1/2$ and $dN_2/d\xi = +1/2$. The result can be expressed in matrix form as

$$[S] = \sum_{m=1}^{M} \begin{pmatrix} \dfrac{\bar{D}}{h} - \dfrac{q}{2} + \bar{\lambda}\dfrac{h}{3} & \Big| & -\dfrac{\bar{D}}{h} + \dfrac{q}{2} + \bar{\lambda}\dfrac{h}{6} \\[2mm] \hline \\[-2mm] -\dfrac{\bar{D}}{h} - \dfrac{q}{2} + \bar{\lambda}\dfrac{h}{6} & \Big| & \dfrac{\bar{D}}{h} + \dfrac{q}{2} + \bar{\lambda}\dfrac{h}{3} \end{pmatrix} (2.89a)$$

$$[O] = \sum_{m=1}^{M} \begin{pmatrix} \bar{\epsilon}\dfrac{h}{3} & \Big| & \bar{\epsilon}\dfrac{h}{6} \\[2mm] \hline \\[-2mm] \bar{\epsilon}\dfrac{h}{6} & \Big| & \bar{\epsilon}\dfrac{h}{3} \end{pmatrix} \qquad (2.89b)$$

$$\{F\} = \sum_{m=1}^{M} \begin{Bmatrix} q_c\Big|^{-} + \bar{w}\dfrac{h}{2} \\[2mm] \hline \\[-2mm] q_c\Big|_{+} + \bar{w}\dfrac{h}{2} \end{Bmatrix} \qquad (2.89c)$$

Using the time discretization Equation 2.73 also for Equation 2.70, the finite element scheme becomes

$$([O]/\Delta t + \gamma[S]) \{c^{t+\Delta t}\} = ([O]/\Delta t - (1 - \gamma) [S]) \{c^t\}$$
$$- \gamma\{F^{t+\Delta t}\} - (1 - \gamma) \{F^t\} \qquad (2.90)$$

where γ is again the time weighting factor.

The summation over all the elements from $m = 1$ to M according to Equation 2.89 by superimposing the partial element matrices [ ]^m and vectors { }^{m} yields Equation 2.91 if a uniformly spaced node pattern and $\gamma = 2/3$ are used. For node i the following characteristic equation is obtained:

$$\left[\bar{\epsilon}\, \frac{h}{6\Delta t} - \frac{2}{3}\!\left(\frac{\bar{D}}{h} + \frac{q}{2} - \bar{\lambda}\, \frac{h}{6} \right) \right] c_{i-1}^{t+\Delta t} \qquad (2.91)$$

$$+ \left[\bar{\epsilon}\, \frac{2h}{3\Delta t} + \frac{2}{3}\!\left(\frac{2\bar{D}}{h} + \bar{\lambda}\, \frac{2}{3}\, h \right) \right] c_{i}^{t+\Delta t}$$

$$+ \left[\bar{\epsilon}\, \frac{h}{6\Delta t} - \frac{2}{3}\!\left(\frac{\bar{D}}{h} - \frac{q}{2} - \bar{\lambda}\, \frac{h}{6} \right) \right] c_{i+1}^{t+\Delta t}$$

$$= \left[\bar{\epsilon}\, \frac{h}{6\Delta t} + \left(1 - \frac{2}{3} \right)\!\left(\frac{\bar{D}}{h} + \frac{q}{2} - \bar{\lambda}\, \frac{h}{6} \right) \right] c_{i-1}^{t}$$

$$+ \left[\bar{\epsilon}\, \frac{2h}{3\Delta t} - \left(1 - \frac{2}{3} \right)\!\left(\frac{2\bar{D}}{h} + \bar{\lambda}\, \frac{2}{3}\, h \right) \right] c_{i}^{t}$$

$$+ \left[\bar{\epsilon}\, \frac{h}{6\Delta t} + \left(1 - \frac{2}{3} \right)\!\left(\frac{\bar{D}}{h} - \frac{q}{2} - \bar{\lambda}\, \frac{h}{6} \right) \right] c_{i+1}^{t} - \bar{w}h$$

where i denotes an internal position, i.e., a node surrounded by elements.

In formulating this equation, use was made of the continuity of fluxes across interelement boundaries where $q_c|^- = -q_c|_+$. Consequently, boundary solute fluxes q_c must only be provided at external boundaries, i.e., at $x = 0$ (node 1) and at $x = L$ (node N) for the example considered. For these nodes the boundary conditions become incorporated in Equation 2.91, which therefore obtains a modified form.

The FORTRAN program ALGO, shown in Table 2.7, was written to execute algorithms A, B, and C [2.43]. Definitions of the variables used in this program are given in Table 2.6. The input data are chosen according to the example presented in Figure 2.1 as follows:

IART = 1
IRAB = 1
TITL = comparison of Equation 2.39e with Equation 2.81
ND = 5
ID(I) = {5, 9, 13, 17, 21}
MD = 16
JD(I) = {2, 4, 6, . . ., 32}
NE = 24
NSTE = 32
DELX = 12.5 m
DELT = 12.5 days
V = 34.56 m²/day
D = 207.4 m³/day

Table 2.6. Definitions of Variables Used in the FORTRAN Program ALGO

FORTRAN program	Meaning of the input data
IART = 1	application of algorithm A
= 2	application of algorithm B
= 3	application of algorithm C
IRAB = 1	$c(x = 0)$ is fixed
= 3	$c_0 q_0 (x = 0) = CR3$ is fixed
TITL	maximum 60 alphanumeric characters
ND	number of elements with result printing
ID(I)	element number with result printing
MD	number of time intervals with result printing
JD(I)	time interval number with result printing
NE	number of spatial steps Δx
NSTE	number of time intervals Δt
DELX	Δx
DELT	Δt
V	q
D	$\overline{D}$
R	$\overline{\epsilon}$
DZER	$\overline{w}$
DONE	$\overline{\lambda}$
CR	$c(x = 0) = CR$
CI	$c(t = 0) = CI$
CR3	$c_0 q_0 (x = 0) = CR3$

$$
\begin{aligned}
R &= 40 \text{ m} \\
DZER &= 0.3456 \text{ g/m}^2/\text{day} \\
DONE &= 0.0691 \text{ m/day} \\
CR &= 10 \text{ g/m}^3 \\
CI &= 2 \text{ g/m}^3
\end{aligned}
$$

The computed results are compared with the analytical solution of Equation 2.39e, and both the numerical and the analytical results are plotted in Figure 2.26.

2.2.4.4 Generalization of Algorithms A, B, and C

For practical applications, it is desirable to generalize the above three algorithms with respect to stream tubes in general flow systems, axisymmetric stream tubes, and multidimensional migration processes. A general stream tube is characterized by the following migration model (the x-axis is used as the tube's curvilinear center line):

$$\frac{\partial}{\partial t} (Ec) = \frac{\partial}{\partial x} DA \frac{\partial c}{\partial x} - \frac{\partial}{\partial x} (Qc) - L^*c + W \tag{2.92}$$

where
$$
\begin{aligned}
E &= \epsilon MB \\
DA &= DMB
\end{aligned}
$$

Table 2.7. FORTRAN Program ALGO

```
      PROGRAM ALGO
C
C     NUMERICAL SOLUTION OF THE ONE-DIMENSIONAL CONVECTIVE-
C     DISPERSIVE TRANSPORT EQUATION WITH SEVERAL ALGORITHMS
C     FOR BOUNDARY CONDITIONS OF 1ST, 2ND, AND 3RD TYPE
C
C     INPUT:   IART - KEY NUMBER TO SELECT THE ALGORITHM
C                     =1 - ALGA
C                     =2 - ALGB
C                     =3 - ALGC
C              IRAB - KEY NUMBER TO SELECT THE BOUNDARY CONDITIONS
C                     =1 - BOUNDARY CONDITION OF 1ST TYPE
C                     =2 - BOUNDARY CONDITION OF 2ND TYPE
C                     =3 - BOUNDARY CONDITION OF 3RD TYPE
C              NSTE - NUMBER OF TIME STEPS
C              NI   - NUMBER OF ELEMENTS
C              DELT - TIME STEP
C              DELX - SPACE STEP
C              V    - FLOW RATE
C              D    - DISPERSION COEFFICIENT
C              R    - STORAGE COEFFICIENT
C              DZER - AREAL INPUT (SOURCE)
C              DONE - DECAY RATE COEFFICIENT
C              CR3  - BOUNDARY CONCENTRATION * FLOW RATE
C              CR   - BOUNDARY CONCENTRATION FOR 1ST TYPE
C              CI   - INITIAL CONCENTRATION IN THE FIELD
C              ND   - NUMBER OF NODES AT WHICH TO PRINT RESULTS (<7)
C              ID   - ELEMENT NUMBERS AT WHICH TO PRINT RESULTS
C              MD   - NUMBER OF TIME INTERVALS AT WHICH TO PRINT
C                     RESULTS (<81)
C              JD   - TIME INTERVAL NUMBERS AT WHICH TO PRINT RESULTS
C     FOR ALGORITHM B THE COMPUTATION FOR 3RD TYPE BOUNDARY
C     CONDITION IS REALIZED ACCORDING TO ALGORITHM A
C     AUTHOR: D.SCHAEFER (DRESDEN UNIVERSITY OF TECHNOLOGY)
      COMMON/BL1/TITL(15),C(61),X(61),ID(10),JD(80),CD(60)
      COMMON/BL2/F(60),U(60),IRAB,DELX,DELT,V,D,R,DZER,DONE,BN,DN2,
     *EN,B0,D2,D1,E0,W1
      OPEN(UNIT=3,FILE='ALGO.IN',STATUS='OLD')
      OPEN(UNIT=4,FILE='ALGO.OUT',STATUS='NEW')
    1 FORMAT(2I5)
    2 FORMAT(60(1H*)/1H*,16X,25HNUMERICAL SOLUTION OF THE,17X,1H*/
     *1H*,12X,37HONE-DIMENSIONAL CONVECTIVE-DISPERSIVE,9X,1H*/
     *1H*,19X,18HMIGRATION EQUATION,21X,1H*/1H*,58X,1H*)
    3 FORMAT(1H*,14X,31HFD-METHOD ACC. TO VAN GENUCHTEN,13X,1H*/
     *1H*,23X,13H(ALGORITHM A),22X,1H*/1H*,58X,1H*)
    4 FORMAT(1H*,17X,24HFD-METHOD ACC. TO STOYAN,17X,1H*/
     *1H*,23X,13H(ALGORITHM B),22X,1H*/1H*,58X,1H*)
    5 FORMAT(1H*,25X,10HFEM-METHOD,23X,1H*/
     *1H*,23X,13H(ALGORITHM C),22X,1H*/1H*,58X,1H*)
    6 FORMAT(1H*,18X,20HBOUNDARY OF 1ST TYPE,20X,1H*/1H*,58X,1H*)
    7 FORMAT(1H*,18X,20HBOUNDARY OF 3RD TYPE,20X,1H*/1H*,58X,1H*)
    8 FORMAT(1H*,14X,29HLINEAR STORAGE BEHAVIOR S=R*C,15X,1H*/
```

Table 2.7 continued

```
     *1H*,11X,36HKINETIC REACTION OF 1ST ORDER C*DONE,11X,1H*/
     *1H*,16X,26HCONSTANT SOURCE TERM DZERO,16X,1H*/
     *1H*,58X,1H*/60(1H*)///)
   9 FORMAT(15A4)
  10 FORMAT(2I5)
  11 FORMAT(10I5)
  12 FORMAT(2I5,2F10.3)
  13 FORMAT(5E13.5)
  14 FORMAT(2F10.3)
  15 FORMAT(/44HNUMBER OF ELEMENTS (MAX 60)                NE = ,I3/
     *44HNUMBER OF TIME STEPS               NSTE = ,I3/
     *44HSPATIAL STEP                       DELX = ,F6.2/
     *44HTIME STEP                          DELT = ,F7.3/
     *44HFLOW RATE                             V = ,E10.3/
     *44HDISPERSION COEFFICIENT                D = ,E10.3/
     *44HSTORAGE COEFFICIENT                   R = ,E10.3/
     *44HAREAL INPUT (SOURCE)               DZER = ,E10.3/
     *44HDECAY RATE COEFFICIENT             DONE = ,E10.3/
     *44HBOUNDARY CONCENTRATION               CR = ,F10.3/
     *44HINITIAL CONCENTRATION IN THE FIELD   CI = ,F10.3/
     *44HBOUNDARY CONCENTRATION * FLOW RATE  CR3 = ,F10.3/)
  16 FORMAT(///50HCONCENTRATIONS AS A FUNCTION OF TIME AND SPACE IN ,
     *6HG/M**3/56(1H-)//
     *7HT I M E,7H (DAYS),1X,1HI,30H    DISTANCE FROM THE BOUNDARY,
     *9H IN METER/15X,1HI,2X,6(F7.2,2X))
  17 FORMAT(72(1H-))
  18 FORMAT(F7.2,8X,1HI,2X,6(F7.3,2X))
C
     READ(3,1) IART,IRAB
C    ***HEADING PRINTOUT***
     WRITE(4,2)
     IF(IART.EQ.1) WRITE(4,3)
     IF(IART.EQ.2) WRITE(4,4)
     IF(IART.EQ.3) WRITE(4,5)
     IF(IRAB.EQ.1) WRITE(4,6)
     IF(IRAB.EQ.3) WRITE(4,7)
     WRITE(4,8)
C    ***READING THE INPUT DATA***
     READ(3,9)(TITL(I),I=1,15)
     WRITE(4,9)(TITL(I),I=1,15)
     READ(3,10) ND,MD
     READ(3,11)(ID(IL),IL=1,ND)
     READ(3,11)(JD(JL),JL=1,MD)
     READ(3,12) NE,NSTE,DELX,DELT
     READ(3,13) V,D,R,DZER,DONE
     READ(3,14) CR,CI
     CR3=CR*V
     IF(IRAB.EQ.3) CR3=CR
     WRITE(4,15) NE,NSTE,DELX,DELT,V,D,R,DZER,DONE,CR,CI,CR3
C    ***FIELD DATA GENERATION**
     NN=NE+1
     DO 100 I=1,NN
       C(I)=CI
       X(I)=(I-1)*DELX
```

Table 2.7 continued

```
  100 CONTINUE
C     ***PRINTOUT OF RESULTS - TABLE HEADING***
      DO 110 IL=1,ND
       IDD=ID(IL)
       X(IL)=X(IDD)
  110 CONTINUE
      WRITE(4,16)(X(IL),IL=1,ND)
      WRITE(4,17)
      IF(IART.EQ.2.AND.IRAB.EQ.3) IART=1
      IF(IART.EQ.1) CALL ALGA
      IF(IART.EQ.2) CALL ALGB
      IF(IART.EQ.3) CALL ALGC
C     ***TIME DEPENDING COMPUTATIONS***
      DO 190 J=1,NSTE
       TIME=J*DELT
C     ***COMPUTATION OF THE KNOWN VECTOR***
       F(1)=D1*C(1)+E0*C(2)+W1+CR3
       IF(IRAB.EQ.3) GOTO 120
        F(1)=CR
        C(1)=CR
  120    CONTINUE
       DO 130 I=2,NE
        F(I)=B0*C(I-1)+D2*C(I)+E0*C(I+1)+DZER
  130    CONTINUE
C     ***COMPUTATION OF THE UNKNOWN VECTOR***
       R=BN/U(1)
       F(2)=F(2)-R*F(1)
       IF(IRAB.EQ.1) R=0.
       U(2)=DN2-R*EN
       DO 140 I=3,NE
        R=BN/U(I-1)
        U(I)=DN2-R*EN
        F(I)=F(I)-R*F(I-1)
  140    CONTINUE
       C(NN)=F(NE)/(U(NE)+EN)
       DO 150 I=2,NN
        K=NN+1-I
        C(K)=(F(K)-EN*C(K+1))/U(K)
  150    CONTINUE
       IF(IRAB.EQ.1) C(1)=F(1)
C     ***PRINTOUT OF RESULTS***
       DO 160 JL=1,MD
        IF(JD(JL).EQ.J) GOTO 170
  160    CONTINUE
       GOTO 190
  170    CONTINUE
       DO 180 I=1,NE
        DO 180 IL=1,ND
         IF(ID(IL).EQ.I) CD(IL)=C(I)
  180    CONTINUE
       WRITE(4,18) TIME,(CD(IL),IL=1,ND)
  190 CONTINUE
      STOP
      END
```

Table 2.7 continued

```
      SUBROUTINE ALGA
C
C     ALGORITHM A - FD METHOD AFTER VAN GENUCHTEN
C
      COMMON/BL2/F(60),U(60),IRAB,DELX,DELT,V,D,R,DZER,DONE,BN,DN2,
     *EN,B0,D2,D1,E0,W1
      V=V*DELT/DELX
      D=D*DELT/DELX**2
      RN=(R+0.5*DELT*DONE)/6.
      RO=(R-0.5*DELT*DONE)/6.
      DZER=DZER*DELT
      W1=DZER/2.
      DN=D-V*V/6./R
      DO=D+V*V/6./R
      EN=-0.5*DN+0.25*V+RN
      E0=0.5*DO-0.25*V+RO
      BN=-0.5*DN-0.25*V+RN
      BO=0.5*DO+0.25*V+RO
      U(1)=0.5*DN+0.25*V+2.*RN
      IF(IRAB.EQ.1)U(1)=1.
      DN2=DN+4.*RN
      D1=-0.5*DO-0.25*V+2.*RO
      D2=-DO+4.*RO
      RETURN
      END


      SUBROUTINE ALGB
C
C     ALGORITHM B - FD METHOD AFTER STOYAN
C
      REAL N1,MY,NY,M1,N2,MAXP,MAXM
C
      COMMON/BL2/F(60),U(60),IRAB,DELX,DELT,V,D,R,DZER,DONE,BN,DN2,
     *EN,B0,D2,D1,E0,W1
C     ***COMPUTATION OF THE WEIGHTING FACTORS***
      S=DELX*(-V)/2./D
      AS=ABS(ALFA(S))
      AB=ALFA(S)
      ER=DONE*DELT/2./R
      AR=ALFA(ER)
      A=(D/DELX+ABS(V)*(0.5+AS))*DELT/R/DELX
      A2=2.*AS
      IF(A.GT.A2) GOTO 120
        IF(A.EQ.A2) GOTO 100
          GAMA=0.
          OMEG=0.
          GOTO 110
 100    CONTINUE
        GAMA=0.5+AR
        OMEG=0.5-AR
 110    CONTINUE
        MY=0.
        N1=1.-(0.5+AS-A)/(2.*ER*(0.5+(1.-2.*OMEG)*AS))
        NY=GAMA
```

Table 2.7 continued

```
         IF(GAMA.LE.N1) NY=N1
         GOTO 130
 120 CONTINUE
     G1=0.5+AR
     G2=1.-(.5-AS)/(A-2.*AS+ER*(1.+2.*AS))
     GAMA=G1
     IF(G2.GT.G1) GAMA=G2
     M1=(A-A2)/4./ER/AS
     MY=GAMA
     IF(M1.LT.GAMA) MY=M1
     N2=1.-0.5/ER
     NY=GAMA
     IF(N2.GT.GAMA) NY=N2
     O1=((1.-GAMA)*A+A2*GAMA)/(4.*ER*(1.-MY)*AS)
     OMEG=0.5
     IF(O1.LT.0.5) OMEG=O1
 130 CONTINUE
     MAXP=0.
     MAXM=AB
     IF(AB.LE.0.) GOTO 140
       MAXP=AB
       MAXM=0.
 140 CONTINUE
C    ***COMPUTING THE MATRIX ELEMENTS***
     X1=(D-V*DELX*(0.5+AB))/DELX
     X2=(D+V*DELX*(0.5-AB))/DELX
     T1=2.*GAMA*DELX*R*MAXP/DELT
     T2=R*DELX*(0.5+AB-2.*GAMA*MAXP)/DELT
     T3=R*DELX*(0.5-AB-2.*CAMA*MAXM)/DELT
     T4=2.*GAMA*R*DELX*MAXM/DELT
     S1=2.*OMEG*DONE*DELX*MAXP
     S2=DONE*DELX*(0.5+AB-2.*OMEG*MAXP)
     S3=DONE*DELX*(0.5-AB-2.*OMEG*MAXM)
     S4=2.*OMEG*DONE*DELX*MAXM
     W1=DELX*(0.5+AB)*DZER
     W2=DELX*(0.5-AB)*DZER
     DZER=(-W1)-W2
     BN=X2*GAMA-T4-S4*MY
     DN2=X1*(-GAMA)-X2*GAMA-T2-T3-S2*NY-S3*NY
     EN=X1*GAMA-T1-S1*MY
     B0=X2*(GAMA-1.)-T4+S4*(1.-MY)
     D2=X1*(1.-GAMA)+X2*(1.-GAMA)-T2-T3+S2*(1.-NY)+S3*(1.-NY)
     E0=X1*(GAMA-1.)-T1+S1*(1.-MY)
     U(1)=1.
     RETURN
     END

     SUBROUTINE ALGC
C
C    ALGORITHM C - FE METHOD AFTER DIERSCH
C
     COMMON/BL2/F(60),U(60),IRAB,DELX,DELT,V,D,R,DZER,DONE,BN,DN2,
    *EN,BO,D2,D1,E0,W1

     EPS6=R*DELX/6./DELT
     EPS3=R*DELX*2./DELT/3.
     DP=D/DELX+V/2.-DONE*DELX/6.
```

Table 2.7 continued

```
        DM=D/DELX-V/2.-DONE*DELX/6.
        DL=2*D/DELX+DONE*2.*DELX/3.
        BN=EPS6-DP*2./3.
        DN2=EPS3+DL*2./3.
        EN=EPS6-DM*2./3.
        BO=EPS6+DP/3.
        D2=EPS3-DL/3.
        EO=EPS6+DM/3.
        DZER=DZER*DELX
        W1=DZER/2.
        U(1)=1.
        IF(IRAB.EQ.3)U(1)=R*DELX/3./DELT+2.*(D/DELX+V/2.+DONE*DELX/3.)/3.
        D1=R*DELX/3./DELT-(D/DELX+V/2.+DONE*DELX/3.)/3.
        RETURN
        END


        FUNCTION ALFA(X)
C       COMPUTING THE ALPHA-VALUES FOR THE STOYAN ALGORITHM
        REAL KA,KAPA
        KA=3.+3.*ABS(X)+3.*X*X+2.*ABS(X)**3
        PA=3.+3.*ABS(X)+2.*X*X
        KAPA=KA/PA
        ALFA=(KAPA-1.)/2./X
        RETURN
        END
     2    1
comperison of eq.2.39e with the numeric solution of eq2.81
     5   16
     5    9   13   17   21
     2    4    6    8   10   12   14   16   18   20
    22   24   26   28   30   32
    24   32   12.5      12.5
  34.56000e+00207.40000e+00  40.00000e+00   0.34560e+00   0.06910e+00
    10.0       2.0
```

```
**********************************************************
*             NUMERICAL SOLUTION OF THE                  *
*        ONE-DIMENSIONAL CONVECTIVE-DISPERSIVE           *
*                MIGRATION EQUATION                      *
*                                                        *
*             FD-METHOD ACC. TO STOYAN                   *
*                  (ALGORITHM B)                         *
*                                                        *
*             BOUNDARY OF 1ST TYPE                       *
*                                                        *
*          LINEAR STORAGE BEHAVIOR S=R*C                 *
*       KINETIC REACTION OF 1ST ORDER C*DONE             *
*            CONSTANT SOURCE TERM DZERO                  *
*                                                        *
**********************************************************

comparison of eq.2.39e with the numerical solution of eq.2.81
```

```
NUMBER OF ELEMENTS (MAX 60)          NE  =   24
NUMBER OF TIME STEPS                 NSTE =   32
SPATIAL STEP                         DELX =   12.50
TIME STEP                            DELT =   12.500
FLOW RATE                              V  =   0.346E+02
```

Table 2.7 continued

DISPERSION COEFFICIENT	D =	0.207E+03
STORAGE COEFFICIENT	R =	0.400E+02
AREAL INPUT (SOURCE)	DZER =	0.346E+00
DECAY RATE COEFFICIENT	DONE =	0.691E-01
BOUNDARY CONCENTRATION	CR =	10.000
INITIAL CONCENTRATION IN THE FIELD	CI =	2.000
BOUNDARY CONCENTRATION * FLOW RATE	CR3 =	345.600

CONCENTRATIONS AS A FUNCTION OF TIME AND SPACE IN G/M**3

T I M E (DAYS) I	DISTANCE FROM THE BOUNDARY IN METER				
I	50.00	100.00	150.00	200.00	250.00
25.00 I	3.274	2.186	2.129	2.127	2.127
50.00 I	5.740	2.751	2.290	2.251	2.248
75.00 I	7.617	3.913	2.600	2.389	2.366
100.00 I	8.640	5.372	3.200	2.590	2.489
125.00 I	9.132	6.696	4.100	2.932	2.639
150.00 I	9.356	7.672	5.153	3.469	2.857
175.00 I	9.455	8.303	6.167	4.187	3.187
200.00 I	9.499	8.677	7.009	5.004	3.653
225.00 I	9.519	8.885	7.637	5.814	4.239
250.00 I	9.527	8.996	8.066	6.531	4.897
275.00 I	9.531	9.054	8.341	7.111	5.562
300.00 I	9.533	9.083	8.509	7.545	6.174
400.00 I	9.534	9.110	8.710	8.267	7.645

$$Q = vMB$$
$$L^* = \lambda MB$$
$$W = wMB$$

After differentiation of the products and rearrangement, one obtains

$$E \frac{\partial c}{\partial t} = \frac{\partial}{\partial x} DA \frac{\partial c}{\partial x} - Q \frac{\partial c}{\partial x} - Lc + W \qquad (2.93)$$

where $L = L^{-*} + \partial E/\partial t + \partial Q/\partial x$. For the general case of a nonuniform temporal and spatial step pattern, and with algorithm A, Equation 2.81a yields

$$\frac{\Delta t}{2} \left[\frac{\partial}{\partial x} \left(DA - \frac{Q^2 \Delta t}{6} \right) \frac{\partial c}{\partial x} - Q \frac{\partial c}{\partial x} - Lc - W \right]^{t+\Delta t} - Ec^{t+\Delta t} \qquad (2.94)$$

$$+ \frac{\Delta t}{2} \left(\frac{\partial}{\partial x} \left(DA + \frac{Q^2 \Delta t}{6} \right) \frac{\partial c}{\partial x} - Q \frac{\partial c}{\partial x} - Lc + W \right)^t + Ec^t = 0$$

where the spatial derivatives

$$\frac{\partial}{\partial x} (\alpha) \frac{\partial c}{\partial x} \text{ and } \beta \frac{\partial c}{\partial x}$$

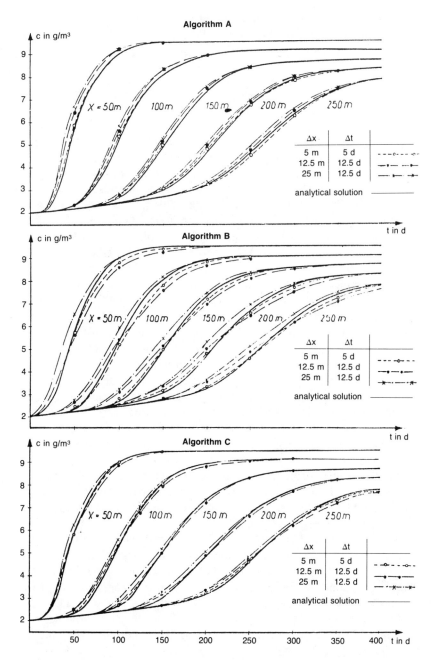

Figure 2.26. Comparison of the analytical solution 2.39e with the numerical solutions of the algorithms A, B, and C.

for variable Δx are approximated according to Table 2.4.

The summary source-sink term $S = E\partial c/\partial t + Lc - W$ is approximated by means of Simpson's rule as

$$S_n \approx \frac{x_n - x_{n-1}}{x_{n+1} - x_{n-1}} \frac{S_{n-1} + 2S_n}{3} + \frac{x_{n+1} - x_n}{x_{n+1} - x_{n-1}} \frac{2S_n + S_{n+1}}{3} \quad (2.95)$$

Using these results, the equation corresponding to Equation 2.81c or 2.81d may again be derived. The coefficients BN_n, $DN2_n$, and EN_n, as well as F_n, now differ for each of the n equations and for each time level. The parameters E, DA, Q, and W, as well as the boundary conditions, can therefore change in space and time. This also allows the use of nonlinear parameters and boundary conditions characterized by their functional dependency on the dependent variable. The resulting algorithm is used in the FORTRAN program SEM1, described in Schäfer [2.86].

An axisymmetric migration process may be considered as a special case of Equation 2.92 with $B = 2\pi r$. The corresponding FORTRAN program, named SROM, is described in Mansel et al. [2.68].

Under certain conditions, multidimensional migration processes may be decomposed into their dimensional components and treated as quasi one-dimensional systems using the ADI method (see, e.g., Luckner and Schestakow [2.63]). Such conditions include stationariness of the flow process and linearity of the transport process. Many applied migration problems are formulated with linear migration parameters. The assumption of steady-state flow conditions are reasonable, for instance, in groundwater management and other long-term problems where migration processes take place over time periods that are much longer than seasonal hydrometeorological variations. An average annual flow regime may therefore often be assumed. In such cases impacts of temporal flow conditions may be represented by means of enhanced hydrodynamic dispersivities (see, e.g., Ackermann [2.2]).

Steady-state flow fields may be characterized by spatially constant stream tubes. To generate these tubes, tracking (walking) points can be used (see §2.2.5), or a Laplace-type equation can be solved directly for stream functions Ψ [2.36]. Suitable computer programs for two-dimensional flow fields are described, for instance, in Gutt and Victor [2.50], Schimmel [2.88], and Victor [2.95]. If curvilinear orthogonal nets are applied in which x coincides with streamlines and y is orthogonal to them, Equation 2.93 takes the following form:

$$E \frac{\partial c}{\partial t} = \frac{\partial}{\partial x} DA_\ell \frac{\partial c}{\partial x} + \frac{\partial}{\partial y} DA_{tr} \frac{\partial c}{\partial y} - Q \frac{\partial c}{\partial x} - Lc + W \quad (2.96)$$

This equation may be split into quasi one-dimensional components and solved by means of the ADI method. With

$$L_x = \frac{\partial}{\partial x} DA_\ell \frac{\partial}{\partial x} - Q \frac{\partial}{\partial x} \qquad \text{and } L_y = \frac{\partial}{\partial y} DA_{tr} \frac{\partial}{\partial y}$$

the two steps of the solution procedure can be written as

Step 1:	$L_x^* c^{t+\Delta t/2} + L_y c^t = 2E(c^{t+\Delta t/2} - c^t)/\Delta t$	(2.97a)
Step 2:	$L_x c^{t+\Delta t/2} + L_y^* c^{t+\Delta t} = 2E(c^{t+\Delta t} - c^{t+\Delta t/2})/\Delta t$	(2.97b)

where * symbolizes the operator relating to the unknowns.

This method is used in the computer program SIMKA-2D [2.48]. In this program the one-dimensional problem of step 1 is solved by the generalized algorithm B, and step 2 is treated explicitly. By contrast, in Frind [2.33] both equations, Equations 2.97a and 2.97b, are solved implicitly achieving second-order accuracy in time over the two steps, and the spatial terms are approximated by finite elements. In this paper the stability criteria are also derived [2.33].

Advantages of algorithm 2.97 include high accuracy and a relatively low computational effort. This assessment, made in Frind [2.33], coincides with our experiences. Figures 2.27 and 2.28 show two typical examples solved with SIMKA-2D: a plume of groundwater pollution originating from a wastewater channel, and the spreading of a tracer injected into an observation well located in the axisymmetric cone of depression of a pumping well.

2.2.5 EFFICIENT ALGORITHMS FOR TWO-DIMENSIONAL MIGRATION PROCESSES

Some of the best-known numerical methods for digital simulation of two-dimensional migration processes are

- the method of characteristics (MOC) (§2.2.5.1)
- the method of random walk (MORW) (§2.2.5.2)
- the finite element method (FEM) (§2.2.5.3)

As a basis for the following discussion, we will use the migration model according to Equation 2.65 and restated here:

$$\frac{\partial}{\partial x_i} D_{ij} \frac{\partial c}{\partial x_j} - v_i \frac{\partial c}{\partial x_i} = \epsilon \frac{\partial c}{\partial t} + \lambda c - w \qquad (2.98)$$

where $i = 1, 2$; $j = 1, 2$; $x_1 = x$; $x_2 - y$; $v_1 = v_x$; $v_2 = v_y$; and D_{ij} is the dispersion coefficient as a tensor of second order (see Table 2.4).

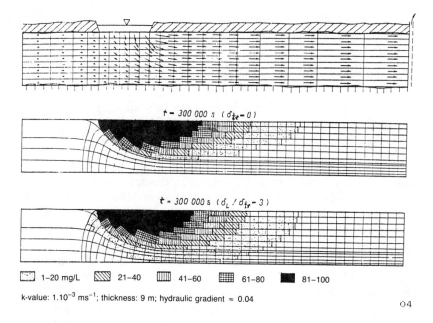

$t = 300\ 000\ s\ (\delta_{tr} = 0)$

$t = 300\ 000\ s\ (\delta_L / \delta_{tr} = 3)$

☐ 1–20 mg/L   ▨ 21–40   ▥ 41–60   ▦ 61–80   ■ 81–100

k-value: $1.10^{-3}\ ms^{-1}$; thickness: 9 m; hydraulic gradient ≈ 0.04

Figure 2.27. Groundwater pollution starting from a wastewater channel simulated by means of the computer program SIMKA-2D. *Source:* Gutt [2.48].

2.2.5.1 Method of Characteristics (MOC)

The method of characteristics subdivides concentration changes $\partial c/\partial t$ (or temperature changes $\partial T/\partial t$) into three components corresponding to convective transport, hydrodynamic dispersion (including diffusion or heat conduction), and conversion processes (including all source and sink terms). Accordingly, Equation 2.98 splits as follows:

$$\partial c/\partial t = (\partial c/\partial t)_1 + (\partial c/\partial t)_2 + (\partial c/\partial t)_3 \qquad (2.99)$$

$$(\partial c/\partial t)_1 = -(v_i/\epsilon)\partial c/\partial x_i = -u_i\partial c/\partial x_i \qquad (2.99a)$$

$$(\partial c/\partial t)_2 = \frac{\partial}{\partial x_i}\left(\frac{D_{ij}}{\epsilon}\frac{\partial c}{\partial x_j}\right) \qquad (2.99b)$$

$$(\partial c/\partial t)_3 = (\lambda/\epsilon)c - w/\epsilon \qquad (2.99c)$$

The boundary and initial conditions must also be split in the same manner.

Equation 2.99a is solved for each of the discrete field elements k,n in a time-discrete way, starting from

$$(\partial c/\partial t)_1 = (c_{k,n}^{t+\Delta t} - c_{k,n}^{t})_1/\Delta t \qquad (2.100)$$

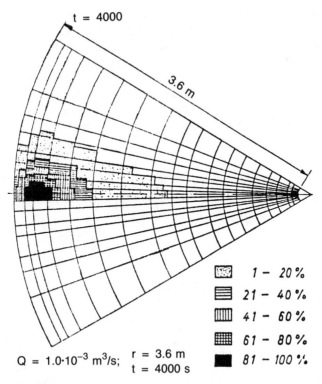

t = 4000

3.6 m

1 – 20 %
21 – 40 %
41 – 60 %
61 – 80 %
81 – 100 %

$Q = 1.0 \cdot 10^{-3}$ m³/s;   r = 3.6 m
t = 4000 s

Figure 2.28. Spreading of a tracer injected in an observation well located in the axisymmetric cone of depression of a pumping well solved by means of the computer program SIMKA-2D. *Source:* Gutt [2.48].

where $(c_{k,n}^t)_1$ is a known value. The unknown value $(c_{k,n}^{t+\Delta t})_1$ is determined by means of the walking point method.

Using this method, marked points are allowed to move with mean migration velocity $\vec{u} = v_x/\epsilon + v_y/\epsilon$ through the flow field, each of them representing the center of gravity of a specified amount of solute or heat. Therefore, the concentration $(c_{k,n}^{t+\Delta t})_1$ results from the number of walking points located in the field element k,n at time $t + \Delta t$, in relation to the fluid volume contained within this element.

Thus, the basic task consists of the determination of the position of each walking point at time $t + \Delta t$ with sufficient accuracy and minimal expenditure. For this purpose, the continuous velocity field $v_x = v_x(x,y)$ and $v_y = v_y(x,y)$ as solution of the flow problem must be known. These velocities are usually determined on the basis of either (1) a continuous potential function, such as piezometric height h(x,y), as an analytical solution of the flow problem, and its partial differentiation $v_x = -k\partial h/\partial x$ and $v_y = -k\partial h/\partial y$ under isotropic permeability conditions, or (2) spatially discrete potentials, such as h_{ij}, and the approximation of the terms $k\partial h/\partial x$ and $k\partial h/\partial y$ by finite differences, or on the basis of spatially discretized velocities $v_{x,m,n}$ and $v_{y,m,n}$, resulting from digital simulations of the flow problem, followed by interpolation. Figure 2.29 illustrates three different forms of spatially discrete velocities resulting from numerical solutions of the flow problem. The bilinear interpolation can be done by means of Equation 2.63c. With the interpolation values of v_x and v_y the spatial positions of the walking points at time $t + \Delta t$ can be determined as

$$x_{WP}^{t+\Delta t} \approx x_{WP}^{t} + \frac{\Delta t}{4\epsilon}[v_x^t(x_{WP}^t, y_{WP}^t) + v_x^t(x_{WP}^{t+\Delta t}, y_{WP}^{t+\Delta t}) \quad (2.101a)$$
$$+ v_x^{t+\Delta t}(x_{WP}^t, y_{WP}^t) + v_x^{t+\Delta t}(x_{WP}^{t+\Delta t}, y_{WP}^{t+\Delta t})]$$

$$y_{WP}^{t+\Delta t} \approx y_{WP}^{t} + \frac{\Delta t}{4\epsilon}[v_x^t(x_{WP}^t, y_{tWP}) + v_y^t(x_{WP}^{t+\Delta t}, y_{WP}^{t+\Delta t}) \quad (2.101b)$$
$$+ v_x^{t+\Delta t}(x_{WP}^t, y_{WP}^t) + v_y^{t+\Delta t}(x_{WP}^{t+\Delta t}, y_{WP}^{t+\Delta t})]$$

where $x_{WP}^{t+\Delta t}$ and $y_{WP}^{t+\Delta t}$ on the right-hand side are treated iteratively.

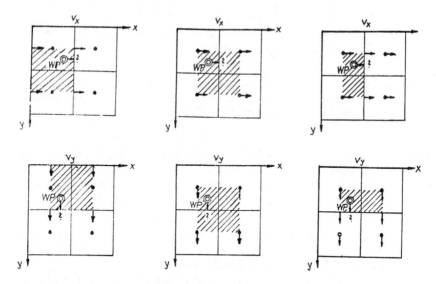

Figure 2.29. Bilinear interpolation of the spatial discretized velocity components v_x and y_y for given velocities at the element edges, given velocities at the element nodes, and given velocities at the element edges and nodes.

The procedure can also be reduced to a noniterative form by setting $x_{WP}^{t+\Delta t} \approx x_{WP}^t$ and $y_{WP}^{t+\Delta t} \approx y_{WP}^t$ on the right-hand side.

In any case, the time step length must be constrained in order to prevent walking points from moving longer distances per time step than the space step width. This constraint can be expressed as

$$\Delta t \leq \frac{\alpha \Delta x}{(v_x/\epsilon)_{max}} \quad \text{and} \quad \Delta t \leq \frac{\alpha \Delta y}{(v_y/\epsilon)_{max}} \qquad (2.102)$$

On the other hand, it can be shown that the paths walked per Δt should be greater than the distances between the walking points at their start position in order to avoid oscillation. For example, with four walking points starting per square element, the value of α in Equation 2.102 should be 0.75, and with nine points it should be 0.50 [2.54].

At heat or solute sources, new walking points must be started continuously; likewise, walking points arriving at sinks must be deleted. Occasionally, after a certain time period, a redistribution of starting positions becomes necessary.

Equation 2.99b is solved using an FD approach (see Table 2.4). At least for convectively dominant transport an explicit solution scheme ($\gamma = 0$ in Equation 2.73) is recommended. If, however, the stability criterion for the explicit solution scheme

$$\Delta t \leq \begin{array}{c} \text{min of} \\ \text{all elements} \end{array} \left(\frac{0.5}{\dfrac{D_{xx}}{(\Delta x)^2} + \dfrac{D_{yy}}{(\Delta y)^2}} \right) \qquad (2.103)$$

results in time steps Δt that are too small, a time-centered approach with $\gamma = 1/2$ may be more effective.

The solution of Equation 2.99c can be performed by an explicit FD scheme. Occasionally, better accuracy is obtained if a time-centered scheme ($\tau = 1/2$) is used. A constraint on Δt can result here because the source-sink term must not withdraw a larger amount of heat or solute than the amount which is stored within the element. This important constraint can be expressed as

$$\Delta t \leq \begin{array}{c} \text{min of} \\ \text{all elements} \end{array} \left(\frac{\epsilon}{|\lambda - w/c|} \right) \qquad (2.104)$$

Provided explicit schemes are used for the solution of Equations 2.99b and 2.99c, the MOC algorithm does not require a solution of simultaneous migration equations. Hence, this algorithm is fast. However, a disadvantage exists in the great organizational expenditure needed, which tends to make the computer code rather lengthy.

The computer programs presented in Konikow and Bredehoeft [2.54] and Victor [2.95] are typical examples of MOC programs. Both programs use implicit FD approximations for representing the flow problem, and the ADI method for solving the system of difference flow equations.

As the test examples in Konikow and Bredehoeft [2.54] show, the best results are achieved with nine walking points starting per square element. The solute or heat balance calculation after each time step has proven valuable in monitoring the digital simulation procedure. The MOC may also be applied to simulations of three-dimensional migration problems as shown, for instance, in Mehlhorn [2.67]. Further refinements of the MOC decomposition based on a combined Euler/Lagrange approach are described in Neumann [2.73] and Neumann and Sorek [2.74], and in modified form in Lever and Rae [2.56].

2.2.5.2 Method of Random Walk (MORW)

The method of random walk is a technique that can be considered as a refinement of the MOC. Figure 1.36 illustrates its basic principle. The MORW, just as the MOC, is based on decomposition of the migration model into components. The convective transport component is simulated just as with the MOC, using Equation 2.101. However, the spatial positions of the walking points determined in this way are now regarded as expected values in the statistical sense. The real positions of these points, which are calculated from the relations given in Figure 1.36, scatter around the expected positions. Thus, with the MORW a stochastic representation of solute or heat transport due to convection and dispersion replaces the deterministic approach of the MOC expressed by the Equations 2.99a and 2.99b.

The impact of exchange and transformation processes as well as the impact of sources and sinks reflected by Equation 2.99c must be taken into account in the MORW procedure by correction of the solute or heat amount carried by each walking point. This can be achieved for walking point k in the simplest form by:

$$n_k^{t+\Delta t} = n_k^t - \frac{\Delta t \lambda n_k^t}{\epsilon} + \frac{\Delta t w \Delta V}{\epsilon K}$$ (2.105)

where n = solute or heat amount carried by the walking point k
   ΔV = volume of the element in which the kth walking point is located at time $t + \Delta t$
   K = number of walking points within the element at that time

On purely computational grounds, the MORW has the following advantages over the MOC:

1. The solution of Equation 2.99b is omitted and replaced by a faster program module based on a random generator.
2. In using the MORW, spatial discretization is only necessary for the

numerical solution of the flow problem and for the solution of Equation 2.105. Therefore, it is subjected to fewer restrictions.

3. The concentration distribution, which in the MOC must be determined at the end of each time step, is calculated only at the target times for printing.

4. The MORW easily allows the growth of the hydrodynamic dispersivity with the distance covered by the migrants to be taken into account (see Figure 1.33).

5. The MORW guarantees a minimum expenditure if the migration of several (also coupled) migrants is simulated.

6. If an example is computed repeatedly, the stochastic character becomes apparent, so that it may be advantageous, for instance, to solve the problem ten times with 200 WP (walking points) instead of once with 2000 WP.

7. Due to the first three reasons the MORW requires less computing time than the MOC.

On the other hand, there are also the following drawbacks:

1. A relatively large number of WP or repetitions is necessary to guarantee a sufficiently smooth concentration or temperature function.

2. Occasionally the simulated concentrations or temperatures may become greater than those of the boundary conditions, especially if a coarse discretization was chosen.

The programs presented in Prikett et al. [2.84] and Gutt [2.49] provide two typical examples of MORW programs. In both cases, an implicit FD approximation based on a rectangular space discretization, and the ADI method were used to solve the flow problem. This approach seems to be very compatible with the MOC and MORW in terms of accuracy.

Figure 2.30 shows a test example selected from Prikett et al. [2.84]. This example, which considers the injection of a pulse of conservative migrants into a steady-state parallel groundwater flow system, was already examined in §1.3.2 (see Equations 1.35 to 1.37) and in Figure 1.35. With Equation 1.36, an analytical solution is given, providing an accurate reference basis to verify the numerical solution of the selected example.

Figure 2.30 shows the distribution obtained at 120 days (20 time steps) with 500 and 2000 walking points, respectively, started at time zero. A discrete grid of 10 m × 10 m and migration parameters of u = 1 m/day, $\delta_\ell = 4.5$ m, and $\delta_{tr} = \delta_\ell/4$ are used in computing this example. The number of WP at time t is shown at each grid point of this figure. Seen from the standpoint of a geologist or engineer, the degree of approximation of the theoretical solution appears to be sufficient for only 500 WP. Because the MORW solution reflects the stochastic char-

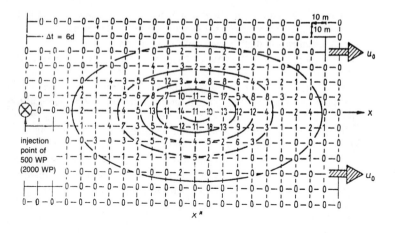

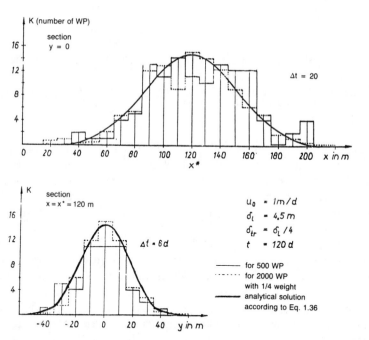

Figure 2.30. Comparison of MORW solutions for a point-shaped and pulse-shaped tracer injection into a steady-state groundwater flow field according to Prikett et al. [2.84] with the analytical solution according to Equation 1.36 with $c_o V_o = 500$ (see also Figure 1.35) .

acter of the migration process, this solution may be considered to be more realistic than a smooth solution, which only gives the appearance of a high degree of accuracy.

2.2.5.3 Finite Element Method (FEM)

The FEM algorithms used today for digital simulation of migration processes are based almost exclusively on Galerkin's formulation (see §2.2.2, Equation 2.68, and the following equations). The computer program FEFLOW [2.22; 2.20; 2.26] is regarded as one of the well-tested and established computing programs and is used here as an example.

FEFLOW was developed as a flexible research computer program. It calculates the flow velocity field as well as the concentration or temperature field by means of the FEM. A choice of element shapes is provided (see Table 2.5), and the variables (h, c, or T) as well as the natural variables (h, v_x, v_y, c, or T) may be used as primary unknowns [2.26].

For the time approximation, FEFLOW offers the options to use a fully implicit scheme ($\gamma = 1$) or the Crank-Nicolson scheme ($\gamma = 1/2$) with $O(\Delta t)$ and $O(\Delta t^2)$ order accuracy, respectively. Since for $\gamma = 1$ significant numerical dispersion can be expected, which may be characterized by $\Delta D = \Delta t v^2/(2\epsilon)$ (compare Equation 2.76a), the time step length Δt must be limited accordingly. Therefore, the less stable but more accurate Crank-Nicolson scheme is used during time periods in which no abrupt changes of the boundary conditions occur.

The coupled set of equations which finally results from spatial and temporal discretization of the migration model (see Equation 2.65) is of the form:

$$[A] \{X(t + \Delta t)\} = [B] \{X(t)\} + \{F(t)\} \qquad (2.106)$$

where the subvector $X_i = (h_i, v_{x,i}, v_{y,i}, c)$ is formulated at point i, [ ] is the matrix symbol, and { } is the vector symbol.

The matrix Equation 2.106 is asymmetrical. This equation is solved by means of frontal and profile techniques [2.20]. The system characteristics entering the matrix or vector elements of [A], [B], or [F] are taken element-wise constant. The double integrals are solved numerically by means of a 3×3 Gaussian quadrature scheme. The computing code of FEFLOW is implemented on a mainframe computer, but an IBM PC AT/XT version has also been developed.

Figure 2.31 shows an example, solved by means of FEFLOW, which involves the characteristic region of a bank filtration system with a double well gallery [2.26]. The upper part of the figure shows the finitization of the region with 96 quadrilateral elements and 347 nodes, resulting in a total of 1168 unknowns of the variables h_i, $v_{x,i}$, $v_{y,i}$, and c_i.

The middle part of Figure 2.31 shows a computer plot of the velocity vectors which characterize the steady-state flow field under the conditions that the wells A and B each are pumping at a constant rate of 36

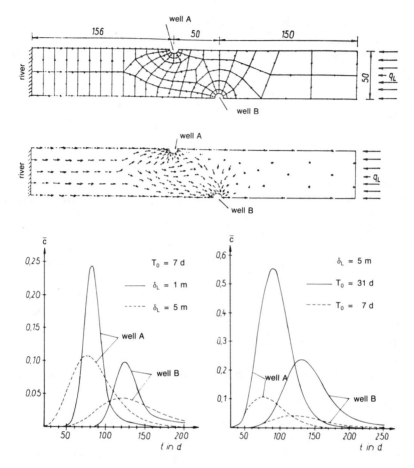

Figure 2.31. Migration of pollutants to a double well gallery of a bank filtration plant.

m³/hr, the water level at the left-hand boundary is kept constant (BC of the first type, river), and that an influx of $q = 1.72 \cdot 10^{-5}$ m²/s occurs at the right-hand boundary (BC of the second type, groundwater inflow from the land side). The aquifer of this example was considered homogeneous and isotropic ($k_o = 2.5 \cdot 10^{-4}$ m/s, $M = 20$ m, $\epsilon = 0.36$).

The lower part of Figure 2.31 shows pollutant breakthrough curves $\bar{c}_A = c_A/c_1$ and $\bar{c}_B = c_B/c_1$ at the two wells A and B, corresponding to inputs at the left-hand (river) boundary with a duration of $T_o = 7$ days or 31 days, respectively. The curves are obtained at a radial distance of 5 m from the wells. The dispersion parameters are $\delta_\ell = 1$ m and 5 m, respectively, and δ_{tr} is used as $\delta_\ell/20$. The storage capacity of the aquifer with respect to the pollutant integrated over the aquifer thickness is $\bar{\epsilon} = 5$ m, and the time step length is $\Delta t = 1$ day in this example.

2.3

Parameter Estimation

The inverse problem of parameter estimation concerns the optimal determination of parameters (constants) which are imbedded in the mathematical process model and in its initial and boundary conditions. Solving an inverse migration problem requires knowledge of the dependent variables c (concentration) or T (temperature) at discrete points in space and time, and also knowledge of certain model parameters. Observations and measurements are the basis of this knowledge.

This chapter focuses on practical methods which estimate lumped parameters of migration models. These same methods are commonly used in today's pump test analysis to estimate lumped parameters of flow models. Starting from a more general approach and algorithm of parameter estimation in §2.3.1, examples are presented applying the straight-line method (§2.3.2), the method of type-curve matching (§2.3.3), and the method of minimizing objective functions (§2.3.4) for estimating dispersive and retardation properties in practical migration problems. The inverse problem of distributed parameter estimation is not considered in this book; the reader is referred to more specialized textbooks, such as Willis and Yeh [2.97, Chapter 8].

2.3.1 FUNDAMENTAL APPROACH AND ALGORITHM

The estimation of unknown migration parameters is based upon

- a "historical" migration process recorded by measuring devices (PRM), which provides the required data for the dependent variables such as solute concentrations c, temperatures T, mass fluxes (c$\overset{\bullet}{V}$), and in

some cases for a few of the migration parameters, such as the coefficients of dilution and degradation λ, dispersivity δ, or Darcy's velocity v
- a **m**athematical migration **p**rocess **m**odel formed by forward modeling (MPM)
- an inverse model derived from this MPM, known as the **p**arameter **i**dentification **m**odel (PIM)

Figure 2.32 shows this triangular relation of the elements PRM, MPM, and PIM which is actively formed and influenced by the model user and builder (MUB), the person or group of persons building and using the PIM under specified aims. Model users and builders, hence, stay at a higher hierarchical level.

Practical solutions of inverse problems are complicated due to the complex structure illustrated in Figure 2.32. Decoupling procedures (algorithms) are required to simplify the structure of the problem. Figure 2.33 shows a flow chart of such an algorithm as established in Beims et al. [2.9]. This algorithm is comprised of six main activities, or steps (see Figure 2.33), which are discussed in the following six sections 2.3.1.1 to 2.3.1.6, and then applied to the solution of the practical examples in §2.3.2 to §2.3.4.

2.3.1.1 Formulation of the MPM (Step 1)

Fundamentals concerning the formulation of the mathematical process model of migration, known as *forward modeling,* and the analytical

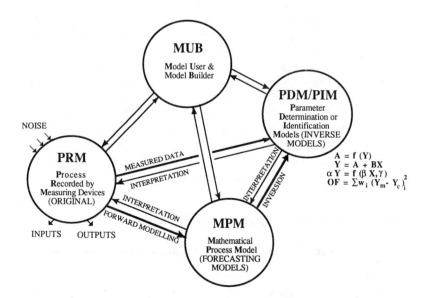

Figure 2.32.   Tetrahedral relation of parameter determination or identification.

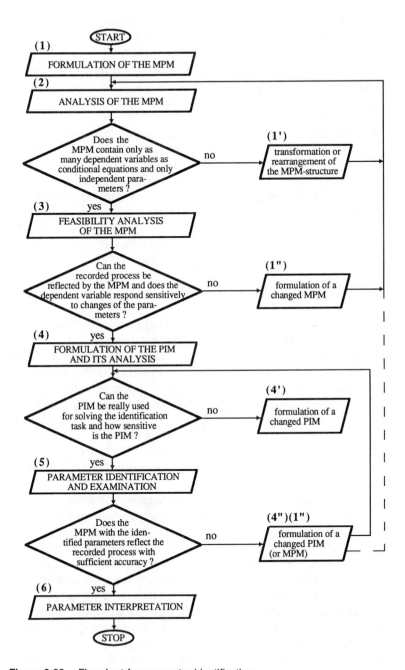

Figure 2.33.   Flowchart for parameter identification.

or numerical solutions of these mathematical models are described in the preceding sections of this book. However, some peculiarities of the MPM formulation must be considered under the objectives of the inverse problem of parameter estimation. Migration processes used in parameter estimation are either (1) generated in laboratory, pilot, or field test plants or (2) proceed as naturally occurring or human-induced processes and are only used for the goal of parameter estimation by the modeler. In each of these cases the estimation of migration parameters is based upon a reliable process-measuring technique providing sufficiently accurate data of the dependent variables used in the MPM. Nevertheless, some noise corruption of observations is unavoidable and must be considered.

The modeler must have sufficient experience in order to formulate a reliable MPM which is as simple in its structure as possible, and which contains a minimum number of parameters. Such formulated MPMs reduce the danger of nonuniqueness and instability of commonly ill-posed inverse problems of parameter estimation.

2.3.1.2 Analysis of the MPM (Step 2)

The MPM formed according to §2.3.1.1 usually contains more natural parameters than determinable or identifiable parameters. Determinable or identifiable parameters are often called *independent parameters*. This required parameter property is given if unique and reversible functional relations exist between individual parameter values a_i and the values of the dependent variable y_j at specified spatial and temporal points xj:

$$y_j = f(a_1, a_2, \ldots, a_n) \text{ direct task} \qquad (2.107a)$$

$$a_i = F(y_1, y_2, \ldots, y_k) \text{ inverse task} \qquad (2.107b)$$

where F should express an explicit or implicit relation for a_i.

If Equation 2.107a does not contain only independent parameters, an infinite number of parameter value combinations may result for the same value of y_j. The reduction of the number of parameters in an MPM which is expressed in form of a differential equation is frequently required by substitution of fictitious parameters for groups of natural parameters. A comparable lumping process takes place when those differential equations are solved analytically. In other words, an analytical solution may have different parameters than its governing differential equation. Independent parameters in differential equations may be obtained by generating at least one parameter-free derivative for each dependent variable of the MPM by division, and replacing occurring combinations of natural parameters, such as sums, products, etc., by fictitious parameters. After transformation (for example, Laplace transformation), the independence of parameters in the image space must again be guaranteed.

2.3.1.3 Feasibility Analysis of the MPM (Step 3)

The feasibility analysis of the MPM must show that the MPM is capable of reflecting the measured data with meaningful parameter values, and that the dependent variable in the domains covered by the measured data is sensitive to parameter changes. Feasibility analyses are performed by scenario analyses in which meaningful and possible variants or scenarios of parameter sets and boundary conditions are specified and the completed MPM is solved. The results obtained are compared with the measured data. Dimensionless variables and parameters often simplify this comparison. The feasibility of a MPM is only assured when at least one of the variants approximately reflects the measured data points. If not, a new MPM must be formulated.

Parameter scenarios are also used to delineate the spatial and temporal ranges or domains in which the dependent variable is sensitive to changes of the independent parameters which are to be estimated. The sensitivity may be expressed quantitatively by sensitivity coefficients, which are defined as the partial derivative of the dependent variable with respect to each of the parameters ($\partial \phi / \partial P$). The feasibility of parameter estimation is only given if the sensitive domains are covered by measured data. If not, new observations are required.

The knowledge of sensitive ranges of an MPM frequently allows simplifications of the parameter estimation problem because only parameters which are involved in each domain may be estimated simultaneously. The parameter estimation method which makes use of such decomposition is known as *step parameter search*.

2.3.1.4 Formulation of the PIM and its Analysis (Step 4)

The simplest parameter estimation model provides a parameter A which is to be estimated explicitly in dependence on the measured data of the dependent variable Y:

$$A_i = f(Y) \tag{2.108}$$

The calculation of the parameter value A in this ideal case is designated as *parameter determination*. The general term *parameter identification* is used if such a direct inversion of the MPM is impossible and an explicit governing equation for A cannot be derived.

In the implicit case, the simplest parameter identification model PIM is the straight line:

$$Y = A + BX \qquad\qquad (2.109a)$$

where Y is the dependent and X the independent variable in the transformed space, and A and B are the independent parameters of the PIM.

The derivation of a straight-line PIM presumes the existence of an analytical closed-form solution of the MPM. Such solutions of migration processes are described in Chapter 2.1. It is of major importance that analytical solutions do not require the knowledge of parameter values for their derivation.

Straight-line PIMs are usually obtained by transformations of analytical solutions. These transformations may include the functional transformation of the dependent variable, the functional transformation of the independent variable, truncations of solution parts, approximate substitutions, etc. However, straight-line PIMs may only be formed if the MPM has only one dependent and one independent variable, and only two independent parameters which pass over to A and B in the straight-line equation. It may be desired from a theoretical point of view that Y contains only the dependent variable of the MPM, and X contains only the independent variable of the MPM.

Type curves are used as a PIM when a straight-line PIM cannot be derived. A type-curve PIM with the three parameters α, β, and γ is the standard case and has the following structure:

$$\alpha\eta = f(\beta\xi, \gamma) \qquad\qquad (2.109b)$$

where $\alpha\eta$ is the dependent and $\beta\xi = \beta^*$ the independent variable of the PIM.

The formulation of a PIM as type curve or type-curve set is advisable if the MPM has one dependent and one independent variable and only two or three independent parameters which are to be identified. Only one of the independent parameters should be linked with the dependent variable, and only one with the independent variable. The identification of the two parameters linked to the corresponding variables is obtained by axis-shifting techniques, and the identification of the third parameter by selection of that type curve from the plotted set that fits the data points best. This procedure of axis-shifting is known as *type-curve matching*.

The formulation of a PIM as an optimization algorithm is obtained in the most simple case by linking an objective function (OF) with a minimization algorithm. An objective function assesses the difference between the observed and calculated values of the dependent variable in the MPM and contains the parameters as independent variables. In the simplest form an objective function PIM may be expressed as follows:

$$OF = \sum_{i=1}^{N} w_i(Y_{m,i} - Y_{c,i})^2 = f(parameters) \stackrel{!}{=} min \qquad (2.109c)$$

where $Y_{m,i}$ = measured data of the dependent variable
$\quad\quad\quad Y_{c,i}$ = value calculated by means of the MPM at the spatial and time point i
$\quad\quad\quad w_i$ = weight, often specified corresponding to the transformed axial intercepts represented by the point i

It must be shown that the objective function OF is sensitive to and responds to changes in the parameters to be determined. The delineation of sensitive parameter domains is also of major importance in this approach. The use of objective functions as PIMs is recommended when the solution of the MPM requires the use of a computer, and three to perhaps six independent parameters are to be identified. Some people prefer objective function PIMs because these PIMs provide solutions which are less subjective than the solutions obtained by straight-line or type-curve matching.

Several algorithms have proven successful in minimizing objective functions (see, e.g., Luckner and Beims [2.58] and Beims [2.7]). A parameter set is regarded as identified if the objective function reaches its "global" minimum. Three-dimensional computer graphics of OF = $f(a_1,a_2)$ are useful tools in illustrating the minimization problem more effectively.

If more than five to six parameters per sensitivity domain have to be identified, a more demanding PIM must be formulated [2.97; 2.51], or the rigorous, tedious, and time-consuming trial-and-error method must be used. However, convincing success has rarely been achieved in solving practical problems with any method if many parameters are to be identified.

2.3.1.5 Parameter Identification and Examination (Step 5)

Parameter determination according to Equation 2.108 requires no further explanation. Unfortunately, parameter determination can be achieved only rarely, and the use of identification procedures is unavoidable.

Parameter identification in this more limited context means the solution of a parameter identification model PIM by corresponding techniques such as regression analysis (e.g., for the PIM 2.109a), matching techniques (e.g., for the PIM 2.109b), and minimization procedures (e.g., for PIM 2.109c). The result of a PIM solution are the PIM parameters, for instance A and B when the PIM 2.109a is considered. From these parameters the values of the independent parameters of the MPM must be recalculated. These MPM parameters should be examined by solving the MPM with these parameter values. The solution obtained may now be compared graphically with the measured data. In graphical representations of discrete measured data and discrete calculated values, the same interpolation technique should be used.

2.3.1.6 Parameter Interpretation (Step 6)

The original physical, chemical, and biological process parameters of the PRM and its MPM must be estimated by interpretation of the identified independent parameters of the MPM. For this task, further information or approximations are usually necessary in addition to the equations defining the independent parameters. Such additional information is usually obtained by laboratory and field tests or by theoretical considerations.

2.3.2 EXAMPLES USING THE STRAIGHT-LINE METHOD

The required assumptions for reasonable applications of the straight-line method were discussed in §2.3.1.4. Two typical examples of illustration and application of the fundamental approach and algorithms discussed in the previous section will be presented in this section.

2.3.2.1 Example A

Formulation of the MPM

In a completely penetrating well, water is injected at a constant rate Q with a constant migrant concentration c_w (e.g., $c_w = 0$) until quasi-stationary conditions are reached in the monitored area characterized by $\partial h/\partial t|_w \approx \partial h/\partial t|_{mw}$ and $c_w \approx c_{mw}$. In this context w symbolizes the injection well and mw the monitoring well. Then a concentration c_o which is significantly different from the initial concentration $c_I = c_w$ is generated in the well where Q remains unchanged. At the well-aquifer interface the concentration $c(r_o) = c_o$ is assumed to be constant over the whole aquifer thickness. If we can assume that the considered migrant is nonreactive, Equation 2.48a would be the MPM and Equation 2.49a its analytical approximate solution. Both equations with the natural parameters δ (dispersivity) and $\bar{\epsilon}$ (migration storage coefficient $\bar{\epsilon} = \epsilon M$, where M is the aquifer thickness) are again given here:

$$\frac{\partial}{\partial r}\left(Q\delta\,\frac{\partial c}{\partial r}\right) - \frac{\partial}{\partial r}(Qc) = 2\pi r\bar{\epsilon}\,\frac{\partial c}{\partial t} \qquad (2.110a)$$

$$\bar{c}(r,\ t) = \frac{c(r,t) - c_I}{c_o - c_I} \approx 0.5\ \mathrm{erfc}\left[\frac{r^2 - r_F^2}{\sqrt{16\delta r_F^3/3}}\right] \qquad (2.110b)$$

where $r_F = Qt/(\pi\bar{\epsilon})$.

Analysis of the MPM

The MPM according to Equation 2.110a has two independent parameters

$$\chi_1 = Q\delta/(2\pi\bar{\epsilon}) \text{ and } \chi_2 = Q/(2\pi\bar{\epsilon}) \qquad (2.111)$$

which result when the derivative $\partial c/\partial t$ in Equation 2.110 is made parameter-free by division. The natural and the independent MPM parameters are then related as follows:

$$\bar{\epsilon} = Q/(2\pi\chi_2) \text{ and } \delta = \chi_1/\chi_2 \qquad (2.111a)$$

Feasibility Analysis of the MPM

For $r = 5$ m the dispersivity δ may be initially estimated to be 0.2 m according to Figure 1.33. The best guess for $\bar{\epsilon}$ is assumed to be $\bar{\epsilon} = M\sigma_f$ = 5 m, where σ_f in this case is the volumetric water content of the aquifer (see Equation 1.78b). For an injection rate of $Q = 3$ L/s, the following parameter scenarios appear reasonable:

$$\bar{\epsilon} = \{3, 5, 7\} \text{ m} \qquad \Rightarrow \quad \chi_2 = \{13.75, 8.25, 5.89\} \text{ m}^2/\text{day}$$

$$\delta = \{0.1, 0.2, 0.3\} \text{ m} \Rightarrow \quad \chi_1 = \{0.1, 0.2, 0.3\} \chi_2 \text{ m}^3/\text{day}$$

from which nine solutions of the MPM result. In Figure 2.34a these solutions are plotted together with the measured data. This graph demonstrates the feasibility of the MPM chosen. Plotting such scenarios prior to the measurements is recommended in order to design reasonable sampling schedules. From the graph the sensitive domains may be assessed also. As shown, in the concentration range $0.4 < \bar{c} < 0.6$, only $\bar{\epsilon}$ can be identified, and in the ranges $\bar{c} > 0.6$ and $\bar{c} < 0.4$, both parameters $\bar{\epsilon}$ and δ influence $\bar{c}$ significantly.

Formulation of the PIM and its Analysis

First it is necessary to transform the MPM according to Equation 2.110 in such a way that the independent parameters and independent variables forming the argument of the erfc function are set free:

$$\text{inv erfc } [2\bar{c}(r,t)] = \frac{r^2 - r_F^2}{\sqrt{\dfrac{16}{3}\delta r_F^3}} = \frac{r^2 - 2\chi_2 t}{3.88 \sqrt{\chi_1} \sqrt[4]{\chi_2} \, t^{3/4}} \qquad (2.112)$$

In the simplest form, the following straight-line equation results as the PIM for $r = R = $ constant [2.80]:

$$Y = A + BX \qquad (2.113a)$$

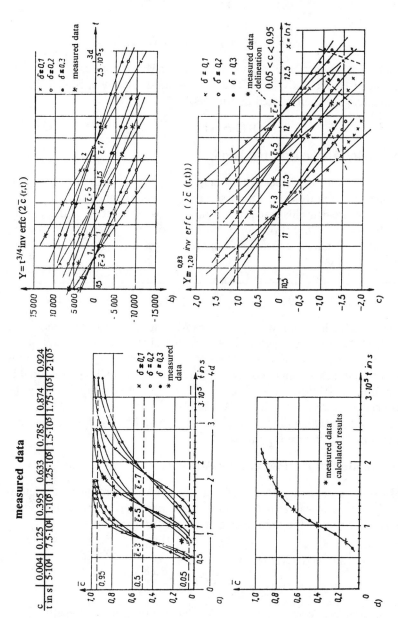

Figure 2.34. Steps of parameter identification in using the straight-line method for Example A.

where

$$Y = t^{3/4} \text{ inv erfc}(2\bar{c}(R,t)) \text{ and } X = t \qquad (2.113b)$$

$$A = \frac{R^2}{3.88 \sqrt{\chi_1} \sqrt[4]{\chi_2}} \quad \text{and} \quad B = -A \frac{2\chi_2}{R^2} \qquad (2.113c)$$

Since this PIM occasionally leads to some confusion due to the natural independent variable t entering both the new independent variable X and the new dependent variable Y, the following approximate equation may be suggested:

$$\frac{R^2 - 2\chi_2 t}{3.88 \sqrt{\chi_1} \sqrt[4]{\chi_2} \, t^{3/4}} = \frac{1}{4} \sqrt{\frac{R}{\delta/3}} \left(\frac{1-\eta}{\eta^{3/4}} \right) \approx$$

$$
\begin{cases}
\dfrac{-1.2}{4} \sqrt{\dfrac{R}{\delta/3}} \ln \eta \text{ for } 0.05 < \bar{c} < 0.50 \\[3mm]
\dfrac{-0.83}{4} \sqrt{\dfrac{R}{\delta/3}} \ln \eta \text{ for } 0.50 < \bar{c} < 0.95
\end{cases}
\qquad (2.113d)
$$

where $\eta = t/t_{50}$
$t_{50} = \pi R^2 \bar{c}/Q = R^2/(2 \chi_2)$

The variables of the PIM 2.113d are then the following:

$$Y = \begin{cases} 0.83 \text{ inv erfc } (2\bar{c}(R,t)) \text{ for } \bar{c} < 0.5 \\[2mm] 1.20 \text{ inv erfc } (2\bar{c}(R,t)) \text{ for } \bar{c} > 0.5 \end{cases}$$

$$X = \ln t \qquad (2.113e)$$

$$B = -\frac{1}{4} \sqrt{\frac{R}{\delta/3}} = -\frac{1}{4} \sqrt{\frac{3R}{\chi_1/\chi_2}} \quad \text{and} \quad A = -B \ln\frac{R^2}{2\chi_2} \qquad (2.113f)$$

For the function inv erfc(2c(R,t)), the following approximate function is very suitable, which may be solved on a programmable pocket calculator or computer [2.89]:

$$Y = \text{inv erfc}(2\bar{c}(R,t)) = \frac{-1}{\sqrt{2}} \left| \frac{c_0 + c_1 P + c_2 P^2}{1.0 + d_1 P + d_2 P^2 + d_3 P^3} - P \right| + \text{res}$$

$$P = \sqrt{-2 \ln (2\bar{c}(R, t))} \quad \text{and} \quad |res| < 0.00045$$
$$c_o = 2.515517 \qquad d_1 = 1.432788$$
$$c_1 = 0.802853 \qquad d_2 = 0.189269$$
$$c_2 = 0.010328 \qquad d_3 = 0.001308 \qquad (2.114)$$

The analysis of the PIM 2.113a with the two different interpretations of the variables and parameters according to Equations 2.113b and 2.113c, and Equations 2.113e and 2.113f, should be done graphically (ordinate Y, abscissa X). The measured data are required to fall approximately along a straight line (in the second case only in the range of $0.05 < \bar{c} < 0.95$ — see Equation 2.113d). The sensitive domains of the natural parameters $\bar{\epsilon}$ and δ in the transformed space of the PIM may become obvious when the nine variants are calculated by means of Equation 2.110b and the resulting straight lines are plotted (see Figures 2.34b and 2.34c).

Parameter Identification, Examination, and Interpretation

From the best-fitting straight line drawn subjectively through the measured data in Figure 2.34b or 2.34c, one obtains the following PIM parameters:

$$A = Y(0) \text{ and } B = (Y_o - A)/X_o \qquad (2.115)$$

where (Y_o, X_o) is an arbitrary point at the fitted straight line. The formal parameters χ_1 and χ_2 are determined from Equation 2.113b or 2.113f as

$$\chi_2 = -BR^2/(2A) \qquad \chi_1 = R^4/(3.88^2 \, A^2 \sqrt{\chi_2})$$
or
$$(2.116)$$

$$\chi_2 = 0.5 \, R^2 \exp(A/B) \qquad \chi_1 = 3R\chi_2/(16B^2)$$

and the natural parameters $\bar{\epsilon}$ and δ are determined from Equation 2.111a as follows:

$$\bar{\epsilon} = Q/(2\pi\chi_2) \text{ and } \delta = \chi_1/\chi_2$$

Thus for the fitted straight lines in Figures 2.34b and 2.34c, one obtains:

	A	B	χ_1	χ_2	$\bar{\epsilon}$	δ
Figure 2.34b	$11 \cdot 10^3$	−0.100	$0.32 \cdot 10^{-4}$	$1.14 \cdot 10^{-4}$	0.28	4.19
Figure 2.34c	20.78	−1.790	$0.33 \cdot 10^{-4}$	$1.14 \cdot 10^{-4}$	0.29	4.15

If the identified values of $\bar{\epsilon}$ and δ are inserted into the MPM 2.110b, a function results which reflects the measured data very well (see Figure 2.34d).

This example shows that both PIMs (Equations 2.113b/c and 2.113e/f) prove to be of equal practical value. Due to its simpler structure, fewer restrictions, and its general nature, the PIM 2.113b/c is recommended for practical applications. In a varied form it also allows spatial and time-integrated parameter estimation where all data points are used in one PIM. In this case, Equation 2.113a is transformed as follows:

$$\text{inv erfc } (2\bar{c}(r, t)) = \frac{1 - 2\chi_2 t/r^2}{3.88 \sqrt{\chi_1} \sqrt[4]{\chi_2} \, t^{3/4}/r^2} \tag{2.117}$$

and the variables and parameters of the straight-line PIM are obtained as follows:

$$
\begin{array}{ll}
Y = \text{inv erfc } (2\bar{c}(r, t)) \, t^{3/4}/r^2 & X = t/r^2 \\
A = 1/(3.88 \sqrt{\chi_1} \sqrt[4]{\chi_2}) & B = -2A\chi_2 \\
\chi_2 = -B/(2A) & \chi_1 = 1/(3.88^2 \, A^2 \sqrt{\chi_2})
\end{array} \tag{2.118}
$$

2.3.2.2 Example B

Formulation of the MPM

A one-dimensional migration process is considered in this example, which can be reflected by Equation 2.39a as MPM and, for instance, by Equation 2.39e as its analytical solution. Both model equations are again given here, where c is concentration, $DM = \bar{D} = q\delta$, and the remaining parameters are defined as in Equations 2.13 and 2.39e (definition of A and B):

$$\bar{D} \, \frac{\partial^2 c}{\partial x^2} - q \, \frac{\partial c}{\partial x} = \bar{\epsilon} \, \frac{\partial c}{\partial t} + \bar{\lambda} c - \bar{w} \tag{2.119a}$$

$$c(x, t) = \frac{\bar{w}}{\bar{\lambda}} + \left(c_1 - \frac{\bar{w}}{\bar{\lambda}}\right) A(x, t) + \left(c_o - \frac{\bar{w}}{\bar{\lambda}}\right) B(x, t) \tag{2.119b}$$

Analysis of the MPM

The analysis of Equation 2.119a reveals four independent parameters, for example,

$$\bar{D}/\bar{\epsilon}, q/\bar{\epsilon}, \bar{\lambda}/\bar{\epsilon}) \quad \text{or} \quad (\bar{D}/\bar{\lambda}, q/\bar{\lambda}, \bar{\epsilon}/\bar{\lambda}, \bar{w}/\bar{\lambda})$$

(see also Equation 2.119b). Therefore, if q is known from flow field analysis, all other natural parameters of the model Equation 2.119a, $\bar{D}$, $\bar{\epsilon}$, $\bar{\lambda}$, and $\bar{w}$, may be estimated.

Feasibility Analysis of the MPM

As the investigations of Equation 2.39 in §2.1.4.4 showed, several approximate solutions hold for specific ranges of the independent variable t. They also characterize the sensitive time domains of the parameters. If, for instance, Figures 2.1 and 2.12 are considered, three sensitive domains may be delineated for the MPM according to Equation 2.39e or 2.119b, which are also shown in Figure 2.35d:

Domain I   MPM: $c(t) \approx \bar{w}/\lambda + (c_I - \bar{w}/\lambda) \exp(-t\lambda/\bar{\epsilon})$
Eq. 2.16b   sensitive parameters: $\bar{w}/\lambda$, λ/ϵ

Domain II   MPM: $c(t) \approx c_I + 0.5(c_2 - c_I) \operatorname{erfc} \dfrac{x_o - tq/\bar{\epsilon}}{2\sqrt{t\bar{D}/\bar{\epsilon}}}$
Eq. 2.33h   sensitive parameters: $\bar{D}/\epsilon$, q/ϵ

Domain III   MPM: $c(t \to \infty) = \bar{w}/\lambda + (c_o - \bar{w}/\lambda) \exp(-x_o\lambda/q)$
Eq. 2.17c   sensitive parameters: $\bar{w}/\lambda$, λ/q
for $t \to \infty$

The "step search" identification procedure starts with domain I, followed by III, and ends up with II. The sensitivity of the four parameters is illustrated by solving Equation 2.39e (identical with 2.119b) for the following five scenarios (see also Figure 2.35a):

$$c_I = 2 \text{ g/m}^3 \quad c_o = 12 \text{ g/m}^3 \quad q = 4 \cdot 10^{-4} \text{ m}^2/\text{s} \quad x = 350 \text{ m}$$

Scenario	0	1	2	3	4	
$\bar{w}$	$4 \cdot 10^{-6}$	$6 \cdot 10^{-6}$	$4 \cdot 10^{-6}$	$4 \cdot 10^{-6}$	$4 \cdot 10^{-6}$	g m^{-2} s^{-1}
λ	$8 \cdot 10^{-7}$	$8 \cdot 10^{-7}$	$6 \cdot 10^{-7}$	$8 \cdot 10^{-7}$	$8 \cdot 10^{-7}$	m/s
$\bar{\epsilon}$	40	40	40	50	40	m
D	$3 \cdot 10^{-3}$	$3 \cdot 10^{-3}$	$3 \cdot 10^{-3}$	$3 \cdot 10^{-3}$	$1 \cdot 10^{-3}$	m^3/s

where scenario 0 reflects the best guess or expected values of the parameters.

Formulation of the PIM and its Analysis

The PIMs of the three domains have the following form:

(I) $\underbrace{\ln (\bar{w}/\lambda - c)}_{\mathbf{Y}} = \underbrace{\ln (\bar{w}/\lambda - c_I)}_{\mathbf{A}} - \underbrace{(\lambda/\epsilon)}_{\mathbf{B}} \underbrace{t}_{\mathbf{X}}$ 　　　(2.120a)

Domain I must be sufficiently large in order to obtain a straight-line representation. Domain I may be expanded by a domain I* if measured data from a monitoring point with $x_m > x_o$ are available (see Figure 2.35d, where in the example $x_o = 350$ m and $x_m = 1000$ m).

Different $\bar{w}/\lambda$ values were used to calculate the function $Y = Y(t)$. The

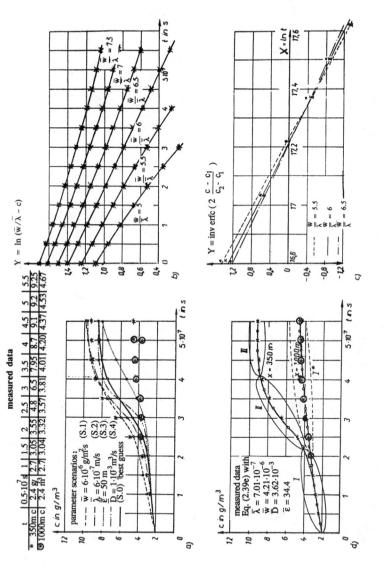

Figure 2.35. Steps *(a)* to *(d)* of parameter identification using the straight-line method within the three different sensitivity domains of Example B.

resulting plots of this function are shown in Figure 2.35b. The straight-line PIM is the function $Y = Y(t)$ or point sequence $Y_i = f(t_i)$ to which a straight line can be fitted. As Figure 2.35b shows, the calculated points (Y_i, t_i) arrange themselves best along a straight line for $\bar{w}/\lambda \approx 6$. The sensitivity for $\bar{w}/\lambda$, however, is not very high. Equation 2.120a provides $\lambda/\bar{\epsilon}$ as $-\Delta Y/\Delta t$ with greater reliability.

(III) $\lambda = \dfrac{q}{x_o} \ln \dfrac{c_o - \bar{w}/\lambda}{c(x_o, \infty) - \bar{w}/\lambda}$ (explicit expression)   (2.120c)

If λ is known, $\bar{w} = \lambda(\bar{w}/\lambda)$ and $\bar{\epsilon} = \lambda/(\lambda/\bar{\epsilon})$ may be calculated, allowing the calculation of c_1 and c_2 from their defining equations given with Equation 2.33h.

(II) $Y = \text{inv erfc} \left(2 \dfrac{c - c_1}{c_2 - c_1} \right) = \dfrac{x_o - (q/\epsilon)t}{2\sqrt{t\bar{D}/\epsilon}} \approx A + BX$

(2.120b)

with $X = \ln t$ $A = \sqrt{\dfrac{x_o q}{4\bar{D}}} \ln \dfrac{x_o}{q/\epsilon}$ $B = -\sqrt{\dfrac{x_o q}{4\bar{D}}}$

$\bar{D} = qx_o/(4B^2)$ $q/\bar{\epsilon} = x_o/\exp|A/B|$ $\bar{\epsilon} = (q/x_o) \exp |A/B|$

This PIM is only valid when the Y_i values in Figure 2.35c are reasonably arranged along a straight line.

Parameter Identification and Examination

For the measured data $c(x_o, t_i)$ given in a table along with Figure 2.35, the following parameters result:

(I) $A = 1.4$ $\bar{w}/\lambda = 6.0$ $B = \lambda/\bar{\epsilon} = 2 \cdot 10^{-8}$

(III) $\lambda = \dfrac{4 \cdot 10^{-4} \text{ m}^2/\text{s}}{350 \text{ m}} \ln \dfrac{12.0 - 6.0}{9.25 - 6.0} = 7.01 \cdot 10^{-7} \text{ m/s}$

$\bar{w} = \lambda(\bar{w}/\lambda) = 7.01 \cdot 10^{-7} \cdot 6.0 = 4.21 \cdot 10^{-6} \text{ g m}^{-2} \text{ s}^{-2}$

$\bar{\epsilon}_1 = \lambda/(\lambda/\bar{\epsilon}) = 7.01 \cdot 10^{-7}/2 \cdot 10^{-8} = 35.05 \text{ m}$

(II) $A = 53.5$ $B = -3.11$

$\bar{D} = 4 \cdot 10^{-4} \cdot 350/(4(3.11)^2) = \underline{3.62 \cdot 10^{-3} \text{ m}^3/\text{s}}$

$\bar{\epsilon} = (4 \cdot 10^{-4}/350) \exp |53.5/-3.11| = \underline{34.4 \text{ m}}$

If these four identified values are inserted into the MPM 2.119b, the function shown in Figure 2.35d results. This function matches the measured data points satisfactorily.

Parameter Interpretation

The hydrodynamic dispersivity is obtained by $\delta = \bar{D}/q = 9.05$ m and is thus related reasonably well to the graphical representation shown in Figure 1.33. When the following additional information is available:

- water saturated thickness of the aquifer $M = 12$ m
- aquifer porosity $\phi = 0.35$
- groundwater recharge rate $w = 5$ L/(s km^2) $= 5 \cdot 10^{-9}$ m/s

the following interpretations could be made:

$$\epsilon = \bar{\epsilon}/M = \phi + K_d \rho_{dry} = 2.9 \text{ (see Equation 1.78b)}$$

where
$$\rho_{dry} = (1 - \phi)2.65 \text{ g/cm}^3 = 1.72 \text{ g/cm}^3$$
$$\phi = 0.35$$
$$K_d = 1.46 \text{ cm}^3/\text{g (distribution coefficient)}$$
$$R = \epsilon/\phi = 8.2 \text{ (retardation coefficient} - \text{see Equation 1.78a)}$$
$$\lambda_o = (\lambda - \bar{w}_o)/M = 5.8 \cdot 10^{-8} \text{ s}^{-1} \text{ (degradation rate} - \text{see Equation 2.12)}$$
$$\lambda_o' = \lambda_o/R = 7.08 \cdot 10^{-9} \text{ s}^{-1}$$
$$\text{(degradation rate in groundwater)}$$

2.3.3 EXAMPLES USING THE TYPE-CURVE METHOD

As previously mentioned in §2.3.1., this method may be used when the MPM has two to three independent parameters and can be expressed in the following form [2.13, p. 274]:

$$\alpha\eta = f(\beta\xi, \gamma) \tag{2.121}$$

$\alpha\eta = \alpha*$ is the dependent variable

$\beta\xi = \beta*$ is the independent variable

α, β, γ are the independent parameters

The formal independent parameters α, β, and γ of Equation 2.121 must contain the parameters to be identified. The variables Y and X are obtained from the terms $\alpha\eta$ and $\beta\xi$ by log transformation:

$$\log \alpha + \log \eta = Y \text{ and } \log \beta + \log \xi = X \tag{2.122}$$

The plot $Y = \log f(X) = \log f(\log \beta^*)$ is designated as a type curve. If γ is a third parameter, $Y_i = \log f(\log \beta^*, \gamma_i)$ represents a type-curve family with the family parameter γ_i. Such type curves usually originate from analytical solutions of the MPM, but may also result from numerical solutions.

The measured data of the dependent variable are plotted on a second transparent sheet in the form of $Y' = \log \eta$ as a function of $X' = \log \xi$. By parallel axis-shifting of both sheets, the optimal fit of a type curve with the measured data is desired (see Figure 2.36). If any reference point (match point) is chosen, the parameters to be identified can be determined from the corresponding values of the variables as follows:

$$\alpha = (f/\eta)^{mp} \text{ and } \beta = (\beta^*/\xi)^{mp} \qquad (2.123)$$

Dimensionless variables are frequently used in both graphs.

As an example of type-curve fitting the identification of the natural parameters c_o and K_{tr} of the MPM according to Equation 2.42c

$$c(x,y) = (c_o/2) \text{ erfc } [y/(4K_{tr}t]^{1/2}]$$

is performed. In this case the terms of Equation 2.121 have the following meaning:

$$\eta = c; \quad \alpha = 2/c_o; \quad f(\beta^*) = \text{erfc}(\beta^*); \quad \beta^* = y/\sqrt{4K_{tr}t};$$

$$\xi = y \text{ or } \xi = 1/\sqrt{t}; \quad 1/\beta = \sqrt{4K_{tr}t} \quad \text{or } 1/\beta = \sqrt{4K_{tr}}/y$$

For further illustration of this method, however, two examples which are somewhat more complex should be examined.

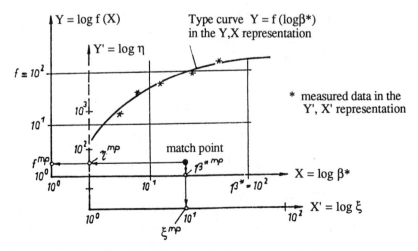

Figure 2.36. Parameter identification using type-curve matching.

2.3.3.1 Example C

Formulation of the MPM

Fluid is pumped from a well gallery along a river bank at relatively constant rate throughout the year, forming a steady-state parallel groundwater flow field normal to the river bank (x-direction) with constant $v = v_o$ (see Figure 2.37). The propagation of heat in the subsurface is therefore subject to convection, dispersion, and storage. In addition, heat exchange normal to the bottom and the upper boundary of the aquifer takes place. Hence, Equation 2.38a should be used as the MPM:

$$K \frac{\partial^2 T}{\partial x^2} - U \frac{\partial T}{\partial x} = \frac{\partial T}{\partial t} - w^* \qquad (2.124)$$

For a sinusoidal annual temperature variation of the water in the river $T(0,t) = T° \sin(2\pi t/t_o)$, and when vertical heat transport in the overlying and underlying strata occurs only through conduction (see Equation 2.38b), one obtains Equation 2.38c as an analytical solution of the MPM:

$$T(x,t) = T° \exp(-\alpha x) \sin(2\pi t/t_o - \beta x) \qquad (2.125)$$

Analysis of the MPM

It can be seen from this analytical solution that the MPM 2.124 has only two independent parameters α and β. The reason is that 2.125 is a periodic or steady-state solution, and the term $\partial T/\partial t$ in Equation 2.124 is, in this case, insensitive and must not be considered in specifying the independent parameters. The formal parameters α and β are related to

- temperature amplitude a: $a(x) = T° \exp(-\alpha x)$
- phase shift of the temperature: $\Delta t(x) = \beta t_o x/(2\pi)$

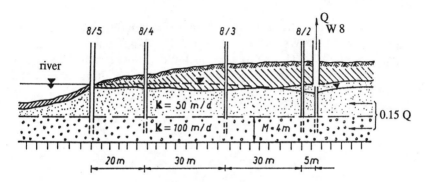

Figure 2.37. Well gallery at a river bank.

From α and β, only two of the three parameters K, U, and w* are identifiable. The functional relations between α, β and K, U, w* may be found by determination of $\partial T/\partial t$, $\partial T/\partial x$, and $\partial^2 T/\partial x^2$ of Equation 2.125, and their substitution in Equation 2.124. This procedure yields

$$K(\alpha^2 \sin \varphi + 2\alpha\beta \cos \varphi - \beta^2 \sin \varphi) + U(\alpha \sin \varphi + \beta \cos \varphi)$$

$$= (2\pi/t_o) \cos \varphi + \omega(\sin \varphi + \cos \varphi)$$

and after rearranging:

$$\underbrace{[2K\alpha\beta + U\beta - 2\pi/t_o - \omega]}_{A} \cos \varphi + \underbrace{[K(\alpha^2 - \beta^2) + U\alpha - \omega]}_{B} \sin \varphi = 0$$

This equation requires A = B = 0. Therefore, two conditional equations are obtained:

$$2K\alpha\beta + U\beta - 2\pi/t_o - \omega = 0 \qquad (2.126a)$$

$$K(\alpha^2 - \beta^2) + U\alpha - \omega = 0$$

which result with known a, β, and K in the following relations with the natural migration parameters:

$$U = \frac{v(\rho c)_w}{(\rho c)_\Sigma} = \frac{2\pi/t_o - K(\beta^2 + 2\alpha\beta - \alpha^2)}{\beta - \alpha} \qquad (2.126b)$$

$$\omega = \frac{2\lambda^*\alpha^*}{M(\rho c)_\Sigma} = \frac{2\pi\alpha/t_o - K(\alpha^2 + \beta^2)\beta}{\beta - \alpha}$$

Feasibility Analysis of the MPM

The initial estimates of the natural MPM parameters may be estimated as follows:

v_o = 1 m/day (from the flow field analysis and the pumping rate)
$(\rho c)_w$ = $4.2 \cdot 10^6$ Ws/(m^3 K) = 48.6 Wd/(m^{3}K) (from Table 1.22)
$(\rho c)_\Sigma$ = $\phi(\rho c)_w + (1 - \phi)(\rho c)_s$ = 30.1 Wd/(m^3 K) (from Table 1.22)
δ = 1 to 3 m (according to Figure 1.33 for x = 30 to 100 m)
λ^* = $\lambda' = \lambda'' = 2$ W/(mK) (from Figure 1.52)
M = 4 m (from well drilling)

which thus become

$$U \approx 1.0 \cdot 48.6/30.1 \approx 1.5 \text{ m/day}$$

$$K \approx (2 + 2 \cdot 1.0 \cdot 48.6)/30.1 \approx 3 \text{ m}^2/\text{day}$$

$$\omega = \frac{-2\lambda^*}{M(\rho c)_\Sigma} \sqrt{\frac{\pi(\rho c)_\Sigma}{\lambda^* t_o}} = \frac{-2 \cdot 2}{4 \cdot 30.1} \sqrt{\frac{\pi 30.1}{2 \cdot 365}} = 0.012$$

The parameters α and β may be now determined iteratively from Equation 2.126a (p = number of current iteration step):

$$\beta^{(p)} = (2\pi/t_o + \omega)/U - 2(K/U) (\alpha\beta)^{(p-1)}$$

$$\alpha^{(p)} = \omega/U + (K/U) (\beta^2 - \alpha^2)^{(p-1)}$$

Using the estimates of U, K, ω and $\alpha^{(o)} = \alpha^{(o)} = 0$, this iteration results in

$$\beta^{(1)} = 0.0195 \qquad \beta^{(2)} = 0.0189 \qquad \beta^{(3)} = 0.0188$$

$$\alpha^{(1)} = 0.0080 \qquad \alpha^{(2)} = 0.0086 \qquad \alpha^{(3)} = 0.0086$$

Hence, $\beta = (0.01, 0.02, 0.03)$ m^{-1} and $\alpha = (0.005, 0.010, 0.015)$ m appear to be reasonable parameter scenarios. Five of these scenarios are shown in Figure 2.38a for the observation point at x = 70 m, $\bar{\theta} = 10°C$, and T° = 8°C, where x comprises the distance between observation well and bank line and an excess length Δx, i.e., the bank line is not at x = 0. However, due to heat conduction normal to the flow lines and transverse dispersion, Δx is much shorter than the hydraulic excess length.

The previous calculations of α and β show that K is without significant influence; in other words, there is no sensitive domain for K. This parameter is unidentifiable by means of the MPM 2.125. Hence, only U and ω are determinable from the identified values $\alpha = 0.0086$ and $\beta = 0.0188$.

Formulation of the PIM and its Analysis

The PIM for the observation point x = x_o is obtained from Equation 2.125 as

$$\Theta(x_o,t) = \bar{\Theta}(x_o) + a(x_o) \sin \left(\frac{2\pi}{t_o} t - \frac{2\pi}{t_o} \Delta t\right) \qquad (2.127a)$$

containing the three parameters: $\bar{\Theta}$ = mean temperature, a = amplitude, and Δt = phase shift. With the transformed variables Y and X:

$$Y = \frac{\Theta - \bar{\Theta}}{(\Theta_{max} - \Theta_{min})/2} \qquad X = \frac{2\pi}{t_o} t - \frac{2\pi}{t_o} \Delta t \qquad (2.127b)$$

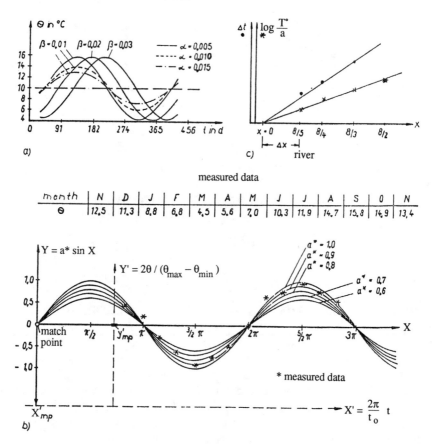

Figure 2.38. Steps *(a)* to *(c)* of parameter identification using type-curve matching for a bank filtration plant according to Figure 2.37 of Example C.

this PIM may be expressed as type-curve family:

$$Y_i = \frac{2a_i}{(\Theta_{max} - \Theta_{min})} \sin (X)$$

with the family parameter $a_i^* = 2a_i/(\Theta_{max} - \Theta_{min})$. Figure 2.38b shows this type-curve set for $a_i^* = \{1.0, 0.9, 0.8, 0.7, 0.6\}$. In the second transparent graph the measured data $Y_j' = 2\Theta_j/(\Theta_{max} - \Theta_{min})$ are plotted as a function of $X_j' = 2\pi t_j/t_o$. After an optimal fit is obtained by parallel axis-shifting, a match point must be chosen. If $X = Y = 0$ is chosen as match point, the parameters to be identified may be determined as follows (see Figure 2.38b):

$$a = a^*(\theta_{max} - \theta_{min})/2, \text{ with } a^* \text{ taken from the most appropriate}$$
curve

$$\bar{\Theta} = Y'_{mp}(\Theta_{max} - \Theta_{min})/2$$

$$t = X'_{mp}t_o/(2\pi)$$

When measured data for several observation points $x_{o,i}$ are available, a function $\bar{\Theta} = f(x)$ must be fitted to the data set $[\bar{\Theta}_i, x_{o,i}]$ (see Figure 2.38c). This function requires careful interpretation! In this case, a straight-line PIM based on the MPM 2.125 may be used to identify a and Δt as follows (see Figure 2.38c):

$$T°\exp(-\alpha x) = a \quad \Rightarrow \quad \underbrace{\log \frac{(T°/a)}{Y}}_{Y} = \underbrace{\frac{(\alpha/2.3)x}{B_\beta^*X}}_{B_\beta^*X} \quad (2.128a)$$

$$\underbrace{\frac{\Delta t}{Y}}_{Y} = \underbrace{\frac{(0.5\beta t_o/\pi)x}{B_\beta^*X}}_{B_\beta^*X} \quad (2.128b)$$

where

$$\alpha = 2.3B_\alpha \quad \text{and} \quad \beta = 2\pi B_\beta/t_o \quad (2.128c)$$

A practical example from a bank filtration system at the Elbe river is presented in Luckner et al. [2.57].

The following steps of the algorithm shown in Figure 2.33 and described in §2.3.1.5 and §2.3.1.6 are omitted for Example C due to the similarities with the previous examples.

2.3.3.2 Example D

Formulation of the MPM

The spreading of a tracer in an aquifer is considered in this example. The tracer is injected via a monitoring well with a constant mass rate $\dot{m}$ (measured in kg/s), and with a constant flow rate of carrier water $V = Q = 0$. A parallel steady-state groundwater flow field prevails in the aquifer (compare Figure 1.35, where a tracer pulse injection is shown). The x-direction should coincide with the direction of Darcy's velocity $\vec{v}$ determined by means of groundwater-level observations. Therefore, Equation 2.46b is chosen as the MPM and shown here again with the following equation:

$$c(x, y, t) = \frac{\dot{m}}{4\pi\sqrt{\bar{D}_x\bar{D}_y}} \exp \frac{qx}{2\bar{D}_x} W(\sigma,b) \quad (2.129)$$

where

$$\sigma = b^2 \frac{\bar{\epsilon}\bar{D}_x}{q^2 t}$$

$$b^2 = \left(\frac{xq}{2\bar{D}_x}\right)^2 + \left(\frac{yq}{2\sqrt{\bar{D}_x\bar{D}_y}}\right)^2$$

$$\bar{D} = DM$$

Analysis of the MPM

The MPM 2.129 has only independent parameters as an analytical solution. The natural parameters are $\bar{D}_x = q\delta_x$, $\bar{D}_y = q\delta_y$, and $q = U\bar{\epsilon}$. They are related to the formal parameters of α, β, and γ of the type-curve Equation 2.121 as follows:

$$1/\alpha = \frac{\dot{m}}{4\pi\sqrt{\bar{D}_x\bar{D}_y}} \exp\frac{qx}{2\bar{D}_x} \qquad \beta = \sigma t \text{ and } \gamma = b$$

Feasibility Analysis of the MPM

The injection well is installed at $x = 0$, $y = 0$. Let $\dot{m} = 100$ kg/day be continuously injected into the aquifer with a water bearing thickness of $M = 10$ m. For the initial estimates $\bar{\epsilon} = 0.3 \cdot 10$ m $= 3$ m, $q = vM = 6$ m²/day, $\bar{D}_x = \delta_t q = 2$ m³/day, and $\bar{D}_y = 0.2$ m³/day, the concentration field shown in Figure 2.39a is obtained after five days.

The observation point is estimated to be located at $x_o = 10$ m and $y_o = 1$ m in the flow field. The function $c(10,1,t) = f(t)$ is shown in Figure 2.39b in order to illustrate the sensitivity of the natural parameters U, $\bar{D}_x$, and $\bar{D}_y$ for the following parameter scenarios:

(1) $U = q/\bar{\epsilon} = 2.0$ m/day $\bar{D}_x = 2$ m³/day $\bar{D}_y = 0.3$ m³/day
(2) $U = q/\bar{\epsilon} = 1.5$ m/day $\bar{D}_x = 2$ m³/day $\bar{D}_y = 0.3$ m³/day
(3) $U = q/\bar{\epsilon} = 2.0$ m/day $\bar{D}_x = 1.5$ m³/day $\bar{D}_y = 0.3$ m³/day
(4) $U = q/\bar{\epsilon} = 2.0$ m/day $\bar{D}_x = 2$ m³/day $\bar{D}_y = 0.5$ m³/day

This figure shows the high sensitivity of the MPM to the transport parameter U and a much lower sensitivity to dispersion ($\bar{D}_x$, $\bar{D}_y$).

Formation of the PIM and its Analysis

The PIM results from the MPM as follows:

$$c(t)/\alpha^* = W^*[(1/\sigma),b] \tag{2.130}$$

where $1/\alpha = \alpha^*$
$W(\sigma,b) = W^*((1/\sigma),b)$
$Y = \log c + \log(1/\alpha^*)$ $X = \log t + \log(1/\beta)$

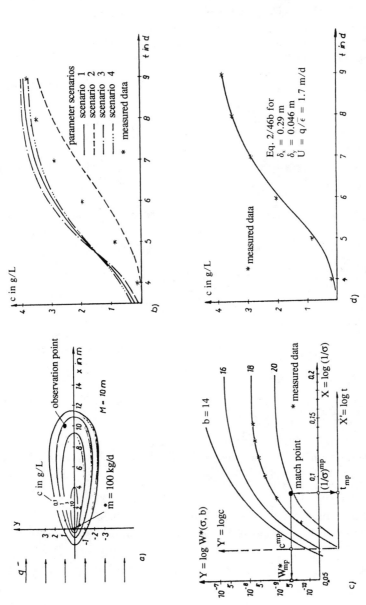

Figure 2.39. Steps (a) to (d) of parameter identification using type-curve matching for a continuous tracer injection into a steady-state parallel groundwater flow field (Example D).

This type-curve family is the well-known Hantush function of well hydraulics [2.13, Figure 5.14], where $Y = \log W^*$ is a function of $X = \log(1/\sigma)$ with the family parameter b. In contrast to well hydraulics, however, the parameter b is in the range of 10 to 20. The transformed, measured data $Y_i' = \log c_i$ must be plotted as a function of $X_i' = \log t_i$. When the two plots are brought to the best fit by parallel axis-shifting, the parameters are calculated from the match point relations as follows (see Figure 2.39c):

$$\alpha = \left(\frac{c}{W^*}\right)^{mp} \qquad \beta = \left(\frac{t}{1/\sigma}\right)^{mp}$$

b = selection of the curve from the family best fitting the data points

The geometrical shape of the type curves is relatively similar, as can be seen in Figure 2.39c. Therefore, the whole range of concentration changes, particularly the initial range, must be monitored carefully.

Parameter Identification and Examination

With the match point data according to Figure 2.39c, the following parameters are obtained:

b = 18

$\alpha = 4.65 \cdot 10^8$ g/L

$\beta = 55$ days

If these identified values are substituted again in the MPM, the measured data are reflected sufficiently well (see Figure 2.39d).

Parameter Interpretation

The natural parameters δ_x, δ_y, and ϵ may be determined iteratively from the identified formal parameters α, β, and b as follows:

$$b^2 = \frac{x^2}{4\delta_x^2} + \frac{y^2(\delta_x/\delta_y)}{4\delta_x^2} \qquad\qquad (2.131a)$$

$$\beta = b^2\epsilon\delta_x/q \qquad\qquad (2.131b)$$

$$\alpha = \frac{\dot{m}}{4\pi q\sqrt{\delta_x\delta_y}} \exp\left(\frac{x}{2\delta_x}\right) \qquad\qquad (2.131c)$$

At the first iteration step p = 1, $(\delta_x/\delta_y)^\circ$ = 1 should be used:

$$\delta_x^P = \frac{1}{2b} \sqrt{x^2 + y^2(\delta_x/\delta_y)^{P-1}} \qquad \bar{\epsilon}^P = \beta q/(b^2\delta_x^P)$$

$$\delta_y^P = \left(\frac{\dot{m}}{\alpha 4\pi q \sqrt{\delta_x^P}} \exp \frac{x}{2\delta_x^P}\right)^2 \qquad (\delta_x/\delta_y)^P = \delta_x^P/\delta_y^P \qquad (2.132)$$

This iteration scheme usually converges. For the identified parameters b = 18, α = 4.65 · 10^8 g/L, and β = 55 days, the iteration results finally in

$$\delta_x = 0.29 \text{ m} \qquad \delta_y = 0.046 \text{ m} \qquad \bar{\epsilon} = 3.6 \text{ m}$$
$$U = q/\bar{\epsilon} = 1.68 \text{ m/days}$$

2.3.4 EXAMPLE OF OPTIMIZATION WITH OBJECTIVE FUNCTIONS

The conditions required for an appropriate formulation of a PIM as an optimization problem in the form of an "objective" were stated in §2.3.1.

2.3.4.1 Example E

Formulation of the MPM

An MPM similar to that in Figure 1.56 is considered with $^{45}Ca^{2+}$ as the only migrant. This MPM describes the migration process in a laboratory column test. This test is characterized by the following conditions:

- lumped parameters v_o (Darcian velocity), D (dispersion coefficient), k_1 and k_2 (velocity constants of first-order transphase exchange between the mobile and immobile phase—see Equation 1.104), θ_o and θ_{im} (volumetric water content of the mobile and immobile phase), $(\theta_o + k') = \epsilon_1$ and $(\theta_{im} + k'') = \epsilon_2$ (storage coefficients of the mobile and immobile phase)
- no transport of migrants within the immobile phase
- thermodynamic equilibrium within the mobile and the immobile phase
- transphase exchange of migrants between the mobile and the immobile phase with different velocity constants k_1, k_2
- no external sources and sinks affecting the migration process

Under these conditions the MPM has the following form (see Equation 2.40):

$$D \frac{\partial^2 c}{\partial z^2} - v_o \frac{\partial c}{\partial z} = (\theta_o + k') \frac{\partial c}{\partial t} + k_1 c - k_2 c_{im} \qquad (2.133a)$$

$$0 = (\theta_{im} + k'') \frac{\partial c_{im}}{\partial t} - k_1 c + k_2 c_{im}$$

where c and c_{im} are the concentrations of the radionuclide $^{45}Ca^{2+}$ in the mobile and immobile mixphase, respectively, and z and t the independent spatial and temporal variables.

Analysis of the MPM

The eight dependent natural parameters of the MPM 2.133a — D, v_o, θ_o, θ_{im}, k', k'', k_1, and k_2 — are combined in five independent formal parameters: $\chi_1 = D/CI$, $\chi_2 = v_o/CI$, $\chi_3 = k_1/CI$, $\chi_4 = k_2/CI$, and $\chi_5 = CI/CI_{im}$ with $CI = \theta_o + k'$ and $CI_{im} = \theta_{im} + k''$, respectively. The corresponding MPM where the independent parameters may be expressed is as follows:

$$\chi_1 \frac{\partial^2 c}{\partial z^2} - \chi_2 \frac{\partial c}{\partial z} = \frac{\partial c}{\partial t} + \chi_3 c - \chi_4 c_{im} \qquad (2.133b)$$

$$0 = \partial c_{im}/\partial t - \chi_3 \chi_5 c + \chi_4 \chi_5 c_{im}$$

If these equations are subjected to Laplace transformation, the independent variable t becomes transformed to the Laplace parameter s as follows:

$$\mathcal{L}[\partial c/\partial t] = s\bar{c}$$

for $c(t=0) = 0$ (see Equation 2.7a), and the two partial differential equations 2.133b may be transformed and combined in the following ordinary differential equation:

$$\chi_1 \frac{d^2\bar{c}}{dz^2} - \chi_2 \frac{d\bar{c}}{dz} = s\left(1 + \frac{\chi_3}{s + \chi_4 \chi_5}\right)\bar{c} \qquad (2.133c)$$

In this MPM only four independent parameters remain. In dimensionless form, this MPM may be expressed as follows:

$$\frac{1}{Pe} \frac{d^2\bar{c}}{d\zeta^2} - \frac{d\bar{c}}{d\zeta} = s^*\left(1 + \frac{\alpha_1}{s^* + \alpha_2}\right)\bar{c}^* \qquad (2.133d)$$

with the independent parameters:

$$Pe = v_o L/D \approx L/\delta \qquad \alpha_1 = k_1 L/v_o \qquad \alpha_2 = \frac{k_2 L}{v_o} \frac{\theta_o + k'}{\theta_{im} + k''}$$

$$1/\tau = s^*/s = (\theta_o + k')L/v_o \quad \text{or} \quad \tau = t^*/t = v_o/(L\theta_o + Lk')$$

and the dependent and independent variables:

$$\bar{c}^* = (\bar{c} - \bar{c}_{min})/(\bar{c}_{max} - \bar{c}_{min})$$
$$\zeta = z/L \text{ and } t^*$$

The analytical solution of Equation 2.133d and its inverse Laplace transformation back to the original space were derived in Luckner et al. [2.62]. This solution has the following form:

$$\underbrace{c^*}_{\substack{\text{dependent} \\ \text{variable}}} = \underbrace{f(Pe, \alpha_1, \alpha_2,}_{\substack{\text{independent} \\ \text{parameters}}} \underbrace{\zeta, \tau t)}_{\substack{\text{independent} \\ \text{variables}}} \qquad (2.133e)$$

The FORTRAN program ALSUB 3 may be used for the numerical evaluation of this MPM with its four independent parameters Pe, α_1, α_2, and τ [2.77; 2.62].

Feasibility Analysis of the MPM

Based on process knowledge, the following expected values were estimated for the $^{45}Ca^{2+}$ migration in a 1.2 m long test column filled with sandy fine gravel. The column was only partially saturated with water, and the flow rate was $v_o = 3.6 \cdot 10^{-5}$ m/s:

$Pe = v_oL/D \approx L/\delta = 50$ (see Figure 1.33; $\bar{\delta}_\ell = 0.02x$ = 0.02L for the homogeneous repacked soil)

$\alpha_1 = k_1L/v_o = 2.0$ (according to Equation 1.68, k_1 is obtained as $k_1 = 0.02 D_M O_{sp}/\overline{\Delta l}$ and with $O_{sp} = 3/d_{10\%} = 3 \cdot 10^3$ m^{-1} and $\overline{\Delta l} = 1 \cdot 10^{-3}$m, e.g., as $k_1 = 0.02 \cdot 10^{-9} \cdot 3 \cdot 10^3/10^{-3} = 6 \cdot 10^{-5}$ s^{-1})

$\alpha_2 = \dfrac{k_2 L}{v_o} \dfrac{\theta_o + k'}{\theta_{im} + k''} = 1.0$ (with $k_2 = k_1$ and $(\theta_o + k')/(\theta_{im} + k'') = 1/2$)

$\tau = \dfrac{v_o}{L(\theta_o + k')} = 1 \cdot 10^{-4}$ s^{-1} (with $\theta = 0.1$ and $k' = 0.2$)

Therefore, the following parameter scenarios appear meaningful for the migrant breakthrough at the end of the column ($\zeta = 1.0$):

Scenario	S1	S2	S3	S4	S5	
Pe	50	50	50	50	50	$\tau = 1.0 \cdot 10^{-4}$ s^{-1}
α_1	2.0	1.0	1.5	3.0	1.0	$\tau = 0.5 \cdot 10^{-4}$ s^{-1}
α_2	1.0	0.5	1.5	0.5	1.5	$\tau = 1.5 \cdot 10^{-4}$ s^{-1}

Figure 2.40 shows these five calculated scenarios and three plots of the data set which correspond to three selected values of $\tau = (1.0 \cdot 10^{-4}, 0.5 \cdot 10^{-4}, 1.5 \cdot 10^{-4})$ s^{-1}:

$$c^* = f(Pe, \alpha_1, \alpha_2, t^* = \tau t)$$

If both log(t*), representing the calculated scenarios, and log(t), representing the measured data, are plotted on the abscissa, the data set can then be represented in one plot (see Figure 2.40). In this case the parameter t may be estimated by the best fit of the plotted data with one of the plotted scenarios obtained by horizontal shifting, where the parameter scenarios or type curves are presented in the graph in the following form:

$$c^* = f(Pe, \alpha_1, \alpha_2, \log t^*)$$

For the example considered, this matching procedure results in (see Figure 2.40):

$$\tau = (t^*/t)^{mp} = 1.4 \cdot 10^{-4} \text{ s}^{-1}$$

The variants plotted in Figure 2.40 and variants with different Pe values show that the sensitive domains may be characterized as follows: In the domain $0 < c^* < 0.1$ and $3 \cdot 10^3 < t < 6 \cdot 10^3$s, Pe and α_1 are sensitive; in the domain $0.3 < c^* < 0.9$ and $1 \cdot 10^4$ s $< t$, α_1 and α_2 are sensitive. These domains must be covered by measured data.

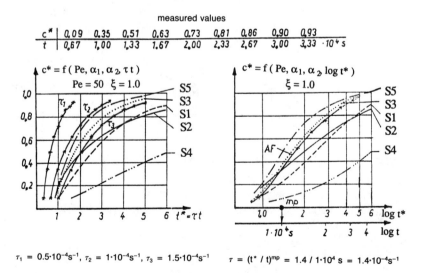

$\tau_1 = 0.5 \cdot 10^{-4} \text{s}^{-1}$, $\tau_2 = 1 \cdot 10^{-4} \text{s}^{-1}$, $\tau_3 = 1.5 \cdot 10^{-4} \text{s}^{-1}$ $\tau = (t^* / t)^{mp} = 1.4 / 1 \cdot 10^4 \text{ s} = 1.4 \cdot 10^{-4} \text{s}^{-1}$

Figure 2.40. Typical parameter scenarios for the MPM of Equation 2.133e and the plotted data, and the estimation of τ by means of type-curve matching.

Formulation of the PIM and its Analysis

The simplest form of Equation 2.108c is used as the PIM:

$$OF = \Sigma \; w_i \; (c_{m,i} - c_{c,i})^2 = f(Pe, \alpha_1, \alpha_2, \tau) \overset{!}{=} \min \qquad (2.134)$$

where $c_{m,i}$ are the measured concentrations at time i and $c_{c,i}$ are the concentrations at time i calculated by means of Equation 2.133e. The weights w_i can be set $w_i = 1$ if i represents horizontal-equidistant points. The objective function OF 2.134 must be sensitive to the parameters to be identified. In the range of meaningful parameters (identification range), OF should have only one minimum and significant gradients in the direction toward this minimum.

This analysis can be performed in illustrative manner for two of the independent parameters by two-or three-dimensional graphs of OF. Based on the assumption that $\tau = t^*/t$ is already identified with a sufficient accuracy by curve matching as shown before, graphs of the functions OF $= f(Pe, \alpha_1)$ and OF $= f'(\alpha_1, \alpha_2)$ seem to be sufficient for this analysis. These graphs are shown in Figure 2.41. It can be seen from the left graph that the sensitivity for Pe is poor, and its identification may therefore cause some problems.

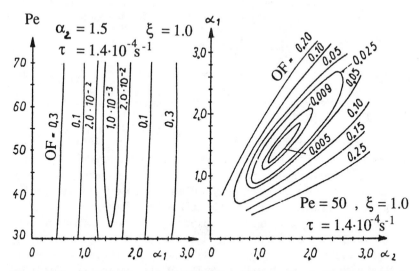

Figure 2.41. Typical 2D graph of the function OF according to Equation 2.134 for Example E.

Parameter Identification and Examination

The identification is executed according to the flowchart shown in Figure 2.42. If τ was not estimated with sufficient accuracy by type-curve

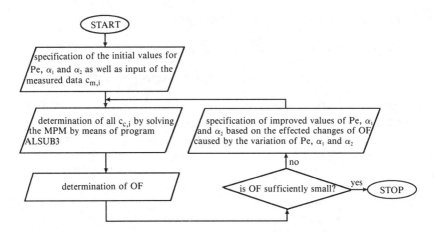

Figure 2.42. Flowchart for parameter identification with the PIM 2.124.

matching based upon Figure 2.40, it must also be included in the objective function as an independent variable.

The initial values may be estimated from Figure 2.40 as Pe = 50, α_1 = 1.0, and α_2 = 1.5. The computer program PIA-1 was used for the minimization procedure. With the measured data and $\tau = 1.4 \cdot 10^{-4}$ s, this program provides the following identification results:

$$\alpha_1 = 1.51 \qquad \alpha_2 = 1.41 \qquad Pe = 53$$

The function AF computed with these parameters is shown in the right part of Figure 2.40.

Parameter Interpretation

The interpretation of the eight natural parameters of the MPM according to Equation 2.133a based on the four identified independent parameters Pe, α_1, α_2, and τ is only possible if four additional pieces of information are available. In the given case, one may proceed from the fact that v_o may be measured to sufficient accuracy and that $k_1 \approx k_2$ because molecular diffusion causing the transphase exchange is a symmetrical process, and one may assume that both θ_o (e.g., from an anion breakthrough curve) and θ_o/θ_{im} (e.g., from laboratory experiments — see Figure 3.32) may be measured independently of the parameter estimation procedure.

With $v_o = 3.6 \cdot 10^{-3}$ m/s, $k_1 = k_2$, $\theta_o = 0.09$, $\theta_o/\theta_{im} = 1.1$, and L = 1.2 m, as well as the identified values $\tau = t^*/t = 1.4 \cdot 10^{-4}$ s^{-1}, $\alpha_1 = 1.51$, α_2 = 1.41, and Pe = 53, the eight natural parameters result as follows:

$$v_o = 3.6 \cdot 10^{-5} \text{ m/s}$$

$$D = v_o L/Pe = 0.82 \cdot 10^{-6} \text{ m/s or}$$
$$\delta = D/v_o = 0.023 \text{ m}$$

$$k_1 = \alpha_1 v_o/L = 4.53 \cdot 10^{-5} \text{ s}^{-1}$$

$$k_2 = k_1 = 4.53 \cdot 10^{-5} \text{ s}^{-1}$$

$$\theta_o = 0.09$$

$$k' = \frac{v}{L_T} - \theta_o = 0.12$$

$$\theta_{im} = \theta_o/1.1 = 0.08$$

$$k'' = (\theta_o + k')k_2 L/(v_o \alpha_2) - \theta_{im} = 0.14$$

Different assumptions lead, of course, to somewhat different results of interpretation. In particular $k_1 = k_2$ has often proven to be a function of the hydraulic gradient, which may be reflected by $\alpha \approx a^*v$. In other words, the transport in dead-end pores is then not only controlled by molecular diffusion but by convection, even at a very low transport velocity.

References for Part 2

1. Abramowitz, M., and I. A. Stegun. Handbook of mathematical functions. New York: Dover Publ., 1970.
2. Ackermann, G. Lehrwerk Chemie (5), *Leipzig:* Dt. Verl. für Grundstoff industrie, 1974.
3. Ameling, W. Laplace-Transformation. Düsseldorf: Bertelsmann Universitätsverlag, 1975, p. 29 (Studienbücher Naturwissenschaften).
4. Barker, J. A. Laplace transform solutions for solute transport in fissured aquifers. Adv. Water Resources 5 (1982) 6, pp. 98–104.
5. Bear, J. Dynamics of fluids in porous media. New York: Elsevier, 1972.
6. Bear, J. Hydraulics of groundwater (Water resources and environmental engineering). New York: McGraw-Hill, 1979.
7. Beims, U. Rechnergestützte Pumpversuchsauswertung auf der Grundlage des KRS 4200. WWT 27 (1977) 11.
8. Beims, U. EDV-gestützte Brunnenberechnung in der DDR. Techn. Univ. Dresden, Diss. B, 1980.
9. Beims, U., L. Luckner, and C. Nitsche. Beitrag zur Ermittlung von Parametern von Migrationsprozessen in der Boden-und Grundwasserzone. Wiss. Z. Techn. Univ. Dresden 31 (1982) 5, pp. 211–217.
10. Botschewer, F. M., and A. E. Oradovskaja. Gidrogeologiceskoe obosnovanie zasily podzemnych vod i vodozaborov ot zagraznenij. Moskva: Nedra, 1972.
11. Bronstein, I. N., and K. A. Semendjajew. Taschenbuch der Mathematik. 19. Aufl. Leipzig: Teubner Verlagsgesellschaft, 1979.
12. Burnett, R. D., and E. O. Frind. Simulation of contaminant transport in three dimensions; 1. The Alternating Direction Galerkin Technique. Water Res. Research 23 (1987) 4, pp. 683–694.
13. Busch, K.-F., and L. Luckner. Geohydraulik. Leipzig: Dt. Verl. f. Grundstoffindustrie, 1972.
14. Cameron, D. R., and A. Klute. Convective-dispersion solute transport with

323

a combined equilibrium and kinetic adsorption model. Water Res. Research 13 (1977), pp. 183–188.

15. Carslow, H. S., and J. D. Jaeger. Conduction of heat in solids, 2nd Ed. London: Oxford University Press, 1959.

16. Chung, T. J. Finite Elemente in der Strömungsmechanik. Leipzig: Fachbuchverlag, 1983.

17. Cleary, R. W., and D. D. Adrian. Analytical solution of the convective-dispersive equation for cation adsorption in soils. Soil Science Soc. of Am. Proc. 37 (1973), pp. 197–199.

18. Crump, K. S. Numerical inversion of Laplace transforms using a Fourier series approximation. Journal of the Association for Computing Machinery 23 (1976), p. 89.

19. Damrath, H. et al. Wasserinhaltsstoffe in Grundwasser-Reaktionen, Transportvorgänge und deren Simulation. Berlin: Schmidt-Verl., 1979 (Berichte Umweltbundesamt, 1979, 4).

20. Diersch, H.-J. Finite-Elemente-Modellierung instationärer zweidimensionaler Stofftransportvorgänge im Grundwasser. Wiss. Konf. der Techn. Univ. Dresden zur Simulation der Migrationsprozesse im Boden-und Grundwasser. Bd. 1, 1979, pp. 26–138.

21. Diersch, H.-J. Primitive variable finite element solutions of free convection flows in porous media. Zeitschr. für Angew. Math. u. Mech. 61 (1981), pp. 325–337.

22. Diersch, H.-J. Finite-Element-Galerkin-Modell zur Simulation zweidimensionaler konvektiver und dispersiver Stofftransportprozesse im Boden. Acta Hydrophysica XXVI (1981) 1, pp. 5–44.

23. Diersch, H.-J. On finite element upwinding and its numerical performance in simulation coupled convective transport processes. Hrsg.: AdW der DDR, Inst. für Mechanik, Berlin, 1982 (P-MECH-04/82).

24. Diersch, H.-J. Modellierung und numerische Simulation geohydrodynamischer Transportprozesse. AdW der DDR, Dissertation B, 1984.

25. Diersch, H.-J., and S. Kaden. Contaminant plume migration in an aquifer. Finite element modelling for the analysis of remediation strategies: a case study. Collaborative paper, Laxenburg: IIASA, 1984.

26. Diersch, H.-J., and P. Nillert. Modelluntersuchungen zu vorbeugenden Maßnahmen gegen Schadstoffhavarien an Uferfiltratfassungen. WWT 33 (1983) 3, pp. 103–107.

27. Dobesch, H. Laplace-Transformation, Einführung von Einschwingvorgängen, 2. Aufl. Berlin: Verlag Technik, 1964.

28. Doetsch, G. Anleitung zum praktischen Gebrauch der Laplace-Transformation und der Z-Transformation. München: R. Oldenburg Verlag, 1967.

29. Doetsch, G. Einführung in Theorie und Anwendung der Laplace-Transformation. Basel: Birkenhäuser Verlag, 1976.

30. Douglas, J., and B. F. Jones. On predictor-corrector methods for nonlinear parabolic differential equations. Jour. of the Soc. for Ind. and Applied Math. 11 (1963) March 1.

31. Dubner, R., and J. Abate. Numerical inversion of Laplace transforms by

relating them to the finite Fourier cosine transforms. Journal of the Association for Computing Machinery 15 (1968), p. 115.

32. Durbin, F. Numerical inversion of Laplace transforms: an efficient improvement to Dubner and Abate's method. Computer J. 17 (1974), p. 371.

33. Frind, E. O. The principal direction technique: a new approach to groundwater contaminant transport modeling. 4. Int. Conference of Finite Elements in Water Resources Proc. Hannover: Springer Verlag, 1982, pp. 13.25-13.42.

34. Frind, E. O. Simulation of long-term density-dependent transport in groundwater. Advances in Water Res. 5 (1982), pp. 73-88.

35. Frind, E. O., W. H. M. Duynisveld, O. Strebel, and J. Boettcher. Modeling of multi-component transport with microbial transformation in groundwater — The Fuhrberg case. Water Res. Research (submitted 1989).

36. Frind, E. O., and G. B. Matanga. The dual formulation of flow for contaminant transport modeling — 1. Review of theory and accuracy aspects. Water Res. Research 21 (1985) 2, pp. 159-169.

37. Frind, E. O., E. A. Sudicky, and S. L. Schellenberg. Micro-scale modeling in the study of plume evolution in heterogeneous media. Stochastic Hydrology and Hydraulics, 1 (1987) 4, pp. 263-279.

38. Gärtner, S. Zur Berechnung von Flachwasserwellen und instationären Transportprozessen mit den Methoden der finiten Elemente. Düsseldorf: VDI-Verlag, 1977, Reihe 4, 30.

39. Genuchten, M. T. van, and P. J. Wierenga. Mass transfer studies in sorbing porous media. I. Soil Science Soc. Amer. J. 40 (1976) 4.

40. Genuchten, M. T. van. On the accuracy and efficiency of several numerical schemes for solving the convective-dispersive equation. Conference of Finite Elements in Water Resources, Finite elements in water resources. Proc London: Springer-Verlag 1977, pp. 1.71-1.90.

41. Genuchten, M. T. van, and W. G. Gray. Analysis of some dispersion corrected numerical schemes for solution of the transport equation. Int. Journal of Numerical Methods in Engineering 12 (1978), pp. 387-404.

42. Genuchten, M. T. van. Analytical solution for chemical transport with simultaneous adsorption, zero-order production and first-order decay. Journal of Hydrology 49 (1981), pp. 213-233.

43. Genuchten, M. T. van, and W. J. Alves. Analytical solutions of the one-dimensional convective-dispersive solute transport equation. U.S. Department of Agriculture (Bulletin No. 1661), 1982.

44. Gershon, N. D., and A. Nir. Effect of boundary conditions of models on tracer distribution in flow through porous media. Water Res. Research 5 (1969), pp. 830-840.

45. Gräber, P.-W. Möglichkeiten zur hybriden Simulation der Migrationsprozesse im Boden-und Grundwasser. Wiss. Konf. der Techn. Univ. Dresden zur Simulation der Migrationsprozesse im Boden- und Grundwasser, Bd. 1, 1979, pp. 184-206.

46. Guymon, G. L., et al. A general numerical solution of the two-dimensional diffusion-convection equation by the finite element method. Water Res. Research 6 (1970), pp. 1611-1617.

47. Guymon, G. L. Note on the finite element solution of the diffusion-convection equation. Water Res. Research 8 (1972), pp. 1357–1360.

48. Gutt, B. Programm SIMKA-2D der FG Grundwasser des IfW Berlin und der TU Dresden. Dresden: Techn. Univ., 1981 (20–01–81).

49. Gutt, B. Anwenderinstruktion zum Programm HORAND. Berlin: Institut für Wasserwirtschaft, 1984.

50. Gutt, B., and N. Victor. Programmsystem STREAM der FG Grundwasser des IfW Berlin und der TU Dresden. Dresden: Techn. Univ., 1981 (20–02–81).

51. Häfner, F. Mathematische Modellierung und Simulation von Strömungsvorgängen in porösen Stoffen unter besonderer Berücksichtigung der Umkehraufgabe. Freiberg: Bergakademie, Dissertation B, 1976.

52. Jensen, O. K., and B. A. Finlayson. Solution of the transport equation using a moving coordinate system. Advan. Water Res. 3 (1980), pp. 9–18.

53. Kaden, S., and L. Luckner. Digitale Simulation der Geofiltration und Migration. Dresden: Techn. Univ., 1980 (Lehrheft im postgr. Studium Grundwasser).

54. Konikow, L. F., and J. D. Bredehoeft. Computer model of two-dimensional solute transport and dispersion in groundwater. Automated Data Processing and Computations 7 (1978).

55. Leismann, H. M., and E. O. Frind. A symmetric-matrix time integration scheme for the efficient solution of advection-dispersion problems. Water Res. Research 25 (1989) 6, pp. 1133–1139.

56. Lever, D. A., and J. Rae. A new computer program for the transport of radionuclides in groundwater. Geow. Vienna: IAEA, 1982, pp. 695–700 (IAEA-SM-257/23P).

57. Luckner, L., et al. Konzeptionelles Blockmodell für die Prognose der Rohwassertemperatur einer Uferfiltrationsanlage. Wiss. Konferenz der Techn. Univ. Dresden zur Simulation der Migrationsprozesse im Boden- und Grundwasser. Bd. 1, 1979, pp. 263–276.

58. Luckner, L., and U. Beims. Beitrag zur digitalen geohydraulischen Parameteridentifikation. Zeitschr. für Angew. Geologie 11 (1976) 10.

59. Luckner, L., A. Eckhardt, and H. Mansel. Handbuch zur Lösung von Migrationsproblemen im Boden und Grundwasserbereich mit dem Taschenrechner Sharp PC-1401. TU Dresden, 1987.

60. Luckner, L., G. Muellerbuchhof, and N. Victor. Nutzung von Uferfiltrat und künstlichem Infiltrat als Wärmequellen für elektrisch angetriebene Kompressions-Wärmepumpen. Berlin: Wasserwirtschaft Wassertechnik 33 (1983) 1, pp. 16–21.

61. Luckner, L., and C. Nitsche. Bildung systembeschreibender Modelle der Migrationsprozesse in der Aerationszone und ihre digitale Simulation. Geodätische und geophysikal. Veröffentl. Reihe IV (1980) 32, pp. 99–109.

62. Luckner, L., C. Nitsche, and H. Wenzel. Analytische Lösung der eindimensionalen Diffusions-Konvektionsgleichung unter Berücksichtigung der linearen kinetischen Austauschreaktion 1. Ordnung. Acta Hydrophysica XXVII (1982) 2, pp. 109–123.

63. Luckner, L., and W. M. Schestakow. Simulation der Geofiltration. Leipzig: Dt. Verlag f. Grundstoffindustrie, 1975.

64. Luckner, L., and K. Tiemer. Mathematische Modellbildung der Geofiltration und Migration. Dresden: Techn. Univ., 1981 (Lehrheft im postgr. Studium Grundwasser, H. 1).

65. Lynch, D. R., and K. O. O'Neill. Continuously deforming finite elements for the solution of parabolic problems with and without phase change. Int. J. Num. Math. Eng. 17 (1981), pp. 81–96.

66. Mansel, H., D. Schäfer, and U. Beims. Beschreibung des FORTRAN-Programms SROM für den BC 5120 zur digitalen Simulation rotationssymmetrischer Migrationsprozesse. Techn. Univ. Dresden, Sektion Wasserwesen, WB Wassererschließung, 1985.

67. Mehlhorn, H. Temperaturveränderungen im Grundwasser durch Brauchwassereinleitungen. Mitteilungen des Institutes für Wasserbau der Univ. Stuttgart 50 (1982), p. 181.

68. Mehlhorn, H., et al. Kurzschluströmung zwischen Schluck-und Entnahmebrunnen—kritischer Abstand und Rückstromrate. Wasser und Boden 33 (1981) 4, pp. 117–174.

69. Mercado, A. Nitrate and chloride pollution of aquifers: A regional study with the aid of a single-cell model. Water Res. Research 12 (1976) 4, pp. 731–747.

70. Muskat, M. The flow of homogeneous fluids through porous media. New York: McGraw-Hill, 1937.

71. Narasimhan, T. N., and P. A. Witherspoon. An integrated finite difference method for analyzing fluid flow in porous media. Water Res. Research 12 (1976) 1.

72. Nestler, W., L. Luckner, and J. Hummel. Die Nutzung von Signalmodellen zur Wasserbedarfsprognose als Basis einer effektiven Steuerung der Wasserversorgungsprozesse. Berlin: Wasserwirtschaft Wassertechnik 32 (1982) 12, pp. 414–418.

73. Neumann, S. P. A Eulerian-Lagrangian numerical scheme for the dispersion-convection equation using conjugate space-time grids. Jour. Comp. Phys. 41 (1981) 2, pp. 270–294.

74. Neumann, S. P., and S. Sorek. Eulerian-Lagrangian methods for advection-dispersion. 4. Int. Conference of finite elements in water resources. Hannover: Springer Verlag, 1982, pp. 14.41–14.68.

75. Nitsche, C. Beitrag zur mathematischen Modellbildung und digitalen Simulation von Stofftransport-, Stoffaustausch-, Stoffspeicher- und Stoffumwandlungsprozessen in der Aerationszone. Dresden: Techn. Univ., Sektion Wasserwesen, Dissertation A, 1979.

76. Ogata, A. Mathematics of dispersion with linear adsorption isotherm. Washington: U.S. Geol. Survey Prof. Paper, 1964 (411-H).

77. Ogata, A., and R. B. Banks. A solution of the differential equation of longitudinal dispersion in porous media. Washington: U.S. Geol. Survey Prof. Paper, 1961 (411-A).

78. O'Neill, K. Highly efficient, oscillation free solution of the transport equation over long times and large spaces. Water Res. Research 17 (1981) 6, pp. 1665–1676.

79. Parlange, J. Y., and J. L. Starr. Linear dispersion in finite columns. Soil Science Soc. of Am. Proc. 39 (1975), pp. 817–819.

80. Piessens, R. A. A bibliography on numerical inversion of the Laplace transforms and its applications. J. Comp. App. Math. (1975) 1, p. 115.
81. Pinder, F. P., E. O. Frind, and S. S. Papadopulos. Functional coefficients in the analysis of groundwater flow. Water Res. Research 9 (1973) 1, pp. 222–226.
82. Pinder, G. F., and W. G. Gray. Finite element simulation in surface and subsurface hydrology. New York: Academic Press, 1977.
83. Prakash, A. Finite element solutions of the non-self-adjoint convective dispersion equation. Journ. Num. Meth. Engin. 11 (1977), pp. 269–287.
84. Prikett, T. A., et al. A random-walk solute transport model for selected groundwater quality evaluations. Illinois State Water Survey 65 (1981), p. 149.
85. Sauty, J.-P. An analysis of hydrodispersive transfer in aquifers. Water Res. Research 16 (1980) 1, pp. 145–158.
86. Schäfer, D. Beschreibung des FORTRAN-Programms SEM 1 für den BC 5120 zur digitalen Simulation von Migrationsprozessen in beliebigen Stromröhren. Techn. Univ. Dresden, Sektion Wasserwesen, WB Wassererschließung, 1985.
87. Schestakow, W. M. Dinamika podzemnych vod. Moskva: izd. Moskovskogo Universiteta, 1979.
88. Schimmel, B. Programm FROB zur Berechnung der freien Grundwasseroberfläche in definierten Strömungsfeldern. Techn. Univ. Dresden, Abschluß arbeit im postgr. Studium Grundwasser, 1984.
89. Schwan, M. Beitrag zur analytischen Beschreibung von Transportprozessen im Grundwasserleiter bei diffusem Stoffeintrag. Acta Hydrochimica et Hydrobiol. 13 (1985) 1.
90. Segol, G., G. F. Pinder, and W. G. Gray. A Galerkin finite element technique for calculation of the transient position of the saltwater front. Water Res. Research 11 (1975), pp. 343–347.
91. Smith, I. M., et al. Rayleih-Ritz and Galerkin finite elements for diffusion-convection problems. Water Res. Research 9 (1973), pp. 593–606.
92. Stoyan, G. On a maximum norm stable, monotone and conservative difference approximation of the one dimensional diffusion-convection equation. Wiss. Konf. zur Simulation der Migrationsprozesse im Boden- und Grundwasser. Dresden: Techn. Univ., 1979, Bd. 1, pp. 139–160.
93. Talbot, A. The accurate numerical inversion of Laplace transforms. J. Inst. Math. Appl. 23 (1979).
94. Technischer Bericht 82/5: Analytische Lösungen der Schadstofftransportgleichung und ihre Anwendung auf Schadensfälle mit flüchtigen Chlorkohlenwasserstoffen. Mitt. des Institutes für Wasserbau der Universität Stuttgart, 1982.
95. Victor, N. Beitrag zur mathematischen Modellbildung und digitalen Simulation des Wärme- und Stofftransportes unter besonderer Berücksichtigung der thermischen Nutzung unterirdischer Wässer durch Wärmepumpen. Dresden: Techn. Univ. Sektion Wasserwesen, Diss. A, 1982.
96. Wengle, H. Numerical solution of three-dimensional and time-dependent advection diffusion equation by collocation methods. 4. Int. Conference on

finite elements in water resources. Proc. Hannover: Springer Verlag, 1982, pp. 14.87–14.97.

97. Willis, R., and W.-G. Yeh. Groundwater systems planning. Englewood Cliffs, New Jersey: Prentice-Hall, 1987.

98. Zienkiewitcz, O. C. The finite element method in engineering science. London: McGraw-Hill, 1977.

Part 3

Techniques of Parameter Estimation

3.1

Sampling Techniques

This third part deals with specific features of migration parameter estimation in the subsurface, without involving the special disciplinary problems of hydrochemical and hydrobiological analysis. Parameter estimation is always based on data obtained from rock, water, or gas samples collected in laboratories or in the field, or on data obtained from sensors installed in the system where the migration process occurs.

For local sampling, special devices and techniques are required to guarantee a representative sampling procedure (Chapter 3.1) and a reliable analysis (Chapter 3.2). However, parameters determined from local sampling may not be applicable at a larger scale of interest, thereby limiting the utility of local sampling. Field testing is therefore generally required (Chapter 3.3), including the onsite analysis of artificially induced migration processes (tracer studies). The representativeness of even field-scale testing may also be limited. This problem can be overcome only by monitoring migration processes taking place in the entire space of investigation (Chapter 3.4).

For migration parameter estimation, it is necessary to collect rock or soil samples (§3.1.1), soilwater or groundwater samples (§3.1.2), and soil air samples (§3.1.3). Devices and methods currently used, and the problems associated with them are presented in this chapter.

3.1.1 SAMPLING SUBSURFACE SOLIDS

Rock samples are collected to estimate physical, chemical, and microbiological parameters. This is necessary for the following reasons:

1. Undisturbed rock samples (cores) allow the easiest estimation of the dry bulk density ρ_b and the porosity ϕ of loose and solid rocks.

2. Soil samples collected from the unsaturated zone allow the identification of migrants, the monitoring of their movement, and the analysis of associated degradation processes. Soil sampling is particularly useful during the first period of contamination or when the migrants are strongly sorbed on the solid phase. Early diagnosis of a contamination problem is of major importance for efficient containment of the contaminants and for effective rehabilitation.

3. In the groundwater zone, rock samples may often be used to estimate the overall content of migrants and thus the contamination potential for groundwater. Such samples are the only means by which the multiple distribution coefficient (see also §1.5.2) can be estimated under real conditions.

4. Microbial populations controlling biochemical transformation processes in the subsurface are primarily bound to the solids in biofilms. It is frequently impossible to identify or quantify them solely from soilwater and groundwater samples. The most reliable method of identifying microorganisms is based on collecting sterile samples from the interior of undisturbed cores of aquifer material.

5. In any type of monitoring well, it is difficult to prevent surface microbes from penetrating into the subsurface and thus contaminating soil- or groundwater samples.

6. It is also frequently necessary to collect monoliths (large undisturbed rock complexes). In such monoliths migration processes are induced using special laboratory or field test equipment. Monolith tests allow the application of controlled boundary conditions and the use of smaller, more easily handled instrumentation than in field tests.

3.1.1.1 Sampling Equipment

Rock samples may be collected from the soil surface, from the bottom of a test pit or construction excavation, from shafts, or from boreholes. Since collection of disturbed samples requires no special equipment or techniques, it will not be considered further here.

Undisturbed, unconsolidated rock samples may be collected from the surface using the device shown in Figure 3.1. Stainless steel or brass sampling cylinders usually have a diameter of d = 5–10 cm and a length of 1–2 diameters. Sampling cylinders are hydraulically forced vertically or horizontally into the formation and successively removed by hand. Small undisturbed rock samples can be obtained as cores from consolidated rocks using hard rock coring machinery. Finally, monoliths are collected by carefully removing the surrounding rock. Figure 3.2 shows an example of preparing a soil monolith for a lysimeter.

In unconsolidated rock boreholes, thin-walled sampling tubes are primarily used for collecting samples. The sampling tubes are forced into the undisturbed rock at the bottom of the borehole by means of the drill

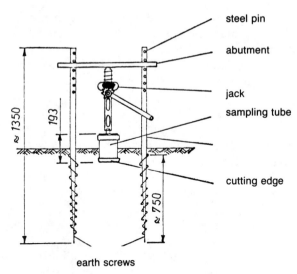

steel pin

abutment

jack

sampling tube

cutting edge

earth screws

Figure 3.1. Device for taking undisturbed unconsolidated rock samples. *Source:* Busch and Luckner [3.12].

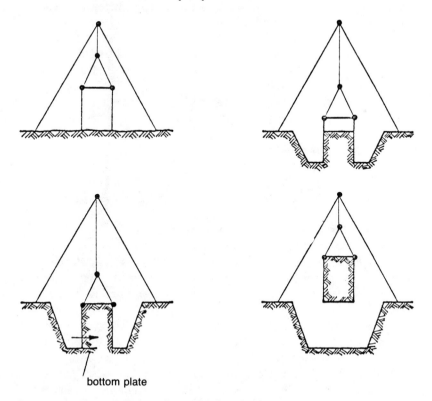

bottom plate

Figure 3.2. Extraction of a soil monolith for a lysimeter.

rod. Thin-wall steel cylinders 2.5–10 cm in diameter and 20–50 cm in length are preferred for this type of sampling (see Figure 3.3a). A cutting edge at the tip of the sampler, with an inner diameter somewhat smaller than that of the sampling tube, is used to reduce the disturbance of the sample and the danger that the sample may slip out when the sampling barrel is removed. This is of great importance in the groundwater zone when noncohesive unconsolidated rock material is present (e.g., loose sands). Borehole sampling devices which force the sampling tube horizontally into the borehole wall have proven highly successful (see Figure 3.3b and Busch and Luckner [3.12]).

Collection of samples from boreholes drilled with mud or other drilling fluids is limited because drilling mud penetrates into the material to be collected. This is practically unavoidable; hence, it is better to use air or at least clean water as the drilling fluid. Recommended techniques for sampling in fluid-drilled boreholes are shown in Figure 3.3c [3.4].

Hollow-stem augers are the optimal tool for drilling and sampling in unconsolidated rocks. Samples may be collected vertically through the hollow stem without removing the auger (see Figure 3.4). If the hollow stem is screened, the water level during drilling can be measured. The use of a casing avoids cross-contamination.

Soil and rock samples obtained in the groundwater zone from boreholes are subject to contamination from drilling mud and drilling cuttings. Therefore, the exterior part of such samples should be removed, and only the interior part should be used for analysis. Borehole geophysics, a nondestructive measurement technique, may make sampling superfluous, depending on the parameter one wishes to determine.

3.1.1.2 Treatment of Samples

Rock samples should be analyzed as quickly as possible, if necessary in situ. Generally, they are transferred to a laboratory in the sampling tubes, which are tightly covered and sealed with wax. The sealed tubes are put into sterile plastic bags, which are also sealed.

Unconsolidated saturated samples should be treated with special care. Recommendations for sample preservation in laboratories depend primarily on the parameters to be investigated. Thus, recommendations range from maintaining the natural rock temperature to freezing at –45°C.

To divide core samples, the material may be slowly forced out of the sampling tube after the cutting edge has been removed (usually affected in situ). During this procedure subsamples for physical and chemical investigations may be collected, and small sterile stainless-steel tubes may be forced into the middle of the sample to take subsamples for microbial investigations (see Figure 3.5) [3.99; 3.23].

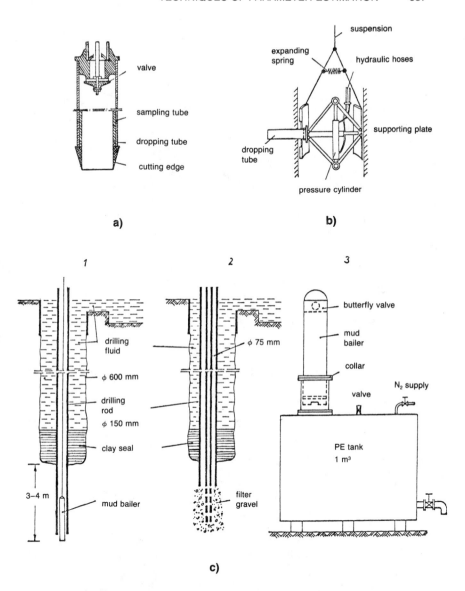

Figure 3.3. Thin-wall sampling tube with a cutting edge for unconsolidated rock sample collection in a borehole. *(a)* Device for vertical sampling. *(b)* Device for horizontal sampling. *(c)* Collection of unconsolidated rock and groundwater samples from fluid drilled boreholes, simplified according to Assmann et al. [3.4]: *1,* placing and forcing the boring rod below the borehole bottom, setting up a clay seal and collecting the rock sample; *2,* installing a screen, casing, and gravel pack and withdrawal of the water sample, subsequent withdrawal of the temporary casing and continuation of drilling; *3,* sample mixing vessel.

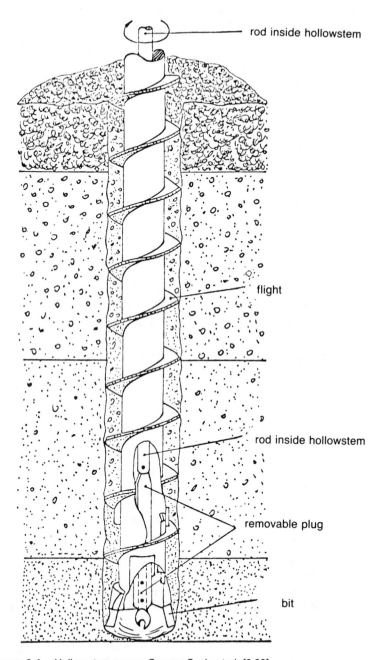

Figure 3.4. Hollow-stem auger. *Source:* Scale et al. [3.99].

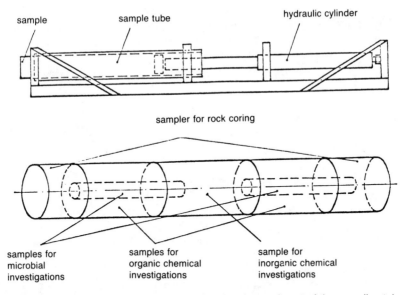

Figure 3.5. Device to force an unconsolidated rock sample out of the sampling tube and example of sample separation. *Source:* Scale et al. [3.99].

The population of aerobic microorganisms may change as a result of airtight seals on samples, which lock out oxygen and prohibit escape of gases. Therefore, samples should either be tested for microorganisms within a few hours of collection or cooled and frozen. Alternatively, they may be preserved in a polyethylene bag, which permits air but not water vapor to pass. However, if anaerobic microorganisms are the investigation objective, samples must be carefully protected against exposure to oxygen. This may be accomplished using sterile glass tubes from which air is evacuated and replaced with nitrogen, and repeating the procedure several times.

Formation samples must be treated in compliance with the regulations effective in the municipality where the overall project takes place. Procedures of chemical and microbiological analysis or the extraction procedures required prior to final analysis may differ between countries, states, and even counties.

3.1.2 COLLECTION OF SOILWATER AND GROUNDWATER SAMPLES

Soilwater and groundwater analyses are strongly influenced by sampling techniques and procedures. Selection and development of sampling sites are important in order to ensure the representativeness of the samples taken. Locations of sampling sites and the design of sampling schedules usually proceed from scenario analyses – in other words, based upon migration simulations with hypothetical parameters and initial and boundary conditions. Simulation methods are discussed in Part 2 of this book. This section focuses on the technically appropriate design and development of sampling sites, the proper selection of sampling devices, and the treatment of water samples in situ and their transport to laboratories.

3.1.2.1 Sampling Sites

A sampling site should not impact the quality of water samples. Changes in water samples should be minimized and controlled [3.123; 3.107; 3.115]. The following principles should be considered:

1. Each installation of a sampling site implies that the natural system of the soil or aquifer has been altered. The changes caused in the immediate environment of the sampling site, including changes in the rock matrix, the microbial ecology, and the migrants, should be forecasted as best as possible and minimized by appropriate construction.
2. The path of the water to be collected in the sample should be anticipated. Provisions should be made to ensure that the sample will be drawn from the desired part of the soil-or groundwater zone. This should be guaranteed by appropriate design and construction of the sampling site.

These principles should be reached with minimum expense.

Springs and Tunnels

Springs, drainage shafts, ghanats, tubes, or tunnels where groundwater flows freely out of the subsurface must usually be developed for regular sampling by construction measures which ensure that the water is collected immediately at the outlet without any mixing with surface water. It must also be guaranteed that surface and rainwater, or animals such as snails, worms, and insects, cannot penetrate into the sampling site [3.123].

Supply Wells

Samples from supply wells are collected by tapping the well pipe above the soil surface and ahead of the well valve. Supply wells represent a strong interference to the natural flow of an aquifer system. They do allow groundwater sampling, but only with some restrictions. Supply wells are usually screened and developed over large sections of the aquifer thickness, sometimes extending over several water-bearing layers.

Therefore, samples collected from supply wells usually represent mixed water samples.

During periods of no pumping, supply wells may cause short-circuit connections between different water-bearing layers. In such cases, these wells allow layers having a lower hydraulic potential to be recharged from those having a higher potential, and thus occasionally cause significant changes to the groundwater quality or contamination of the recharged layer. Vertical flow in idle wells may be verified by use of a highly sensitive flow meter, tracer, or temperature log measurements. Flow meter logs should be supplemented by caliper logs when the wells are not cased. Flow and temperature measurements can also be used in wells to verify the location of discharge areas during pump operation. Figure 3.6 shows two types of such flow meters.

Tracer transport measurements to detect vertical flow in well screens have proven to be effective. Analysis may be performed either with the

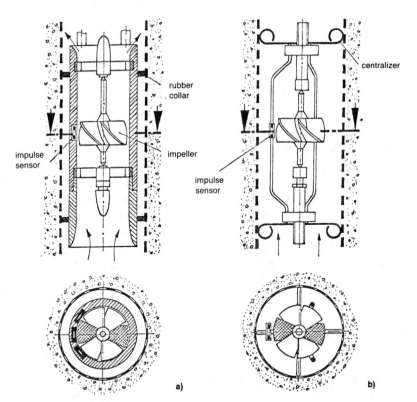

Figure 3.6.   Sensitive impeller flow meters for measuring vertical flow velocities in well tubes: (a) flow meter according to Folkens and Miersch [3.29] with the ability of reducing the cross section to increase the flow velocity; (b) flow meter, minimally influencing vertical flow velocity.

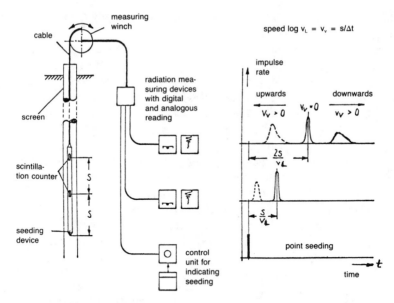

Figure 3.7.   Geophysical probe for measuring vertical spread of a radioactive tracer in the well tube. *Source:* Drost [3.19].

peak-peak technique or the total-count technique. For either technique, the equipment shown in Figure 3.7 may be used. The evaluation of the measured data should be based on a simulation of the vertical tracer transport in the well, that is, in both the screen and the filter pack.

Well head construction for supply wells must prevent surface water, rainwater, and organisms from penetrating into the subsurface. In pumping tests, the pumped water should be discharged at a sufficient distance from the well head to avoid geohydraulic short circuits. To evaluate groundwater analyses, the geological well log, the documentation of the well construction, and the details of water sampling, such as pump specifications and position of its intake, must be known.

Monitoring Wells

Monitoring wells may therefore be thought of as groundwater gauging stations. Their location, design, and construction should guarantee the measurement of representative groundwater levels and the collection of representative groundwater samples. They should meet the following requirements.

Sectional Design

The diameter of monitoring wells should not be larger than necessary. A two-inch casing, without section-reducing parts, generally allows reliable sampling with bailers and pumps, and a maximum 1–2 in. annular space is sufficient to permit the placement of a grout seal. Thus, borehole diameters of 4 in. (to a maximum 6 in.) are recommended. Larger borehole diameters are unnecessarily expensive and cause undesired increases of both the quantity of stagnant water within the well and the disturbance of the natural hydrogeological, hydrochemical, and biological conditions. The zones around the screen of monitoring wells should be backfilled with washed sand or gravel.

Grouting and gravel packs may be avoided if minimal annular spaces are guaranteed. In this case, the in situ unconsolidated aquifer material should move into the narrow annular space and be prevented from penetrating the slotted openings of the screen by wrapping the perforated sections with plastic screen gauze, if necessary. Glued gravel filters are also occasionally inserted into narrow boreholes.

Refilling the annular space with rock cuttings derived from drilling can lead to artificially induced disturbances. This is due primarily to oxidation and contamination of the cuttings during and after the drilling process.

Monitoring wells must be developed—that is, an intensive cleaning procedure of the borehole walls and the adjacent formation is required. The drilling process may cause crusting and clogging of the borehole walls and clogging of the adjacent rock formation. Well development will help to ensure that no artificial geological, chemical, or ecological local environment is created through well installation in the zone from which water samples are to be collected.

In developing monitoring wells, forceful up-and-down motion of a surge block (round plunger) or pulsing of a submersible pump (start-stop operation), usually with double packers to limit the development space, are used. If submersible pumps are utilized, they must be capable of withstanding the sand transport they induce. Sediments should not be allowed to remain inside the casing of monitoring wells. They should be removed after completing the development and periodically thereafter.

If air lift pumps are used in well developing, the air intake should always be above the well screen to avoid the penetration of air into the adjacent rock formation. The importance of careful development of monitoring wells should never be underestimated (see also Scale et al. [3.99]).

Depth of the Well Screen

In general, monitoring wells should be constructed to allow ground-water sampling from the appropriate depth. The screen length or the length of water inflow should be limited to 0.5–1 m or 2 m. In cases where the annular space around the screen is refilled with sand or gravel pack, the remaining sections should be carefully sealed to avoid any possibility of vertical communication or short-circuiting between the water-bearing layers. Improperly constructed and sealed monitoring wells often cause serious cross-contamination or saltwater intrusion into freshwater-bearing layers from deeper saltwater horizons. It is also important to ensure that the well bottom and the connections of the well casing are tightly sealed. For deep monitoring wells, such leaks may cause serious problems [3.3].

Monitoring wells should only be put into operation after proving that no vertical flow occurs by the use of tracers. Flow meter measurements are often not sensitive enough to prove this, because the highest sensitivity available in a flow meter is approximately $v_v = 1$ mm/s, which is equivalent to a transport distance of nearly 100 m/day. If the ground-water quality is affected by a vertical short-circuit flow, the monitoring well should be reconstructed [3.3], or partially or completely grouted to stop the undesired communication between different water-bearing layers or aquifers.

If groundwater is investigated for pollutants which have a density less than that of water ($\rho < \rho_w$), water samples should be collected from the upper zone of the aquifer. In the case of unconfined aquifers, this is the fluctuation zone of the groundwater table. In the case of semipermeable confining layers, this zone includes both the variation zone of the groundwater table and the zone near the interface between the aquifer and the top confining layer.

Because migrants often travel along preferential paths, it is essential to place screens at the proper depth. Otherwise, the danger arises that migrants might pass the monitoring well in a zone of solid casing without being detected. This danger should be reduced by careful sampling during drilling (see §3.1.1.), supplemented by geophysical logging of the borehole, if necessary.

Material Selection

The material to be used for constructing monitoring wells should cause the least possible interference with the local environment. In particular, the subsurface water should not leach out any substances which could change water quality or impact microbial activity. Further, construction materials should not adsorb substances nor catalyze transformation processes either chemically or biologically. The material selection problem is most commonly driven by considerations of cost versus accuracy.

In most cases, PVC can be used. In the United States, PTFE (Teflon) is sometimes recommended since PVC casing may bleed organic components, although only to a minor extent, or PO_4^{3-} may be adsorbed onto PVC surfaces (see Salacz [3.96]). PVC surfaces also may aid biofilm generation.

The installation of metal casings or parts (e.g., spacers) should be avoided unless justifiable. They should be used only as the exception since they usually corrode in groundwater, releasing metal ions, thereby changing the redox potential and consuming oxygen. If a few monitoring wells exist with metal casing, one PVC well should be installed to quantify these interferences by parallel monitoring.

In any critical application, it is recommended that the influence of the available materials on subsurface chemistry be predicted a priori. If this prognosis demonstrates that a representative sample cannot be obtained, because no appropriate materials are available, the construction of a monitoring well serves no purpose.

The well casing should be carefully cleaned with a purifying agent and rinsed with clean water before installation. This also applies to other items to be used in the well. Such cleaning will help remove grease, oil, and other contaminants that may stick to the equipment and be carried into the subsurface.

Multiple Wells

A multiple monitoring well is a gauging station containing more than one measuring probe arranged vertically in a borehole. Such a well allows groundwater sampling from several depths. The best-known construction of a multiple monitoring well is shown in Figure 3.8a, where two or three observation tubes are installed in the same borehole.

A far larger number can be installed in one borehole when the observation tubes are bundled before their installation and the gaps between them are sealed with a watertight foam to avoid preferential vertical flow paths. The drilling diameter should be approximately 5 inches larger than the bundle diameter, to provide a temporary space for a 2-inch tremmie pipe to be used for grouting with inert sand or fine gravel mixed with cement slurry. Careful measurements of the depth of grouting in the borehole should be collected by geophysical logging using one of the observation tubes.

For very deep boreholes, the construction shown in Figure 3.8b is recommended if the various water-bearing layers are to be sampled once. After completion of drilling, the annular space between the centered casing and the borehole wall is grouted. After that, perforations which

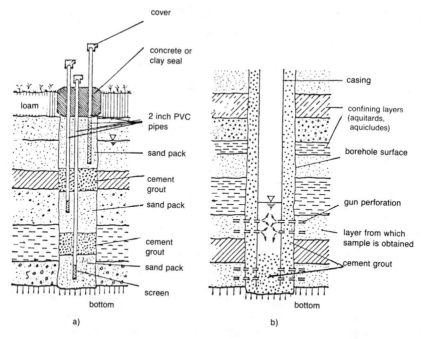

Figure 3.8. Multiple monitoring well: *(a)* multiple completion well for periodic sampling according to Busch et al [3.12]; *(b)* multiple completion well for one-time sampling according to Scale et al. [3.99].

allow water to enter the casing from the lowest permeable layer are shot by a gun. Following cleaning and sampling from this layer, the casing is grouted with cement up to the next deepest layer. That layer can then be gun-perforated after the concrete has cured. It is not appropriate, however, to sample migrants which react chemically with the cement grout. Further types of multiple monitoring stations are shown in Figures 3.9, 3.20, 3.21, and 3.23.

Minifilters

The term *minifilter sampling technique* became a synonym for a whole group of various devices and methods allowing representative subsurface water sampling from a vertical profile. Minifilters are small elements collecting soil-or groundwater. The water collected in these filters is pulled or pressured up to the soil surface. Minifilters are commonly attached to the outside of a carrying pipe with the delivery hoses for all minifilters protected inside this pipe (see Figure 3.9).

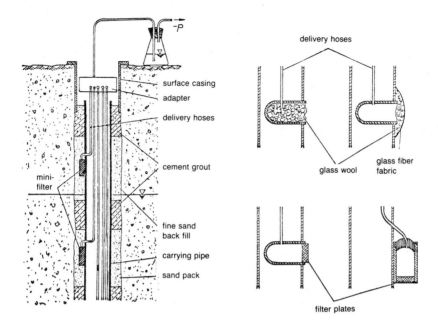

Figure 3.9. Minifilter gauging station with typical sampling equipment.

Such carrying pipes, equipped with minifilters, should be installed in boreholes drilled by cable tools or flight augers. If the annular space is small enough, no gravel pack is required. By removing the drill pipe or stem, formation material will collapse into the annular space, enclosing the filter. However, washing filter sand into the annular space surrounding the minifilters and sealing the sections between them by clay cement grouting is a more difficult installation procedure.

In the groundwater zone, the water pressure is by definition $p_{gw} = p_w \geq p_{atm}$. Therefore, groundwater enters a space free of solids if the air pressure in this space is less than p_{gw}. In the vadose zone, the water pressure is $p_w < p_{atm}$. There, water is either the only fluid phase in the rock matrix (in the closed capillary fringe), or air and water are both present as immiscible fluid phases. In this zone, subsurface water enters a hollow space (e.g., the minifilter) only if there is a separating membrane which allows water but not air to pass, in addition to the vacuum which is to be generated in the hollow space. Measuring probes using such membranes are called *tensiometers*.

Separating membranes are also required for the selective collection of other immiscible fluids. Membranes made of ceramics, sintered materials, plastics, or glass permit water to pass and prevent air from passing up to the air breakthrough point, a specific capillary pressure characterizing the membrane material [3.15; 3.74]. They should be highly permeable to water, permit air to pass only at high capillary pressures, and should have minimal impact on the quality of the sampled water. These requirements, particularly the first two, are contradictory, and compromises are required.

Figure 3.10 shows typical designs of tensiometers with porous ceramic cups. Before installation, the membrane bodies should be tested in the laboratory for the three properties mentioned above. Figure 3.11 shows some laboratory tests reflecting serious changes of water quality caused by passage of model water through a ceramic membrane body. Tensiometers should be installed with special care to ensure that the porous membrane makes a good contact with the backfill material, for which silt or silty fine sand is usually used. Impairment of this contact is often the cause for tensiometer failures in permanent field gauging stations [3.120].

3.1.2.2 Sampling Devices

The following explanations focus on sampling from monitoring wells and minifilters. In general, the devices to be used should be

- economical and sufficiently rugged to be used in situ
- chemically inert (choice of material)
- easy to clean and sterilize
- operable with portable power sources available in the field
- capable of developing the material immediately surrounding the monitoring point as well as removing stagnant water
- capable of sampling the soil- or groundwater after such procedures

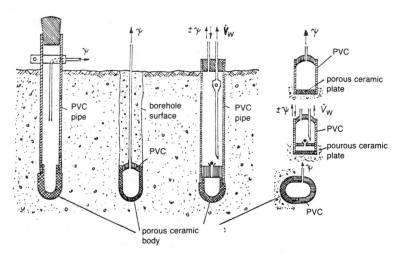

Figure 3.10. Typical designs of tensiometers with ceramic membrane cups as soilwater collectors.

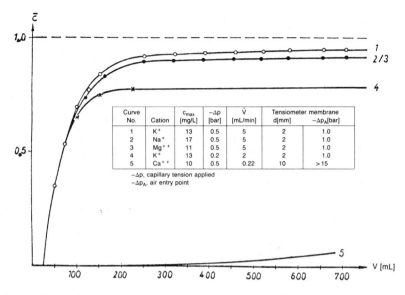

Figure 3.11. Example of a water sample changing quality while passing a porous ceramic cup. *Sources:* Nitsche [3.76] and Raquet [3.85].

A combined use of different devices is usually necessary.

External Pressure and Vacuum Devices

The simplest sample extraction method is to apply pressure or vacuum to an established hollow space (collection chamber) in order to force the collected water sample to the soil surface. Figure 3.12 shows some typical examples.

Vacuum Pumps

Vacuum extraction is restricted to a suction lift of about 7 m or 0.7 bar. When water is vacuum extracted it is exposed to low pressure, which may have a strong influence on the content of dissolved gases.

Therefore, the integrity of a subsurface water sample drawn by vacuum is often not guaranteed. Concentrations of O_2, CO_2, the easily oxidizable constituents Fe^{2+}, NH_4^+, NO_2 or the volatile hydrocarbons are often altered. When suction pumps are used, samples should be collected prior to entering the pump body to prevent further quality alterations. Electrically driven centrifugal pumps may also produce significant water quality alterations. They create high turbulences and pressure gradients in the impeller zone and often create a significant electromagnetic field around the engine, which may polarize ions in subsurface water. Figure 3.13 shows an example of alterations resulting from use of a centrifugal

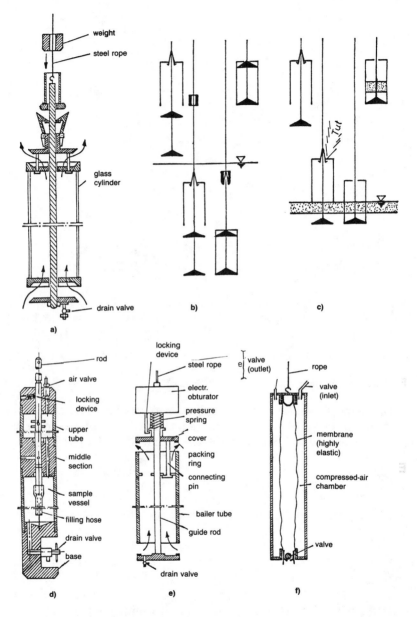

Figure 3.12. Recommended pressure and suction equipment for subsurface water sampling: (a) ram probe with withdrawable steel ram pipe; (b) 1½-to 2-inch monitoring well; (c) tensiometer with return valve and reversible suction/pressure hand air pump; (d) manually operated suction piston pump; (e) pneumatic suction pump with sampling vessel (arranging of two vessels is better, where the first has no water-air interface); (f) sampling from the suction stream, avoiding a gas-water interface.

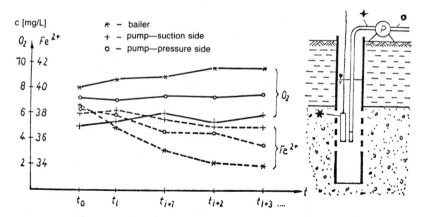

Figure 3.13. Example of a water sample being influenced by the sampling point and the sampling equipment. *Source:* Schmidt [3.102].

pump. It is advisable to test sampling equipment with artificial water (a blank sample of known quality) in the laboratory prior to applying the equipment in the field [3.108].

Hoses in which vacuums are used should not exceed a diameter of 1/2 inch and a length of 12–13 m and should be equipped with a check valve to prevent return flow.

Pressure Pumps

For the most part, using pressure fluids to push water to the surface leads to less alteration in water chemistry than vacuum extraction — particularly, if an inert gas, such as N_2, is used as the pressure fluid. Even so, the use of pressurized gases may also disturb the gas balance of samples and may transform migrants. Squeezing water samples against a membrane with a comparatively low hydraulic conductivity or against rock material of low permeability is also often used. Check valves are commonly applied in conjunction with pressure to increase the operational reliability and efficiency.

Water-Driven Pumps

Recently, the use of jet pumps has become widespread. Jet pumps are water-operated suction pumps equipped with driver and collector nozzles combined with a diffuser (see Figure 3.14b). Jet pumps are installed inside the well casing. The driving jet is generated by a pump installed above ground. This pump should cause only minor changes to water quality. Peristaltic tubing pumps, roller or vane pumps [3.83], or membrane pumps are best suited for this purpose because they cause minimal turbulence and local pressure gradients.

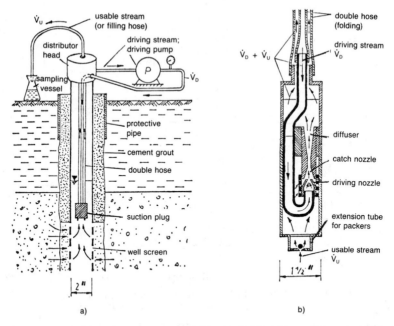

Figure 3.14. Jet pump after Nitsche and Kritzner [3.77]: *(a)* overall plant and *(b)* body of the pump.

In general, jet pumps attain lifting heights of up to approximately 30 m, and in 2-in. well casings they attain pumping rates of about 1 to 5 L/min with a driving jet of about 5 to 10 L/min. In combination with a double packer, they are thus suited for developing a monitoring well and removing stagnant water. Jet pumps are less suited for sampling. Nevertheless, with properly designed jet pumps, water quality changes may be kept within allowable limits in the nozzle-diffuser combination, in the hose system, and in the pump generating the driving jet. Therefore, jet pumps may be a suitable compromise for many tasks in groundwater quality monitoring (see Figure 3.14 for examples).

Bailers

Bailers are comparatively inexpensive sampling devices. In general, they allow fairly reliable sampling from monitoring wells with casings greater than 1½ in. in diameter. Usually, bailers for 2-in. casings have a diameter of about 4 cm, are about 1 m long, and permit sampling of about 1 L of water.

Bailers are lowered by hand into the well on a plastic or steel rope marked with depth indicators at 1-foot intervals. Figure 3.15 shows some typical examples (see also Löffler [3.57]). Bailers can be made from

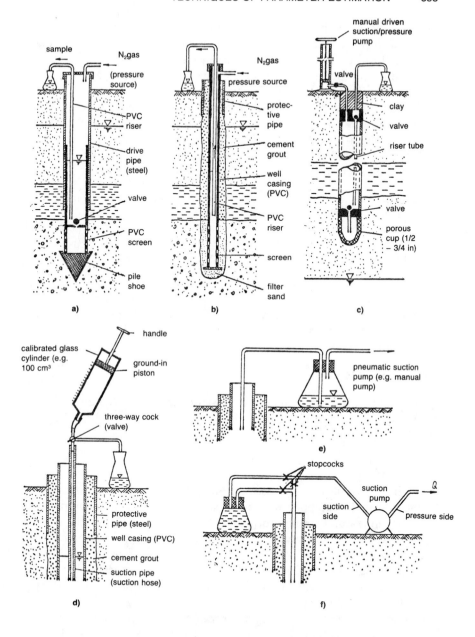

Figure 3.15. Examples of bailers: *(a)* Ruttner bailer for 2-inch well casings; *(b)* samplers after Blasy [3.132]; *(c)* sampler after Blasy [3.132] for withdrawing floating substances (e.g., oils); *(d)* bailer after Nestler et al. [3.73]; *(e)* valve bailer after Jordan et al. [3.41]; *(f)* squeeze bailer (the drain valve is opened by drawing the rope with a jerk, or the drain valve is operated by means of a pressure hose from the soil surface).

different materials that are adaptable to the water quality to be sampled; thus, for instance, brass components should be avoided when sampling for Hg. However, the greatest advantage in using bailers is that they do not require any special power source onsite for their operation. In addition, due to their low air-water interface volume ratio, they scarcely permit degassing such as escape of volatile halogenated hydrocarbons. Bailers may also be used to sample very deep wells. Examples of bailers for sampling from depths of 500–1000 m are presented in Hofreiter [3.39] and Repsold and Friedrich [3.87].

Use of a bailer has the following limitations:

1. In practice, they do not allow the removal of stagnant water from the monitoring well prior to sampling.
2. Mixing processes take place in the well casing above the sampling point, making additional sampling from the same or a smaller depth difficult; therefore, a subsequent sample can only be taken from a depth 1 to 2 m deeper than the previous sampling point.
3. Bottling from bailers may lead to aeration.

In practice, bailers are often used only for sampling, while removal of stagnant water before sampling is best accomplished with special pumps, such as airlift pumps.

Internal Pumps

Internal pumps are herein defined as sampling devices that are permanently installed in well casings below the groundwater level or in minifilters.

Airlift pumps

Airlift pumps blow compressed air through a hose or pipe either open at the bottom or equipped with a special pump head. These pumps are lowered into a water column formed by the well casing or a special tube. The injected air creates small bubbles which disperse into the water. The generated air-water mixture is less dense than the surrounding groundwater, which causes it to rise in the well casing or in a special tube up to the soil surface. Efficiency of air lifting is significantly increased by properly designed pump heads.

The water-air mixture can rise directly in well casings if their diameter is less than 3–4 inches. In larger casings, a special riser tube (1½-2 inches) is necessary to ensure effective operation. Concentric hose constructions, with the air injection hose contained within a rubber-lined textile hose, such as a fire hose, have proven successful. Such double hoses only require small hose winches [3.116].

Figure 3.16 shows some typical examples of airlift pump heads. The

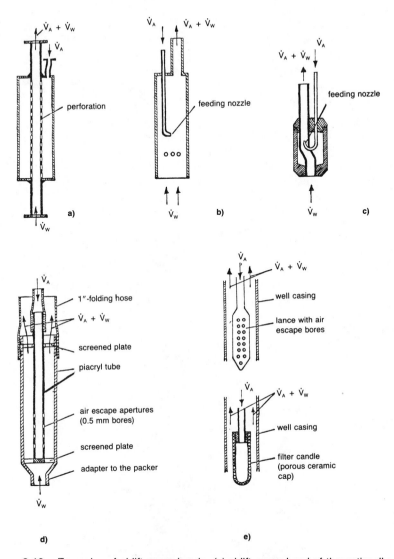

Figure 3.16. Examples of airlift pump heads: *(a)* airlift pump head of the nationally owned enterprise VTK Dresden; *(b)* airlift pump head of the Water Authority Potsdam [3.117]; *(c)* airlift pump head of the GDR Academy of Sciences, Institute of Geography and Geoecology; *(d)* airlift pump head of the Dresden University of Technology, Water Sciences Division [3.116]; *(e)* pressurized air, fed directly into the well casing.

capacity characteristics of airlift pumps may be expressed as $H_F = f(\dot{V}_W, \dot{V}_A)$, where H_F is the lift height, $\dot{V}_W$ is the pumping rate, and $\dot{V}_A$ is the consumption of compressed air per unit time. This function is strongly affected by the immersion depth (depth of air injection below the water table), the size of the air bubbles generated in the water-air mixture (design of the pump head), and the cross section of the riser (see Turnow [3.116] and Schurig [1.108]).

When set at a sufficient immersion depth, airlift pumps are very well suited for developing monitoring wells and removing stagnant water. By injecting air, however, water quality is strongly affected (air-stripping effect). Only alkaline earth (e.g., Ca, Mg), alkalies (e.g., K, N), chloride, and nonoxidizing water constituents, such as nitrate, sulfate, and phosphate, are not affected. Airlift pumps should be used in combination with sensors installed below the air injection point or in combination with a pneumatic squeeze pump for sampling [3.43].

Submersible Centrifugal Pumps

Submersible centrifugal pumps are offered by many pump manufacturing firms for well casings of diameter greater than 4 inches. Figure 3.17 shows an example of a typical mobile, electrically operated pump

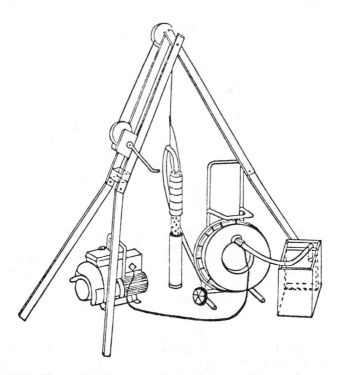

Figure 3.17. Electrically operated, mobile 4-inch submersible pumping equipment.

assembly with a collapsible aluminum tripod. This assembly contains a rope winch, an electrically operated submersible 4-in. centrifugal pump for 20–100 m lifting height and 0.5–1 L/s output, a mobile combination hose/cable winch for a 1-in. hose with electric cable, a gasoline engine- -driven electric generator, and the measuring instrumentation [3.132]. Submersible pumps have also been developed for small casing diameters (see, e.g., Jordan et al. [3.41]); however, their capacity is usually insufficient.

For sampling purposes, submersible pumps are suitable only on a limited scale. The high turbulence and pressure gradients occurring in the pump impeller zone, as well as the electromagnetic field generated by the engine, can frequently lead to significant water quality changes.

Submersible Squeeze Pumps

Submersible squeeze pumps consist of a hollow pump body for collection of subsurface water. Water is squeezed from this body and expelled above ground by displacing a flexible diaphragm by means of compressed air. These pumps are normally equipped with two check valves. The outer diameter of the pump body is typically 1 or $1^1/_2$ in. for $1^1/_2$-or 2-in. well casings. The compressed air required for the pump should be generated by an electric or gasoline-driven compressor because compressed gas bottles permit only short periods of operation.

For a $1^1/_2$-in. pump body, a 200–400 cm^3 displacement is typical, permitting pumping rates of 1–10 L/min at lifting heights of 20–100 m under manual control (approximately 10 displacement cycles/min) [3.99; 3.117; 3.91; 3.127]. Figure 3.18 shows some characteristic constructions. The efficiency $\dot{V}_w/\dot{V}_a$ will be significantly increased if two separate lines for compressed-air inflow and air exhaust are installed, and control valves in the pump body are switched automatically by the diaphragm position (see also Figure 3.20b).

For sampling, squeeze pumps are optimal: they cause minimal turbulences and only minimal water pressure depletions or pressure gradients. The impact of the pump body materials and its built-in components (particularly the diaphragm) on water quality should be investigated carefully before sampling. For field use, equipment such as shown in Figure 3.17 has proven effective, with tripod, rope winch, gasoline engine compressor, hose winch for the compressed air and discharge hoses, measuring instrumentation for the discharged water, and controlling device above ground. Submersible squeeze pumps are also suited for displacing stagnant water, at least in combination with double packers. However, squeeze pumps should not be used in developing monitoring wells; for this purpose, airlift or jet pumps should be employed (see Figure 3.20c).

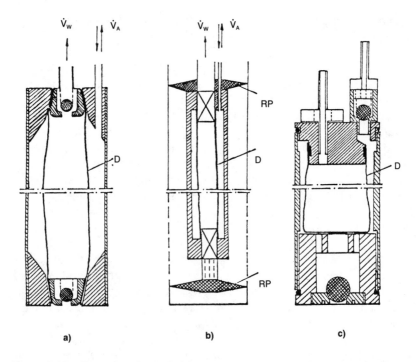

Figure 3.18. Examples of submersible squeeze pumps (*D*, highly elastic diaphragm; *RP*, rubber packers): *(a)* pump with an above-ground control of inflow and outflow air; *(b)* pump after Riha [3.91]; *(c)* pump after Urban and Schettler [3.117].

Submersible Piston Pumps

Submersible piston pumps are produced and used in three versions:

1. mechanically driven cable piston pump
2. electrically driven vibrating membrane pump
3. gas-driven, double-action piston pump

Figure 3.19a shows an example of a cable piston pump. Here, an engine above ground drives a crank with a 30–40 cm lift, where the crank lift is equal to the piston stroke. The pump cylinder is fixed by a stiff riser pipe (1–2 in.), and the piston is forced up and down with a rope and a spring, where the rope draws the piston up and the spiral spring draws it back into its initial position. With the pump described in Schenk [3.100], using a piston diameter of 30 mm and a stroke of 400 mm, capacities of 15 L/ min at a pumping height of 30 m are obtained. However, the stiff riser tube and the crank drive make the pump inconvenient. Therefore, it is not widely used.

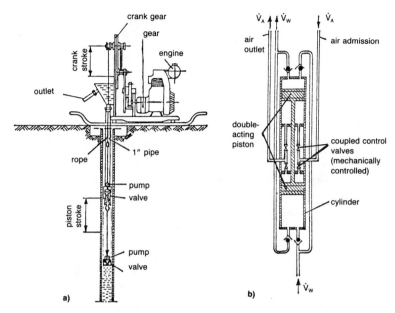

Figure 3.19. Examples of submersible piston pumps: *(a)* cable piston pump after Schenk [3.100]; *(b)* double-acting piston pump with internal mechanical automatic control.

Electrically operated vibrating membrane pumps, used in 2- to 3-in. well casings, are presented in DVWK Merkblätter zur Wasserwirtschaft [3.123] and Käss [3.42]. They require an electric power supply of 25–65 W and have an output of 0.1–1 L/min at pumping heights of up to 50 m. However, these pumps probably will not be important tools in the future.

Gas-driven, double-acting piston pumps are regularly used for sampling. Though more expensive, they generally have a higher capacity [3.79]. They are equipped with an internal self-controlling valve system, and therefore have a high efficiency ratio $\dot{V}_w/\dot{V}_a$.

Double-acting piston pumps require a portable field compressor to be installed above ground. A bundle of three hoses containing the discharge hose and hoses for inflow and exhaust gas is also required. The equipment assembly closely resembles that shown in Figure 3.17.

The device described in SEBA-Hydrometrie [3.132] for pumping from 2-in. monitoring well casings has a capacity of 5–7 L/min and requires a 300 L/min (2.5 kW) compressor to guarantee pumping heights between 10 and 50 m, and a 500 L/min (3.7 kW) compressor pumping at heights of up to 120 m. Such pumps are described in Obermann [3.79], Scale et al. [3.99], Pneumatik-Unterwasserpumpe für Grundwasserprobenahme

[3.131], and Bianchi et al. [3.11]. The materials used for their construction are commonly PVC, brass, and stainless steel.

Packers

Packers permit the local sealing of a borehole or well casing. Straddle packers are among the more successful varieties. They allow sampling from specific sections of the vertical subsurface profile, assuming short-circuit vertical flows around the packers have been eliminated. Use of packers in the well screen only makes sense if there is no zone of preferential permeability around the screen—for instance no sand or gravel pack. Thus, when packers are to be used, the monitoring well should be designed accordingly. Straddle packers not only permit sampling, but also development of the well screens in isolated sections [3.78].

As a rule, packers consist of rubber sleeves which are inflatable with air or water. Figure 3.20a shows a straddle packer for a 2-in. well casing

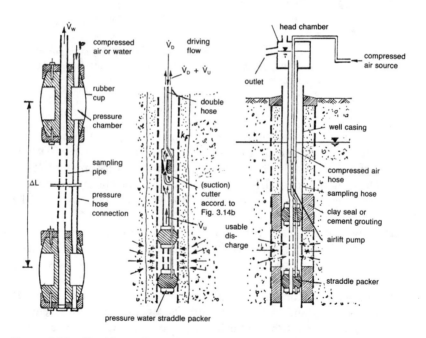

Figure 3.20. Straddle packers: *(a)* straddle packer for 2-inch well casings and boreholes after Turnow [3.116]; *(b)* packer combined with an injector pump; *(c)* packer combined with an airlift pump for well screen development.

consisting of a replaceable sampling pipe, which allows an adjustable sampling length. Packer sleeves may be inflated by a hand or power pump. Typical straddle packers are automatically inflatable by water or air, as represented in Figures 3.20b and 3.20c.

To reduce the sampling errors caused by short-circuit flow around straddle packers, quadruple packers may be used. Three screen sections are simultaneously pumped while samples are collected from only the central section, similar to a double infiltrometer [3.1].

The single special packer shown in Figure 3.21a serves a similar purpose [3.1]. The discharge flow of the submersible pump is supplied from the well screen sections above and below the packer; the sample is drawn from the comparatively stagnant groundwater zone in the middle. The multiple collector shown in Figure 3.21b represents another suitable technical solution for sampling from comparatively long screens [3.1]. A further useful solution is shown in Figure 3.21c for sampling at specific depths during pumping with any kind of pump.

Minifilter Sampling

Groundwater flows freely into minifilters, while a specific vacuum is required to collect water samples in a minifilter when installed in the vadose zone. In both cases, the water sample obtained may be sucked from the minifilter to the soil surface [3.113]. However, this method has the following disadvantages:

1. The depth of a vacuum operated minifilter for sampling in the vadose zone should be smaller than $7m - p/\rho_w g$, where p is the water pressure at the sampling point relative to atmospheric pressure $p = p_{abs} - p_{atm}$.
2. The application of suction, in general, disturbs the gas balance of a water sample. Degassing of such as O_2 and CO_2 cause concentration changes of other water constituents, particularly of the gas-sensitive constituents [3.36].

The first disadvantage is avoidable when the sample is pushed out of the minifilter by means of a pressure fluid [3.11]. Figure 3.22a shows a typical design of such a minifilter. Air or nitrogen gas commonly serves as pressure fluids. These compressed gases seriously interfere with the gas balance of the water sample and may cause serious changes of the various water constituent concentrations. This can only be avoided if a gas-tight, highly elastic interface, such as a rubber or Teflon membrane, is placed between the sample and the pressure fluid, as shown in Figure 3.22b.

Efficiency of sampling, when applying this pressure extraction method, may be greatly increased by employing check valves. However, check valves often fail due to solids or chemical incrustation. Proper filters must ensure that no solids get into the valves. Prior to in situ use, each minifilter construction should be tested in a laboratory for efficiency and its impact on water quality.

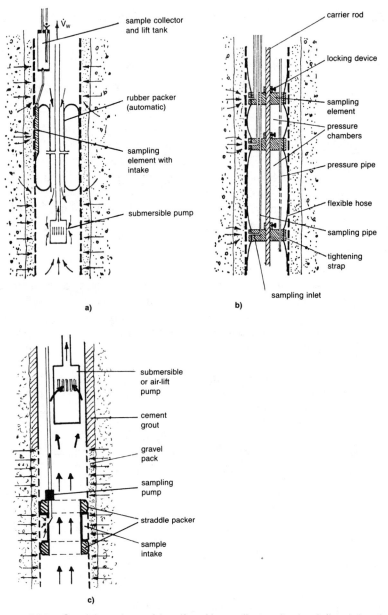

Figure 3.21. Special packers: *(a)* self-packing collector (to be inflated by the withdrawal flow of the submersible pump); *(b)* multiple collector for continuous well screen sections; (c) sampling device in an operated well.

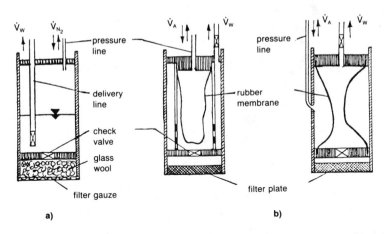

Figure 3.22. Minifilters: *(a)* typical construction with gas-water contact; *(b)* typical construction without gas-water contact.

Minifilters do not allow development of the observation point; i.e., they do not allow washing out drilling mud and other fine materials in their immediate vicinity. Therefore, it has proven successful to install the minifilters on a double tube, with the first containing the connecting hoses and the second designed as screens in the minifilter ranges. This second tube serves the development of each single minifilter surrounding by means of packers. To avoid vertical flow through this tube, it is later filled with fine sand or silt in the groundwater zone. This allows the removal by means of air lifting and a repetition of the development of minifilter surroundings. Figure 3.23 shows such a monitoring well gauging station installed at the well test field near the Nabeshima Hall of the Dresden University of Technology [3.59].

Special Sampling Equipment

For water sampling to analyze organic water constituents and microbes, special equipment is necessary [3.81; 3.86]. Equipment for microbial investigations must meet specific requirements ensuring that all implements are efficiently sterilized. Only glass, stainless steel, and PTFE (Teflon) materials should be used [3.16].

Figure 3.24a shows a sterilizable, simple sampling device in the bypass of a suction lift pump. The bypass line should only be connected after the stagnant water in the monitoring well has been removed by pumping. If the concentration of the organic water constituents is insufficient for applying a specific analysis procedure, it is advisable to enrich the constituents in situ, for example, onto resins. Figure 3.24b shows such a

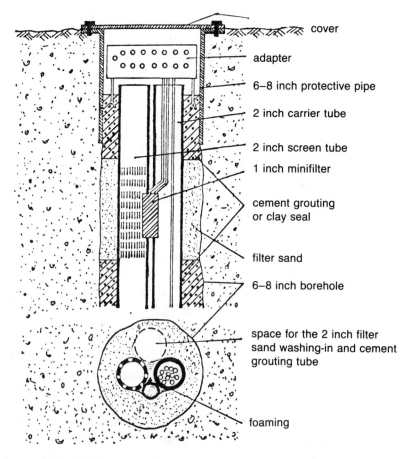

cover

adapter

6–8 inch protective pipe

2 inch carrier tube

2 inch screen tube

1 inch minifilter

cement grouting
or clay seal

filter sand

6–8 inch borehole

space for the 2 inch filter
sand washing-in and cement
grouting tube

foaming

Figure 3.23. Double-tube minifilter gauging station after Luckner et al. [3.59; 3.60].

device which allows a sampling discharge rate of 10–30 mL/min. The resin with the adsorbed constituents can then be backwashed and analyzed in the laboratory.

3.1.3 SOIL AIR SAMPLING

Gas sampling from the unsaturated soil zone is not a special technical problem. Soil air is the nonwetting phase of the subsurface (§1.2.3), thus being immediately withdrawable from soils without using a fluid-separating membrane (§3.1.2). Only a normal screen element is required for the solid-gas separation to retain the solid particles from the gas sample during sampling. Soil air is usually extracted either by manually or electrically operated vacuum pumps. Short-term and long-term mea-

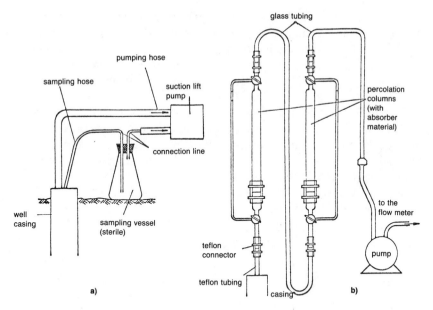

Figure 3.24. Special equipment for water sampling to analyze organic and microbial water constituents (see Scale et al. [3.99]): *(a)* sampling in the bypass; *(b)* continuously operating enrichment system.

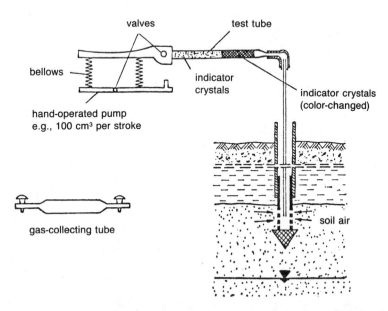

Figure 3.25. Soil air sampling and analysis by means of special test tubes (instead of the test tube, a gas-collecting tube could be used, the content of which is then analyzed in the laboratory with a gas chromatograph).

surements are distinguished. Often it is best to determine the constituents in the air samples immediately in situ using various techniques (see Dässler [3.17]). Due to the limited accuracy usually attributable to methods applied in situ, it may be advisable to collect additional soil air samples in gas-collection tubes for gas chromatographic analysis in the laboratory (see Figure 3.25 and Runge [3.94]).

In analyzing soil air, the constituents usually sampled include O_2, CO_2, H_2O, H_2S, CH_4, and other gases associated with specific weathering and biological transformation processes. In addition, halogenated hydrocarbons such as tri- and tetrachloroethylene may be sampled to monitor pollution processes in the subsurface [3.125].

3.2

Techniques for Sample Analysis

Measurement of the dependent variables of migration process equations such as concentrations, activities, and temperatures for all considered phases in the subsurface appears to be unattainable even in the future. Currently, measuring these variables is only possible in the mobile fluid phases, and then only after their separation from the immobile phases (see Figure 1.12). It is important to keep in mind that such fluid samples allow only the analysis of the intensive state variables of the subsurface. The estimation of the extensive subsurface state variables requires the analysis of the whole rock or soil sample.

Each separation induces quality changes by transferring the fluid sample from its natural environment to an environment which is characterized by different temperature, partial pressures, light, and interface conditions. Therefore, sample analysis in situ as well as sample preservation, transport, and analysis in the laboratory are of great concern in subsurface water quality monitoring.

Section 3.2.1 is devoted to water and air sample analysis, but not addressing the special problems encountered in chemical and biological analysis. Section 3.2.2 considers the estimation of migration parameters based on laboratory measurements of the state variables of fluid phases such as activity $a_{j,i}$ or concentration $c_{j,i}$. Restrictions associated with use of the subsurface migration parameters estimated from rock, water, and gas samples in the laboratory also are discussed. Section 3.2.3 assesses the validity of transferring such estimated parameters to field conditions.

3.2.1 ANALYSES OF WATER AND AIR SAMPLES

3.2.1.1 Field Analyses

Sampling errors associated with chemical concentration in the mobile fluid phase are minimized when the measurements are done soon after the phase separation has occurred and if the temperature, partial pressures, and light conditions are unchanged or only slightly changed compared to the conditions of the natural environment. These requirements inevitably lead to the use of well probes.

The measuring heads of most well probes can be equipped with different sensors [3.114; 3.130]. Frequently, the necessary turbulence of the groundwater in front of the "measuring window" is obtained by an electric driven impeller or by generating a water jet. The measuring head is attached to a cable and lowered into the well. The sensors are connected through this cable to a field measuring device equipped with a microcomputer displaying the measured values. The microcomputer also monitors the transmission of the signals from the sensors in the well, decodes the signals, compensates for measuring errors (such as the influence of temperature [3.93]), and stores the measured data.

The probe described in Stegmann [3.114] permits measuring of the dissolved oxygen concentration, pH, electrical conductivity, and temperature. It is 65 cm long and has a diameter of 51 mm. The pH sensor can also be replaced with a redox sensor.

Another type of probe (see [2.124, probe XIV]) uses a temperature-compensated silicium pressure transducer for depth-measuring up to 100 m. It has a diameter of 1.75 in. and permits the measurement of pH, dissolved oxygen, temperature, and electrical conductivity or total salinity adjusted for temperature.

A further miniaturization of such well probes (see Grünke and Hartmann [3.34], Arche [3.2], Hoffmann et al. [3.38], and Janata [3.40]) and the use of other ion selective electrodes, such as Na^+, Ca^{2+}, Cu^{2+}, Cd^{2+}, Cl^-, and NO_3^- would increase their application field. However, their practical use is still restricted. The impact of interfering ions, known as cross-sensitivity of ion selective electrodes, is still difficult to control. Nevertheless, the use of microprocessor ionometers for computing and correcting the impacts by means of models may help to overcome these restrictions in the near future [3.98].

After the soilwater or groundwater sample has been collected and transported to the soil surface, it is advisable to immediately determine certain state variables. Two techniques are commonly used:

1. flow measurement cells, equipped with ion selective electrodes
2. cuvette photometers or other quick analyzers suited for in situ use

Flow measurement cells guide the flow through a small nozzle forming a jet, which flushes the membrane or the measuring window of an ion selective electrode. This flow causes the turbulence necessary for accurate measurements. The electrodes are usually replaceable, which makes calibration and maintenance easier.

A flow cell frequently contains a set of electrodes. In such a cell, flow is successively directed to each electrode. Cells with one replaceable electrode have also proven successful. Flow measurement cells are used primarily to monitor short-term quality changes and make immediate measurements of selected quality parameters.

Flow cells are also used to establish bottling times for collection of laboratory samples. This time is reached when the quality parameters monitored in the pump discharge reach steady-state values. It is commonly assumed that water withdrawn after this time is representative of formation fluids and is not influenced by the monitoring well. In general, the temperature T, conductivity χ, dissolved oxygen O_2, and pH, and possibly the redox potential and turbidity, are monitored.

Flow cells thus ensure minimal pumping time during sampling. They contribute significantly to a reliable and effective sampling procedure. The values measured at the time of sampling provide information with regard to the representativeness of the subsequent laboratory analysis. Some examples of flow cells are described in Riha [3.90] and Garvis [3.31].

Determination of free carbon acid content and the organoleptical tests, such as color—or with proper precautions smell and perhaps taste—may be performed in situ. The organoleptical assessment may again be of value for evaluating the representativeness of the subsequent laboratory analysis.

It is often advisable to determine additional groundwater quality characteristics in situ [3.124; 3.45]. For this purpose, quick field analyzing devices including microprocessor-equipped field cuvette photometers are used. Such devices are illustrated in Küvetten-Test-Photometer für CSB und Wasseranalysen [3.129] and Kurzmann [3.50]; they permit the rapid analysis of COD, NH_4^+, PO_4^{3-}, NO_3^-, NO_2^-, Cr^{3+}, CN^-, Ni^{2+}, Cu^{2+}, SO_4^{2-}, Fe^{2+}, and Cd^{2+}.

For the in situ determination of migrants in soil air, the test-tube method is commonly used as a quick and relatively inexpensive method [3.53; 3.68]. With the aid of the device presented in Leichnitz [3.54], a variety of migrants contained in soil air may be detected in the ppm range, including ammonia, ethylene, petroleum hydrocarbons, benzol, chlorine, carbon dioxide, phenol, phosgene, oxygen, sulfur dioxide, hydrogen sulfide, and water vapor. When using this method, a prescribed soil air volume is pulled through small glass tubes, which are filled with sorbent showing a color change reaction (see Figure 3.25). The color of the sorbent changes progressively beginning from the inlet side, correlating with the quantity of the substance present in the soil air. From a scale printed on the tube, the amount of substance contained in the gas sample may be read. In many tubes the measuring error is in the range of 10 to 20% (relative standard deviation. Figure 3.25 shows such a test device. Gas chromatographs are also recommended for field use (see, e.g., Halogenkohlenwasserstoffe in Grundwässern [3.125]). They permit an effective determination of halogenated hydrocarbons present in the soil-air, soilwater, and the soil solids.

When using these field test methods, the accuracy of analysis is often not so important as the fact that the results are immediately available. They allow decisionmaking at the site of the investigation, without delay. This helps to determine how to further proceed with sampling or how to stop further subsurface pollution.

3.2.1.2 Preservation

Water samples to be analyzed in the laboratory must be treated or preserved in the field [3.35; 3.33; 3.121] if the constituents to be analyzed may change in chemical composition prior to reaching the laboratory. This preservation is aimed at delaying chemical and biological conversions of the migrants of concern and decreasing their volatility. Complete preservation of samples taken from the subsurface is not possible; only the conversion rate can be reduced. Options for preserving water samples usually involve pH control, conversion to stable substances, and cooling or freezing. Examples of preservation methods, their resulting effects, and their use are presented in Table 3.1 (see also Scale et al. [3.99]).

Table 3.1. Preservation Methods

Preservative	Effect	Use
nitric acid (HNO₃)	avoiding metal precipitation	metals (Fe, Mg, Ca ...)
sulfuric acid (H₂SO₄)	forming salts with organic bases	ammonium, amines
sodium hydroxide (NaOH)	forming salts with volatile compounds	cyanides, organic acids, phenols
sulfuric acid (H₂SO₄)	bacterial inhibitor	organic substances, (COD, organic C, oil, and grease)
mercury chloride HgCl₂	bacterial inhibitor	nitrogen and phosphorous compounds
cooling or freezing	bacterial inhibitor	acidity, alkalinity, organic substances, organic P, organic N, organic C, organisms

To ensure proper laboratory analysis, the sample treatment methodology, the amount of sample to be collected, and the conditions for sample transport should be discussed with the lab supervisor in advance. This will help ensure compatibility of the preservation method and the laboratory equipment and avoid scheduling conflicts. Nevertheless, the following general guidelines may be helpful [3.99; 3.123].

Cations

Only alkali ions (Na^+, K^+) do not require preservation. Samples analyzed for heavy metals (Fe^{2+}, Mn^{2+}), alkaline earths (Ca^{2+}, Mg^{2+}), and NH_4^+ should be brought to a pH of 1.5 by adding mineral acid.

Anions

Only chloride need not be preserved. Nitrite and nitrate should be preserved if the analysis is not to be performed on the same day. The HCO_3^- concentration changes rapidly in soft water, and therefore HCO_3^- should be immediately analyzed in situ.

The use of glass bottles and preservation with NaOH to a pH $\geq$ 12 are advisable for phosphate. Phosphate losses have been observed when polyethylene bottles were used. However, fluorides should not be preserved in glass bottles. Waters with a low redox potential, such as sulfuric waters, require special preservation. Samples from clay rocks and volcanics, in general, make it necessary to determine alkalinity up to pH = 8.2 in situ as they easily take up CO_2.

Gases

Dissolved gases should either be analyzed immediately in situ or the samples should be carefully preserved. Dissolved oxygen is determined with the Winkler method or by means of an O_2-electrode. Carbonic acid $H_2O \cdot CO_3$ must be determined immediately. Dissolved H_2S, CH_4, and inert gases such as N_2, ^{39}Ar and ^{85}Kr (see also §3.4.1) should also be determined immediately in situ.

Organic substances

The most important groups to be analyzed are

- hydrocarbons originating from accidental pollution by gasoline, diesel, and mineral oil
- phenols
- chlorinated hydrocarbons such as solvents which are often volatile
- polycyclical aromatic hydrocarbons
- agrochemicals and detergents

Each of these organic migrants requires a specific preservation procedure. Many of the organic constituents in water are susceptible to bacterial conversion, oxygen, and light, and some are also volatile, due to their high vapor pressure. It is therefore recommended that samples be stored in brown bottles, cooled, and analyzed as quickly as possible.

Isotopes

The most important isotopes present in subsurface waters include ^{18}O, ^{2}H, ^{3}H, ^{13}C, ^{14}C, ^{15}N, ^{39}Ar, and ^{85}Kr. The stable isotopes ^{18}O and ^{2}H and the unstable ^{3}H do not require sample preservation and, in general, only a small amount of sample (50–100 cm^3) needs to be collected. For analyzing ^{13}C and ^{14}C as well as ^{39}Ar and ^{85}Kr, larger quantities are required (see DVWK Merkblätter zur Wasserwirtschaft [3.123]). ^{15}N, ^{39}Ar, and ^{85}Kr also do not require sample preservation.

Microorganisms

In addition to disinfection or sterilization before and during sampling, the analysis of samples for microorganisms requires numerous special preservation measures (see, e.g., Scale et al. [3.99] and DVWK Merkblätter zur Wasserwirtschaft [3.123]). Hence, they are not discussed here.

Summary

Table 3.2 gives a summary of the recommended field treatment procedures for soil- and groundwater samples. A compilation of methods is also contained in Gudernatsch [3.35, Table 3].

3.2.1.3 Laboratory Analysis

The chemical analysis of soil- and groundwater samples is usually trace analysis (see, e.g., Schönborn [3.105]). Even though the concentrations of soilwater and groundwater constituents are quite different from those commonly observed in surface or sewage water, the same analytical methods are applied for many of the criteria [3.55; 3.118]. There are numerous analytical methods currently available. Hence, methods known as "uniform methods" [3.122], "selected methods" [3.121], and "standard methods" [3.127] have been developed to ensure consistency among data sets. It is cautioned, however, that even such standardized methods may lead to faulty results if applied by inexperienced researchers. Therefore, it is always advisable to consult a specialist prior to initiating analysis. The complex nature of chemistry encountered in today's subsurface water problems frequently requires application of

Table 3.2. Recommendations on the Treatment of Soil- and Groundwater Samples In Situ

Constituents to be Determined	Method of Preservation	Holding Time Before Analyzing	Container/Sample Volume in mL
Sensory perception	I	—	G/100
Temperature	I	—	—
Free CO_2	I	—	P,G/100
HCO_3	I in the case of soft waters	—	P,G/100
pH value	I	4 h	—
Dissolved O_2 and	I	—	G/300
other dissolved gases	Winkler fix.	4–8 h	G/300
Sulfite	I	—	P,G/50
Conductivity	4°C	1 d	P,G/100
Acidity	—	1 d	P,G/100
Alkalinity	4°C (I if clayey rock)	1 d	P,G/100
NH_4^+	4°C, H_2SO_4	1 d	P,G/400
NO_3^-/NO_2^-	4°C	1 d	P,G/100
Phosphate	4°C, filt. off, NaOH	1–2 d	G/50
Total dissolved solids	4°C, H_2SO_4, filt. off	1 d	P,G/50
Sulfide	2 mL Zn acetate	1 d	P,G/5000
BOD	4°C	1 d	P,G/1000
Oils and grease	4°C, H_2SO_4	1 d	G/1000
Org. C	4°C, H_2SO_4	1 d	P,G/25
Phenolics	4°C, H_3PO_4 up to pH<4, Cu SO_4 1 g/L	1 d or deep cooling	G/500
Settleable solids	—	1 d	P,G/100
Cyanide	NaOH	1 d	P,G/500
Chloride	—	7 d	P,G/50
Fluoride	—	7 d	P/50
Silicate	4°C	7 d	P/50
Sulfate	4°C	7 d	P/50
COD	H_2SO_4	7 d	P,G/50
Filterable solids	4°C	7 d	P,G/100
Nonfilterable substances	4°C	7 d	P,G/100
Hardness	4°C, HNO_3	6 mon	P,G/100
Dissolved metals	Filt. off, HNO_3	6 mon[x]	P,G/200
Metals, total	HNO_3	6 mon[x]	P,G/100
Hg, dissolved	Filt. off, HNO_3	1 mon	G/100

I = Immediate determination
P = Plastic
G = Glass
Fix. = Fixation
Filt. off = Filtering off
6 mon[x] = only 6 months, if HNO_3 is applicable, otherwise freezing, adding of mineral acid until a pH ≤ 1.5 is reached, or adding of sodium lye until a pH ≥ 12 is reached

analytical techniques more complicated than the classical analytical methods of volumetry (titration) and colorimetry. Among the techniques which are being used with increasing frequency for inorganic constituents are photometry, potentiometry, conductometry, amperometry, and mass and atomic spectroscopy. For the determination of organic water constituents, it is appropriate to distinguish between summation methods used to determine the total organically bound carbon, and methods used to determine single organic substance groups or single compounds, such as chromatographic analysis.

Photometry (radiation measurement) is one of the methods most frequently applied in water analysis. It allows quick and highly sensitive determination of water quality parameters such as the concentration of chlorine, aluminum, iron, manganese, phosphate, nitrate, and sulfide. The concentration of the dissolved substance is determined from the reduction of monochromic light intensity when passing through the colored sample solution. The color intensity of the sample is produced by specific chemical reactions.

Depending on how monochromic light is generated, filter and spectral photometers are distinguished. To ensure high measuring accuracy, a proper wave length adjustment or filter choice is essential. Further, the sample must be optically clear since turbidity adsorbs light and leads to anomalously high concentration data. In addition, the color of the water before adding the chemical substance causing the color reaction should be determined as a blank.

These explanations show that the use of instrumental methods always requires appropriate sample preparation. Further, it is important that the operator has a detailed knowledge of the measuring principle, the sampling procedure, and instrument handling [3.52; 3.80; 3.129].

Potentiometry is an electrochemical method widely applied in water analysis. It is based on the measurement of the potential generated by a chemical system established as a measuring chain. To ensure an appropriate functional interdependence between the potential measured and the chemical criterion to be investigated, measuring electrode chains have been developed allowing relatively high selectivity. To date, potentiometry has worked best when determining proton activity. Therefore, pH measurement is a key application of potentiometry. The pH of water samples is measured by a combination of a glass electrode (measuring electrode) and a reference electrode, usually a calomel or silver chloride electrode. Theoretical problems of pH measurement are described for instance in Schwabe [3.109].

Potentiometric activity determination by means of ion selective electrodes is used with increasing frequency. Ion selective electrodes currently exist for several ions, including calcium, bromide, iodide, fluoride, nitrate, and sulfide. However, using such electrodes is not yet as easy as using glass electrodes since these electrodes may be subject to interference from other ions in solution (see, e.g., Salacz [3.96]). Terms such as steepness, linearity, measuring range, calibration stability, durability, required free-stream velocity, transverse sensitivity, and response time all characterize important properties of such ion selective electrodes, and their quantities should be known and taken into consideration [3.13; 3.14]. Modern microprocessor-controlled ionometers are described, for instance, in Schuler [3.106] and Horiba Ion Meters N-8 Series [3.126]. The comparatively small sample size required by this technique is frequently a great advantage, for instance in soilwater monitoring.

Conductometry (conductivity measurement) is a summation method. Electrolytic conductivity gives information on the summary concentration of the dissociated substances. The conductivity of a solution depends upon concentration of solutes, dissociation of the contained electrolytes, the electrochemical valency of ions, their mobility, and temperature. Therefore, it is impossible to isolate information on one of the electrolytes by this method.

Amperometry is also based on electrode reactions. Basically, it is used to measure the dissolved oxygen content in water. The membrane-covered electrodes play a decisive role in this measurement technique. Because both the solubility of oxygen in water and its diffusion through the membrane are dependent on temperature, modern oxygen-measuring devices are equipped with automatic temperature compensation [3.37].

Among atomic spectroscopic methods used in groundwater analysis, atomic absorption spectroscopy (AAS) and atomic emission spectroscopy (AES) should be mentioned. Due to its high sensitivity, AAS permits direct determination of metals in a concentration range of mg/L and has gained wide acceptance. Modern AAS devices are computer-aided.

There has been a recent trend towards multielement measuring techniques. Atomic emission spectroscopy with inductively coupled plasma (ICP) as the stimulator is the prime example which directly competes with AAS. The speed of sample analysis by ICP techniques is high. After calibration, a simultaneous determination of up to 25 elements can be run in approximately 20 seconds [3.67]. The capital expenditure on such devices is, however, also very high.

With respect to the chemical analysis of organic constituents in groundwater, the primary analytical problem arises from the great variety of organic compounds which may be present in a sample. As a result, summation methods are of special importance. A widely used summation method is the determination of organically bound carbon, known as TOC (total organic carbon). This direct analysis technique allows complete oxidation of organic substances and the measurement of carbon dioxide thereby produced. With the aid of modern instruments, such analyses may be done with high accuracy in a relative short time [3.56].

Chromatographic techniques are used for the determination of organic substance groups and single compounds. Chromatography permits separation of substances through relative transport rates in porous materials. Any substance can be identified by the rate of transport, and its concentration by the magnitude of response to its arrival at the installed sensor.

Thin-layer chromatography is a technique which can be applied in any laboratory because of the modest expenditure for instruments and material. This technique is especially suited for investigations of such important substance groups as chlorinated hydrocarbons (e.g., insecticides) and polycyclic aromatic hydrocarbons [3.119]. Modern high-efficiency, thin-layer chromatography has evolved from conventional methods [3.30]. It is characterized by new plate systems, sample application techniques, and a direct photometric analysis of these plates.

With the aid of gas chromatographic techniques, mixtures in gaseous or vaporized state may be separated. Only in exceptional cases is the direct injection of liquid water samples into a gas chromatograph beneficial. A preliminary treatment of samples, such as isolation and enrichment of organic substances present in the water and purification of the extract, is indispensable.

Chromatographic analysis provides useful results if reference substances are available. However, this is not always the case. Therefore, special identification techniques must often be applied in conjunction with the chromatographic separation. However, due to the high cost of such equipment, it is only justified for central laboratories to purchase and to maintain modern gas chromatographic–mass spectrometry equipment (GS-MS system) [3.44].

Due to a variety of factors, water analysis may be very complex. The analysis of all constituents or of a large number of constituents present in a water sample is not possible with current techniques. Compromises are hence necessary. Figure 3.26 shows an example of steps necessary for preparing a groundwater analysis for investigation of the migration of radionuclides in a fluvial sandy aquifer in Canada.

3.2.2 MIGRATION TESTS IN LABORATORIES

The laboratory tests considered hereafter are classified as batch tests, column tests, and special tests.

3.2.2.1 Batch Tests

In batch tests storage and transformation processes proceeding in soil and rock samples are analyzed. Transport processes in these samples are

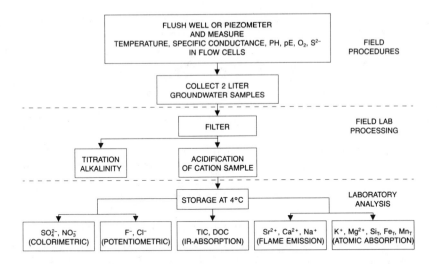

Figure 3.26. Flowchart of groundwater analysis. *Source:* Jackson and Inch. [1.41].

not considered (see Figure 1.14). Tests may be performed on samples which are disturbed or undisturbed, consolidated or unconsolidated, and saturated, unsaturated, or oversaturated. One distinguishes, in general, between static and dynamic batch tests, depending on whether the solid and fluid mixphases are in motion during the batch test. The movement in both cases serves only to accelerate the transphase exchange or the averaging process, which allows mathematical lumping in the representative elementary volume (REV). Nonequilibrium exchange processes occurring in the multiphase system of the soil or rock sample therefore may scarcely be investigated with batch tests. Batch tests finally result in a systems state function according to eq. 1.1a.

Evaluation of Tests

In accordance with the conditions mentioned above, the batch-test migration model consists simply of one node with a lumped storage and a lumped equilibrium transformation (reaction) element (see also Figure 1.14c). Thus, the mathematical model of solute migration in a batch follows as

$$0 = ds_\Sigma/dt + IR_\Sigma + \Delta R_e \qquad (3.1)$$

where S_Σ is substance quantity stored in a batch in kg
  IR_Σ is lumped internal reaction rate in a batch in kg/s where $IR_\Sigma < 0$
    means decay or decomposition
  ΔR_e is external net substance rate feeding to a batch in kg/s

After incorporating a specific storage model, for instance Henry's Law (Equation 1.101), Freundlich's Law (Equation 1.102), or Langmuir's Law (Equation 1.103), and a specific reaction model (e.g., Equations 1.46a-c or 1.46e-g) into the differential Equation 3.1, it must be solved analytically or numerically. To illustrate some typical cases, as well as to demonstrate the procedure, the following examples are considered:

Example 1

With a negligible substance conversion rate $IR_\Sigma = 0$, and a substance added ΔR_e independent of S_Σ, Equation 3.1 provides with $S_\Sigma = S$

$$S_{i+1} - S_i = c_{in} \, \Delta V_{fl,in} - c_i \, \Delta V_{fl,0} \qquad (3.2a)$$

where S_i is the quantity of substance stored and c_i the concentration of the migrant in the mobile fluid phase both at time i, with $\Delta V_{fl,0}$ first taken out of the batch and $\Delta V_{fl,in}$ later filled in, and c_{in} symbolizing the concentration of the migrant in $\Delta V_{fl,in}$. The determination of S_{i+1} is only meaningful when $t_{i+1} - t_i >$ RET (representative elementary time), i.e., if the average (lumping) conditions for c have been established in the batch. Under these conditions S results from the following equation:

$$S = S_m + S_{im} = cV_m + s_{im}m/\rho \qquad (3.2b)$$

where m, im = mobile and immobile mixphase in the batch
    s = specific substance quantity stored in kg per dm³ space
    c = migrant concentration in g per L mobile phase
    m = mass of rock or soil sample in a batch in g rock
    ρ = bulk density of the subsurface sample in kg per dm³ space
      in the natural environment
  V_m = volume of the mobile fluid mixphase in L ($V_{m,i+1} = V_{m,i}$
      $+ \Delta V_{fl,in,i} - \Delta V_{fl,o,i}$)

When the initial values S_1 and $V_{m,1}$ are known, $s_{im,i}$ can be determined cumulatively, the storage function $s_{im,i} = f(a_i) = f'(c_i)$ can be plotted, and a mathematical model—for instance, according to Equation 1.101, 1.102, or 1.103—can be optimally fitted to the calculated and plotted storage (or systems state) function (see Chapter 2.3). If the initial values S_1, or V_{m1}, are unknown a priori, a system of equations must be formulated which is characterized by more equations than unknowns:

$$s_{im,i+1} = s_{im,i} + (\rho/m)(c_i V_{m,i} - c_{i+1} V_{m,i+1} + c_{in,i} \Delta V_{fl,in,i} - c_i \Delta V_{fl,o,i}) \quad (3.2c)$$

After putting meaningful terms for s_{im} into Equation 3.2c (e.g., according to Equation 1.101, 1.102, or 1.103), both the unknown parameters of the storage model and the initial values for $s_{im,i}$ and $V_{m,i}$ may be determined from the system of equations. In Nitsche [3.75], for instance, the SPIB computer program is presented for this purpose. This program provides an optimum storage process model based on a best fit of the test data.

Example 2

When no substance is added to the batch or withdrawn from it ($\Delta R_e = 0$), $s_{im} = \rho K_d c$ is used as the storage model (see Equations 1.78b and 1.101), and $r_m = k_m c^2$ and $r_{im} = k_{im} c^2$, the reaction rate model follows as

$$(V_m + K_d m) \, dc/dt = - (k_m V_m + k_{im} K_d m) \, c^2 \quad (3.3a)$$

with the solution

$$c(t) = c_o/(1 + \alpha t c_o) \text{ or } 1/c(t) = 1/c_o + \alpha t \quad (3.3b)$$

$$\alpha = (k_m V_m + k_{im} K_d m)/(V_m + K_d m) \text{ and } c_o = c(t=0)$$

Batch tests with $IR_\Sigma \neq 0$ are commonly carried out in saturated samples where V_m is $\theta_o V_b$, m is ρV_b, and V_b is the volume of the batch. With the assumption that $k_m = k_{im} = k$, the following equations result:

$$V_m + K_d m = V_b(\theta_o + \rho K_d)$$

$$k_m V_m + k_{im} K_d m = k V_b(\theta_o + \rho K_d)$$

$$\alpha = k$$

When applying more complicated storage and reaction models, closed analytical solutions of Equation 3.1 are rarely possible. Equation 3.1 will then be solved numerically, and the parameters of the storage and reaction models will be estimated by identification (see §2.3.4).

Static batch tests

During the static batch test, solids and fluids are not in motion. In such a test, spreading and averaging processes are caused only by molecular diffusion. Therefore, the batch has to be considered as a representative elementary volume (see §1.2.1) where the representative elementary time interval may become long, which frequently causes long test periods.

Example

The PO_4^{3-}-P sorption and storage process model of a soil sample is to be determined for the case of increasing phosphate concentration in the soil solution where PO_4^{3-}-P conversion is neglected ($IR_\Sigma = 0$). The test is conducted with a water-saturated sample having the volume V ($V_m = \theta_o V$), the mass m $= \rho V$, the specific storage s $= S/V$, and an external exchange of $\Delta V_{fl,in} = \Delta V_{fl,o} = \Delta V_{fl}$. Thus, Equation 3.2c adopts the following form:

$$\text{sorption model:} \quad s_{im,i+1} = s_{im,i} + \theta_o(c_i - c_{i+1}) + (c_{in,i} - c_i)\,\Delta V_{fl}/V$$

$$\text{storage model:} \quad s_{\Sigma,i+1} = s_{\Sigma,i} + (c_{in,i} - c_i)\,\Delta V_{fl}/V = s_{\Sigma,i} + \Delta m_i/V$$

The values c_k ($k = 1, 2, \dots$) are measured according to the test procedure shown in Figure 3.27.

Preliminary tests have shown: $\theta_o = 0.25$ cm^3_W/cm^3_V and $s_{im,1} = 1.4 \cdot 10^{-4}$ mg PO_4^{3-}-P per cm^{3}V where the indices W symbolize water and V space volume. Thus the procedure yields

	measured c	specific substance s_Σ	quantity stored $s_{im} = s_\Sigma - \theta_o c$
i = 1	$1.6 \cdot 10^{-4}$ mg/cm^3_W	$1.8 \cdot 10^{-4}$ mg/cm^3_V	$1.4 \cdot 10^{-4}$ mg/cm^3_V
i = 2	$1.4 \cdot 10^{-3}$	$10.2 \cdot 10^{-4}$	$6.7 \cdot 10^{-4}$
i = 3	$4.0 \cdot 10^{-3}$	$21.1 \cdot 10^{-4}$	$11.1 \cdot 10^{-4}$
i = 4	$7.2 \cdot 10^{-3}$	$32.1 \cdot 10^{-4}$	$14.1 \cdot 10^{-4}$
i = 5	$1.0 \cdot 10^{-3}$	$40.4 \cdot 10^{-4}$	$15.4 \cdot 10^{-4}$

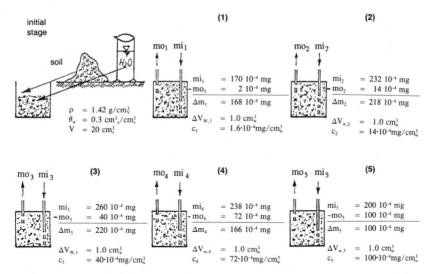

Figure 3.27. Procedure for a static batch test.

Figure 3.28a represents the plots

$$s_\Sigma = s_\Sigma(c)$$

$$s_{im} = s_{im}(c)$$

The data are not arranged along straight lines, except for small activities. Therefore Equation 1.101a cannot be considered as a valid storage process model.

If the storage process model 1.102a were appropriate, the laboratory data would have to satisfy a straight line in a log-log representation (see Figure 3.28b). This is observed for the measured data in the range of $1 \cdot 10^{-3} < c < 1 \cdot 10^{-2}$ mg/cm$_w^3$.

However, if the storage process can be defined by Langmuir's model (see Equation 1.103a), the measured data shown in Figure 3.28c should approximately follow a straight line. As the figure shows, this applies for the entire range of collected data.

Dynamic batch tests

Moving fluid

A forced fluid movement accelerates the spreading and averaging processes within the batch and shortens the test period. The time constant t* for a static batch test may be estimated as follows (see Equation 1.69a):

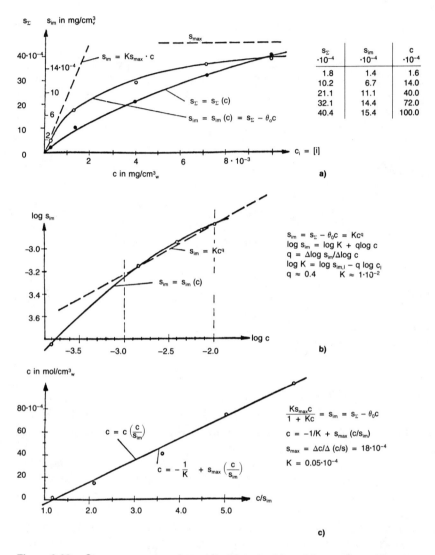

Figure 3.28. Storage process analyses of batch tests: *(a)* model according to Equation 1.101a; *(b)* model according to Equation 1.102a; *(c)* model according to Equation 1.103a.

$$t^* = \frac{S}{c} \; R \approx (0.5 \; Vs_\Sigma/c)/(D_o\omega/\overline{\Delta L})$$

For the dimensions shown in Figure 3.26, this time constant yields

$$t^* \approx 0.5 \cdot 20 \text{ cm}^3 \cdot 0.5/(10^{-6} \text{ cm}^2/\text{s} \cdot 7.37 \text{ cm}^2/2.71 \text{ cm}) \approx 2 \text{ days}$$

In other words, a few days are necessary for one time step of the static test. This illustrates the importance of shortening the representative elementary time interval. Fluid recycling by means of peristaltic pumps, in conformity with Figure 3.29a, has proven very successful. The measuring vessel shown in this figure should be as small as possible, and the additional volume ΔV, including return hose and distribution chambers, should also be small relative to $\theta_o V$ to avoid the formulation and fitting of special reaction and storage models. Otherwise, reaction/storage models will need to be estimated for the return hose, distribution chambers, and measuring vessel from preliminary tests. The desired equilibrium state in dynamic batch tests will usually be attained within a few hours after several cycles of the fluid phase are completed. The flow

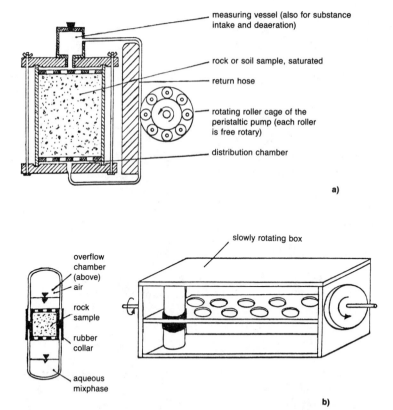

a)

b)

Figure 3.29. Dynamic batch test equipment with forced advection of the liquid mixphase: *(a)* recycling of the mobile liquid phase by a peristaltic pump; *(b)* gravity-forced flow through the rock sample in a slowly moving rotation shaker.

velocity should not exceed the velocity commonly used in column flow tests; that is, the hydraulic gradient should not exceed 1 or 2.

A further dynamic batch test device is presented in Lang [3.51]. In a device as described in Klotz and Oliv [3.48], the fluid motion is gravity forced. A very slow-moving shaker rotates the test vessels slowly (see Figure 3.29b). It should be ensured that the process to be investigated is not affected by the continuous mixing of test liquid and gas (usually air) in the two overflow chambers. The rotation speed should be rated to allow about 50 to 80% of the aqueous mixphase to pass the sample for each half-rotation.

The supporting plates of the rock samples in the two experimental setups shown in Figure 3.29 should prevent fine rock material from being washed out. Textile filters — in particular, glass fiber filters — are also well suited for this purpose.

Moving Solids (Classical Batch Tests)

Classical batch tests mix solid and fluid phases. Experimental layouts shown in Figure 3.29b are common practice. The test vessels (batches) are filled with unconsolidated soil or rock material falling through the fluid in the batch during each half-rotation (see also Lang [3.51]). Common practice is 10–20 rotations per minute. These laboratory tests are very simple and do not demand high technical skills. However, their major disadvantage is that practically no stable equilibrium state is attained. The storage capacity is constantly increasing during this test due to continuous abrasion, which leads to an extension of the internal surface area and causes new fracture surfaces (see, e.g., Klotz and Oliv [3.48] and Moser et al. [3.69]). Therefore, classical batch tests are no longer recommended.

3.2.2.2 Column Tests

In column tests, transport, storage, exchange, and transformation processes are investigated. Disturbed or undisturbed soil or rock samples can be tested. The evaluation of column test data is based on the complete one-dimensional mathematical migration model. With few exceptions, the length-diameter ratio of these columns should be greater than 5:1 (3:1 at a minimum).

Prior to performing a column experiment, an initial concentration of the migrant under consideration often must be attained in all phases of the test column. For this purpose, fluid of a specified concentration is passed through the column until the output concentration is equal to the input concentration $c_{in} = c_{out} = c_I = $ constant. In general, a constant flow rate $v = \dot{V}/\phi = $ constant is generated, where ϕ is the column cross-sectional area.

At the column input a third-type boundary condition is usually applied:

$$\dot{V}c_o = (- \delta_\ell \dot{V} \partial c / \partial x + \dot{V}c)_{|0,t}$$

or   (3.4a)

$$vc_o = (- \delta_\ell v \, \partial c / \partial x + vc)_{|0,t}$$

However, this boundary condition is frequently simplified mathematically to a first-type boundary condition:

$$c(0,t) = c_o \qquad (3.4b)$$

The error resulting from this simplification has been assessed in analyzing Equations 2.36–2.39 (see Figures 2.9–2.11). These figures also show that for typical practical Peclet numbers of 20–30, this approximation should not be used unless necessary. Elaborate instrumentation would be required to technically ensure a first-type boundary condition as in Equation 3.4b. This presupposes a turbulent fluid body contacting the inflow interface, the concentration of which must be kept constant.

Occasionally, a concentration pulse is applied:

$$c(0,t) = \delta(t)m/\dot{V} \quad \text{with} \int \delta(t)dt = 1 \qquad (3.4c)$$

where m is the mass or amount of substance of the pulse. Because such a pulse frequently causes a concentration in the vicinity of its application which is not consistent with the natural systems state, this type of boundary condition should only be used after careful preliminary tests.

At the outflow end of the column (free outflow), it is difficult to define the boundary condition. As explained previously in Chapter 2.1, the test evaluation should be based on a second-type boundary condition $\partial c / \partial x_{|\infty,t} = 0$ or a third-type boundary condition $\dot{V}c_L = (\delta_\ell \dot{V} \partial c / \partial x + \dot{V}c)_{|L}$.

Data from such column tests are usually limited to a measure of the outflow concentration versus time $c(L,t)_i$ recorded as breakthrough curves. The spatial distribution of the concentration $c(x,t_k)$ at selected time points t_k is only rarely recorded (see, e.g., Luckner et al. [3.62]) because the experimental setup is much more complicated. The withdrawal of fluid at the sample points x_i often disturbs the fluid flow in the column. Exceptions are given only by heat migration column tests and tests with radioactive tracers which allow remote sensing and do not require sampling.

Discrete time measurements must be collected if continuously monitoring the state of the system is impractical or too expensive. The time of breakthrough should be forecasted, and a sampling schedule may be devised based on a predicted breakthrough curve. This schedule should then be verified and improved when the first measured data are available.

Figure 3.30 shows two typical column test devices. The evaluation of column tests is practical when based on space-lumped parameters. To ensure stable parameter identification results, efforts must be made to formulate the mathematical migration model as simply as possible. Starting with the simplest possible model is recommended, supplementing the model step-by-step when the accuracy needs to be improved.

It is also advisable to use a mixture of migrants in a column test. The migrants in this mixture must be mutually nonreactive. They are applied simultaneously (see Figure 3.31). In this method, the Cl⁻ ions are classically used for estimating θ_o, ϕ, and δ_ℓ. These parameters are then further applied for identifying parameters of the migration process which are subject to exchange and transformation reactions. The "preliminary determination" of δ_ℓ in a heat migration test can also be useful. In the majority of cases, it is best to carry out laboratory tests with physically, chemically, and biologically undisturbed or minimally disturbed soil, rock, and fluid samples. Using natural soilwater or groundwater as a basis for generating the required "model waters" in laboratory tests has been proven to be a reliable means for conducting laboratory tests (see also Ryan [3.95]).

Analytical solutions have played a dominant role in parameter estimation of migration processes in column studies (see Chapter 2.1). However, it is also common to apply fast digital models with lumped parameters.

Column tests often produce estimates of the global retardation coefficients of the migrants i, relative to migrating Cl⁻ ions or tritium-marked water molecules, as derived from the integrated breakthrough curves (see Equation 1.78a):

$$R = \frac{u_{Cl}}{u_i} = \frac{c_{\infty,Cl} \int [c_{\infty,i} - c_i(t)]dt}{c_{\infty,i} \int [c_{\infty,Cl} - c_{Cl}(t)]dt} = K_d\rho/\theta_w + 1 \qquad (3.5)$$

where $c_\infty = c(t \to \infty)$ and $\int = \int_{dry}$

The mathematical model forming the basis of this solution includes convection as the transport mechanism, linear storage processes in the mix-phases, and transphase exchange processes in the equilibrium state.

As compared to batch tests, column flow tests require much more

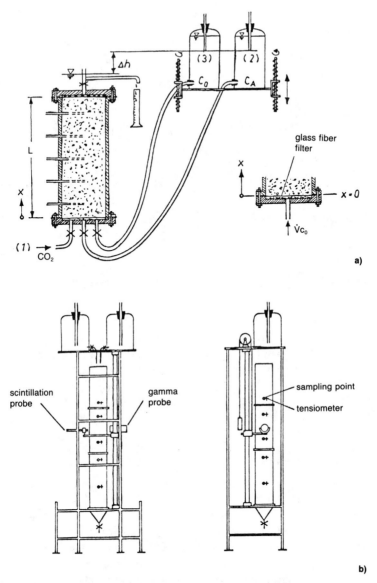

Figure 3.30. Two laboratory test facilities for column flow tests. *(a)* Column tests of saturated unconsolidated rock according to Luckner and Reiβig [3.62]. Test procedure: *1,* displacing of the soil-air by CO_2-gas; *2,* generating the initial state $c(x,o) = c_I$ and the saturated flow conditions by dissolving the residual CO_2 gas; *3,* implementing the migration test with Vc_o at $x = 0$ as the input; *4,* concentration measurement $c = c(x_i, t_k)$. *(b)* Column test in saturated unconsolidated rock according to Luckner and Nitsche [3.61] with γ-radiation and tensiometer measurements to monitor the moisture movement.

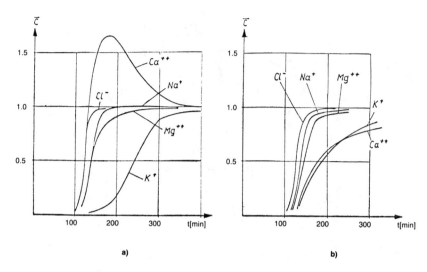

Figure 3.31. Breakthrough curves of a mixture of tracers: *(a)* simultaneous migration; *(b)* single migration.

capital expenditure. In return, the migration parameters determined by replicates are often not so scattered, and fewer repetitions are required. The distribution coefficients K_d or retardation coefficients R estimated from column tests are typically smaller than those estimated from batch tests. The bigger coefficients derived from batch tests with disturbed or repacked samples result from the destruction of preferential flow paths and the extension of the interfaces for the migrant exchange between the different mixphases.

Example

Results of a column test described in van Genuchten [3.32] with boron as the migrant, using a clayey soil with $\rho = 1.22$ g/cm³ and $\phi = 0.45$ in a column of L = 30 cm, is considered. The collected data are plotted in Figure 3.32. Boron was applied at the top of the column in the form of H_3BO_4 for 5.06 days. It was dissolved in the aqueous phase, which was flowing at a rate of q = 17.1 cm/day. This process may be reflected by the following migration model (see also §2.3.4):

$$\sigma_m D \frac{\partial^2 c_m}{\partial x^2} - \sigma_m v_m^* \frac{\partial c_m}{\partial x} = \sigma_m \frac{\partial c_m}{\partial t} + f\rho \frac{\partial s_m}{\partial t} + \alpha \ (c_m - c_{im}) \quad (3.6a)$$

$$0 = \sigma_{im} \frac{\partial c_{im}}{\partial t} + \rho(1 - f) \frac{\partial s_{im}}{\partial t} - \alpha(c_m - c_{im})$$

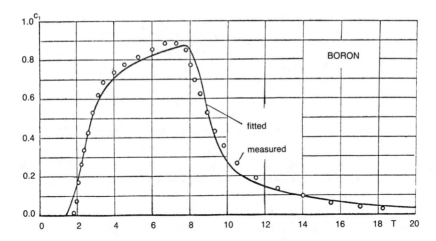

Figure 3.32.   Measured data and fitted breakthrough function for boron when passing a clayey soil in a lab column flow test. *Source:* van Genuchten [3.32].

This mathematical model includes the following linear storage model:

$$s_\Sigma = \rho \, fkc_m + \rho \, (1 - f)kc_{im}$$

where $\sigma_\Sigma = \sigma_{im} + \sigma_m = \phi$ is the porosity
   $\sigma_m = \theta_m$ is the mobile fluid-filled porosity
   $\sigma_{im} = \theta_{im}$ is the immobile fluid-filled porosity
   $v_m^* = q/\sigma_m$ is the flow velocity of the mobile fluid phase
$0 \le f \le 1$ is the fraction coefficient

With the dimensionless variables

$$T = v_m^* t\theta_m/L; \quad z = x/L; \quad Pe = v_m L/D; \quad \omega = \alpha L/(\sigma_m v_m);$$

$$\beta = (\sigma_m + \rho fk)/(\sigma_\Sigma + \rho k) = \theta_m R_m/R; \quad R = 1 + \rho k/\sigma_\Sigma;$$
$$R_m = 1 + f\rho k/\sigma_m$$

$$c_1 = (c_m - c_1)/(c_o - c_1); \quad c_2 = (c_{im} - c_1)/(c_o - c_1)$$

$$c_1 = c(x,o); \quad c_o = c(o,t); \quad \phi_m = \sigma_m/\sigma_\Sigma$$

the following mathematical process model results:

$$\frac{1}{Pe} \frac{\partial^2 c_1}{\partial x^2} - \frac{\partial c_1}{\partial x} = \beta R \frac{\partial c_1}{\partial T} + \omega(c_1 - c_2) \qquad (3.6b)$$

$$0 = (1 - \beta) R \frac{\partial c_2}{\partial T} - \omega(c_1 - c_2)$$

By means of the FORTRAN computer program CFITIM [3.32], using minimization of squared errors for parameter identification, the following independent parameters were found with $T^* = 6.4$ days calculated from $t = t^* = 5.06$ days:

$$Pe = 23.55 \qquad R = 4.120 \qquad \beta = 0.600 \qquad \omega = 0.408$$

If these parameters are put into Equation 3.6b, this model sufficiently fits the breakthrough data plotted in Figure 3.32. The natural parameters must now be interpreted based on the four independent estimates and additional information (see §2.3.4., Example E).

Problems associated with parameter identification in this example arise because the CFITIM program uses an analytical solution of the mathematical model which presupposes a linear storage model. But, as shown by batch tests, the storage process would often be more appropriately defined by a nonlinear model, such as the Freundlich or Langmuir model. Therefore, it will be necessary to utilize a numerical solution for the identification procedure instead of the analytical solution.

3.2.2.3 Special Tests

Only a few typical laboratory tests shall be presented here, as a complete overview would go far beyond the framework of this textbook.

σ_m Determination

It is appropriate to determine the effective porosity, $\sigma_m = \theta_m$, in relatively short columns, where the length-diameter (L/D) ratio should be $\approx$ 3, using Cl$^-$ ions or another tracer migrant at a comparatively large flow rate. Estimations of σ_m are only based on the first measured value of the breakthrough curve data to avoid influence of the transphase exchange with the immobile fluid phase. The laboratory device shown in Figure 3.33 permits the determination of σ_m and $\sigma_m + \sigma_{im}$ for saturated and unsaturated flow conditions. Unsaturated conditions are implemented by controlling the pressure of the nonwetting phase, for example, by controlling the soil air pressure (see §1.2.3).

During step I, a tracer solution (e.g., NaCl) passes through the soil sample until [Cl$^-$] is in the equilibrium state between the mobile and immobile fluid phases. In this case, the concentration of [Cl$^-$] will be equal in the inflow and outflow fluids ($c_{out} = c_{in} = c_l$), and a tracer quantity $m_l = c_l(\sigma_m + \sigma_{im})V$ will be contained in the sample. The sample volume V, c_{in}, and c_{out} must be measured.

During step II, a tracer-free fluid is added to the column until the tracer concentration in the outflow reaches a value of approximately 0.5

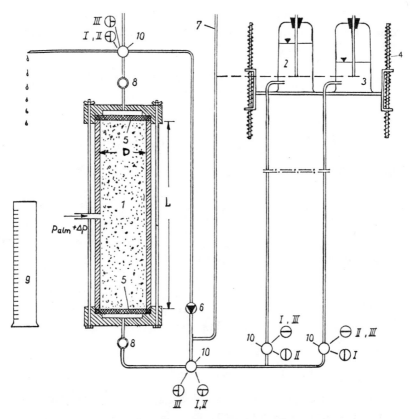

Figure 3.33. Laboratory test device to determine the effective porosity σ_m. *I, II, III,* reference symbols for the procedure steps. *1,* soil sample; *2,* tracer-free water; *3,* tracered water; *4,* elevation control mechanism; *5,* ceramic plates; *6,* peristaltic pump; *7,* standpipe; *8,* measuring cell; *9,* burette; *10,* two-or three-way valve.

of c_{in}. In the outflow, the concentration $c_{II} = c(L,t_{II})$, the tracer quantity m_{II}, which is washed out, and the discharge $\dot{V} = Q$ must be measured. The time point $t_{II} = 0$ is defined as the time when the inflow of tracer-free fluid starts.

In step III, a fluid recycling is implemented. It is operated until the thermodynamic equilibrium has been reestablished between the mobile and immobile mixphase ($c_{in} = c_{out} = c_{III}$). The sample now contains $m_{III} = c_{III}(\sigma_m + \sigma_{im})V$. The concentrations c_{in} and c_{out} must be remeasured. From the balance $m_I = m_{II} + m_{III}$, the following equation results:

$$\sigma_m + \sigma_{im} = m_{II}/((c_I - c_{III})\,V) \qquad (3.7)$$

where σ_m is determined from the approximate solution of Equation 2.36d: $c(L,t) = 0.5\ c_o$ erfc ξ. If the following approximation is used (see also §2.3.2.1, Example A) [3.62]

$$\xi = (L - v^*t)/(2\sqrt{Kt}) \approx - (v^*\sqrt{t_{50}}/\sqrt{4K})\ 1nt/t_{50}$$
$$= - (L/\sqrt{4DL/v})\ \ln t/t_{50}$$

with $\Delta c_{II} = c_I - c_{II}$ and $c_o = \Delta c_{II}/c_I$, the following solution results:

$$\text{inv erfc } (2\Delta c_{II}/c_I) = - (L/\sqrt{4\ DL/v})\ (\ln t - \ln t_{50}) \qquad (3.8a)$$

By graphing, the smallest measured data Δc_{II} follow a straight-line pattern in an x,y-coordinate system with

$$Y = \text{inv erfc } (2\Delta c_{II}/c_I) \quad \text{and} \quad X = \ln t_{II} \qquad (3.8b)$$

The best-fit straight line then crosses the $Y = 0$ line at $t_{II} = t_{50}$. The value σ_m is calculated from these values as

$$v^* = v/\sigma_m = L/t_{50} \qquad \sigma_m = vt_{50}/L \qquad (3.8c)$$

Figure 3.34 shows an example [3.75]. The inv erfc ξ values are determined according to Equation 2.114. The determination of the functional relation

$$\sigma_m/(\sigma_m + \sigma_{im}) = f(\theta_w) \qquad (3.8d)$$

for unsaturated soil or rock samples may also be carried out in the laboratory device described previously and shown in Figure 3.33.

δ_{tr} Determination

The estimation of the transverse dispersivity, δ_{tr}, is only possible by means of a two-dimensional migration model. Estimates are nearly always based on an analytical solution according to Equation 2.42b:

$$c(x,y,t \to \infty) = \frac{c_o}{2} \text{ erfc} \left(y\sqrt{\frac{1}{4\delta_{tr}x}} \right) \qquad \text{for } y \geq 0 \quad \text{and} \quad x \geq 0 \quad (3.9)$$

where $c(0, +y) = c_o$ and $c(0,-y) = 0$.

Thus, equipment is needed which produces a steady-state flow field parallel to the x-direction, with tracer input occurring only on the boundary line $x = 0$ and $y > 0$. Together, the measured data $c_m(L, y > 0)$ and the measured data $c_m(L, y < 0)$, transformed to $c_m^*(L, y > 0) = c_o - c_m(L, y < 0)$, form a single data set called c_m'. The data in c_m' are fitted by a regression line in the x,y-system according to Equation 3.10b. Based on Equation 3.9, we have

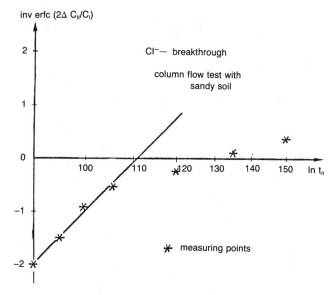

Figure 3.34. Determination of σ_m for the values measured in the test device shown in Figure 3.33.

$$[\text{inv erfc } (2c_m'/c_0)]^2 = y^2/(4\,\delta_{tr}L) \qquad (3.10a)$$

$$Y = [\text{inv erfc } (2c_m'/c_0)]^2 \text{ and } X = y^2/(4L) \qquad (3.10b)$$

$$\delta_{tr} = \Delta X/\Delta Y \qquad (3.10c)$$

It is advisable to use a laboratory test device which permits vertical two-dimensional flow. This device will also permit investigation of two fluids with different densities mixing transversely to their interface. Figure 3.35 illustrates such a laboratory test model according to Rinnert [3.92]. It was constructed to investigate transverse mixing of a freshwater-saltwater interphase.

CEC Determination

The cation exchange capacity (CEC) is nearly always determined in batch tests. Three process steps are normally implemented. During step I, all the exchange sites of the solid phase are occupied by a given cation A. The cations exchanged (expelled) in contact with a control fluid are determined by analyzing their content in the extracted fluid phase. In step II, the cations which are not bound to the exchange sites are washed out. This is generally done by a special washing liquid. Finally, in step III, the cations A are reexchanged by cations B. By determining the amount of expelled cations A, the cation exchange capacity is deter-

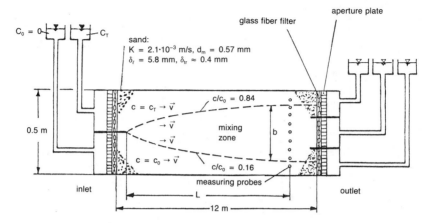

Figure 3.35. Sand model for the determination of transverse dispersion and dispersivity. *Source:* Rinnert [3.92].

mined. Each of the three process steps should proceed at a definite pH value.

There are several variations to this method of determining the CEC. In Richards [3.89] a 1-molar aqueous sodium acetate solution with a pH of 8.2 is recommended for initial occupation (step I), a 95% ethanol solution for washing (step II), and a 1-molar aqueous ammonium acetate solution with a pH of 7.0 for the reexchange (step III). The liquid extraction should be performed by centrifuging the multiphase sample.

The procedure introduced by Mehlich is also used widely [3.72; 3.88]. Here, in step I an aqueous barium chloride solution buffered with triethanolamine (pH between 7 and 8) is used for occupying the exchange sites with Ba^{2+}. Step II involves natural desorption of Ba^{2+}. In step III, barium is desorbed by means of a magnesium chloride solution. The test is performed in a percolation tube ($\phi = 2.5$ cm, L = 11.5 cm) in which the fluids pass the soil or rock sample by gravity drainage. The fluid level is kept constant by utilizing a Marriott bottle.

It is not advisable to use methylene blue dye as an exchange cation in CEC determination. This cation is occasionally used because it is easy to measure colorimetrically; however, its exchange has been shown to be incomplete.

3.2.3 VALUE OF PARAMETERS DETERMINED FROM SAMPLES

The determination of special constituents in soilwater and groundwater, as well as in soil gas samples, is imperative for all investigations of migration processes. This does not imply, however, that all constituents considered are of equal value for a given investigation. The representativeness of the data determined, and thus its usefulness, is endangered by the great variety of potential errors in sample analysis, including all manipulations of sample extraction from the natural environment in the multiphase system subsurface, up to competing analyses in situ or in the laboratory. Because analysis of nonrepresentative samples is useless, applications of the appropriate sampling technique recommended in Chapter 3.1 and use of analysis methods and equipment according to §3.2.1 are essential. Three major factors that directly influence sample integrity are

1. the level of knowledge and competence of the staff
2. the availability of well-documented lab procedures and methods
3. the reliability of the technique for collecting and analyzing samples

The value of parameters determined by means of migration tests in laboratories (see §3.2.2) should be assessed under different conditions. First it seems to be an undisputed fact that laboratory testing provides several distinct advantages for the investigation and study of migration processes and partial processes. Therefore, the batch, column flow, and special tests described above may generate important information for a great variety of migration process investigations. Without such laboratory migration tests, many investigations of complex transport scenarios may not be possible.

However, problems arise regarding the transfer of parameters from the laboratory to the field. It is especially difficult to transfer the dispersivities δ_ℓ and δ_{tr} determined in the laboratory (see also Silliman et al. [3.110] and Silliman and Simpson [3.112]), except, perhaps, from tests with soil monoliths for investigating the soil zone (see §1.3.3). An extrapolation by means of Equation 1.30 or Figure 1.33 may often be unreliable without any supporting field data. Storage, exchange, and conversion model parameters determined in the laboratory are also not easily transferable to the field scale. Undoubtedly, the dependency of these parameters on the scale of application is much smaller than that of the hydrodynamic dispersion. Nevertheless, their transfer to field conditions should be supported by field tests. Field and labortatory tests should always be planned and run together.

3.3

Field Test Methods

In situ test methods provide the most representative characterization of groundwater transport phenomena. Geophysical methods (§3.3.1) can be of exceptional importance, but their application may be limited by geologic, land-use, or regulatory constraints. Therefore, tracer methods which utilize either the undisturbed flow field (§3.3.2), a flow field influenced by pumping or injection (§3.3.3), or a flow field involving groundwater recharge (§3.3.4) often occupy center stage in field-scale tests. The reliability of parameter estimates obtained by various in situ methods is final qualitatively assessed in this chapter (§3.3.5).

3.3.1 GEOPHYSICAL TEST METHODS

Geophysical test methods involve measurement and analyses of physical force fields or processes for purposes of interpretation. Both naturally occurring and artificially induced electromagnetic fields, as well as naturally occurring geothermal signatures, are useful for investigating groundwater transport processes. Surface electromagnetic, downhole electromagnetic, and geothermal borehole surveying are the most widely applied methods. Downhole flowmetric and tracer methods, as discussed in §3.3.2 (see Figures 3.6 and 3.7), may also be characterized as geophysical test methods.

Artificially induced electric fields are used to identify heterogeneities in the subsurface by measuring relative differences in the resistivity (ρ_s) of the subsurface materials. For instance, this method is useful for the differentiation of saltwater, brackish water, and freshwater zones in coastal aquifers affected by saltwater intrusion. Similarly the method may be applied in contaminant migration studies where the concentration of one or more constituents creates a measurable difference in resistivity values, such as in leachate plumes emanating from landfills.

Electrical resistivity surveys are most commonly employed in groundwater investigations. Resistivity surveys are made by introducing a very low frequency current into the ground via electrodes in the soil surface. Additional probes are used to measure the potential. The most important resistivity methods include

- horizontal profiling, by which measurement of ρ_s is made utilizing a linear or areal array of probes arranged in a fixed spatial pattern
- electrical sounding, by which depth investigations are made by varying the distance of the electrode spacing
- directional profiling, by which a series of radial arrays of fixed spacing, arranged along different directions from a central point, are used to detect directional variations in ρ_s

The profiling or sounding curves obtained by measurement can be compared to model curves determined theoretically (see Dohr [3.18], Militzer et al. [3.66], and §2.3.3). Computational methods for identification of parameters from the recorded data are gaining widespread use (see §2.3.4). A series of instructive examples, which detail resistivity methods for identification of localized saltwater intrusion zones within coastal aquifers, is presented in Schneider [3.103].

The measurement of natural electric potentials, known as *spontaneous potentials,* may also be applied to investigation of subsurface transport processes. These potentials reflect

- electrochemical activity, e.g., caused by redox processes (see §1.4.4)
- streaming potentials generated by water moving through a porous medium
- electrofiltration activity, such as the diffusion or membrane potential

The redox potential is of significant importance to investigation of migration processes. Identification of redox potential allows for initial assessment of the system's biochemical state and relevant transformation processes (see Table 1.12).

For the analyses of migration processes, borehole geophysical methods may also provide important information for interpretation of subsurface bedding and structure. As previously discussed in §1.3.3, a better understanding of the vertical heterogeneity within the aquifer system is very important for the estimation of hydrodynamic dispersivity and, therefore, for definition of plume formation. Resistivity and gamma surveys are most often utilized in downhole geophysical tests. However, the groundwater specialist should be aware of regulatory constraints (different in each state) when utilizing gamma methods in aquifers designated for use as a drinking water supply.

Downhole measurement of electrical conductivity and temperature, as well as the use of downhole flow meters (§3.1.2), provide additional information regarding transport mechanisms (see also Silliman and Robinson [3.111]). Borehole thermometry allows for effective determination of the vertical inflow-velocity distribution within a well. Temperature measurements in semipermeable layers which are affected by vertical leakage allow for estimation of the otherwise elusive vertical permeability coefficient $k_v = v_z/\text{grad } h$ (see Figure 3.36).

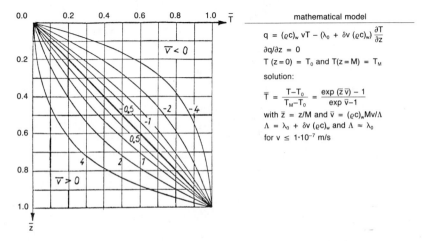

Figure 3.36. Typical curves of the temperature distribution in a subsurface layer passed vertically by water with the Darcy velocity $v_z = v$ of the thickness M under steady-state conditions. *Source:* Schestakow [3.101].

In closing, this brief introduction to geophysical test methods should point out that in situ investigation of transport mechanisms can be carried out with comparatively simple equipment. Of particular interest to groundwater specialists are portable borehole geophysical devices such as the KAT 150 [3.10].

3.3.2 TESTS IN THE NATURAL GROUNDWATER FLOW

Heat and solute transport processes are often investigated by means of tracer tests. Tests are commonly conducted using conservative tracers which are not subject to sorption/retardation processes or chemical reaction with subsurface materials. Checking the degree to which a tracer might interact with subsurface materials may be measured by the percentage of recovery in column or batch tests and by calculation of the observed retardation coefficient (R), which is equal to $1 + \rho K_d/\phi$ (see §3.2.2).

A more detailed discussion of tracer test methodology is presented in Klotz [3.47]. This discussion illustrates that the extent of tracer recovery and retardation is primarily dependent upon the properties of the subsurface material. The following general conclusions may be drawn:

1. $^{82}Br^-$, $^{131}J^-$, and Cl^- exhibit nearly ideal tracer behavior, and have been observed to migrate even more readily than tritium $^3H^1HO$ due to anion exclusion near the pore walls (see Figure 1.41).
2. NO_3^- and ^{51}Cr-EDTA migrate only slightly slower than tritium in many types of aquifer materials.
3. Of the tracer dyes such as uranin, eosin, and pyranin, uranin is transported most readily and exhibits properties of a nearly ideal conservative tracer in sand and gravel aquifers.

Photolytic decomposition should be taken into consideration when using fluorescent dye tracers. Uranin, eosin, and pyranin, for instance, are decomposed in distilled water by light according to a first-order reaction with half-lives of 11.6, and 47 hours, respectively [3.6].

Tracer tests conducted in natural groundwater flow systems are most useful for estimation of the flow velocity, hydraulic conductivity, flow direction, hydrodynamic dispersion, and the partitioning coefficient K_d.

3.3.2.1 Single Borehole Dilution Method

The single borehole test method for estimating groundwater flow velocities and directions is described in more detail in Moser et al. [3.70]. Application of the method has been perfected and has proven to be a valuable tool. The method permits evaluation of the average groundwater flow rate $|v|$ by measuring the rate of dilution of a conservative tracer in a monitoring well. The direction of groundwater flow may be determined from measurement of gamma radiation being emitted by the outflowing tracer. The vertical monitoring well should completely penetrate the aquifer and should be screened and completed with a filter pack over the entire thickness of the aquifer or permeable strata. The method is limited due to errors caused by vertical flow gradients within the filter pack or well casing, even when these gradients are small. Application of the method in the presence of vertical flow components in an aquifer may therefore produce unreliable results.

The functional relationship between the measurable tracer concentration c in the borehole, and the average flow rate $|v|$ of the groundwater in the undisturbed formation around the well bore is dependent upon the relationships between

1. $|v|$ and the discharge $\dot{V}^*$, passing the traced volume of the monitoring well
2. $\dot{V}^*$ and the measurable reduction of the tracer concentration dc/dt within this volume

Assuming a steady-state horizontal flow regime and continual mixing within the traced volume, the following relationship holds:

$$|v| = \frac{A}{B} \cdot \frac{1}{t} \ln \frac{c_o}{c} \qquad \text{with } B = 2 \alpha r \qquad (3.11)$$

where c_o = initial tracer concentration at $t = 0$
   r = radius of the screen
   A = vertical cross-sectional area of the screen pipe
   α = parameter to account for the actual inflow width of the monitored volume, which is dependent upon well construction

The cross section of the well bore illustrated in Figure 3.37 shows a representation of groundwater flow line distortion in the vicinity of the well bore due to the hydraulic influence of well bore and filter pack. The adjustment parameter α is thus a reflection of well completion and related hydraulic factors. This parameter α is dependent upon

- well screen geometry, including perforation size and location of casing joints
- the filter pack material, causing an α range of approximately 2.0 to 3.5 for unpacked versus gravel pack annulus, respectively
- the amount of formation materials removed from the borehole annulus in the course of drilling [3.49; 3.7; 3.22]

The parameter α must also take into account any vertical distortions of the idealized steady-state horizontal flow conditions [3.49; 3.71, Figure 123].

The tracer dilution method does not necessarily require the use of radioactive γ-tracers as do tracer tests designed for evaluating the groundwater flow direction (see below). The use of salt as a tracer, in conjunction with a downhole conductivity meter, allows for efficient and simple measurement of tracer dilution.

The plume of a radioactive gamma tracer, preferably $NH_4^{82}Br$, departing from a monitoring well allows for the determination of time variable location of the plume and, hence, for estimation of the groundwater flow direction. A collimated scintillation detector is used in the well to measure the intensity of gamma radiation caused by the departing tracer cloud, as a function of azimuth, the angle from true north.

Therefore, measurement of the Darcy velocity and groundwater flow direction requires

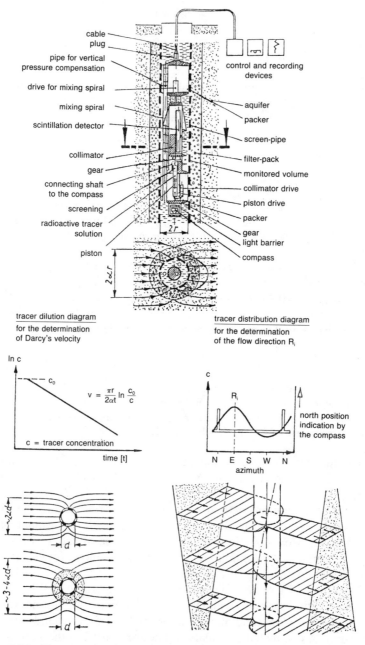

Figure 3.37. Tracer probe and principle of measurement applied in determining the Darcy velocity and its direction according to Drost [3.20], Moser and Rauert [3.71], and Bergmann [3.9].

- a dosed tracer input
- prevention of vertical migration within the borehole or filter pack
- continual mixing within the borehole test volume
- time-dependent measurement of concentration in this volume
- measurement of the azimuth of the outflow tracer cloud based upon gamma emission

A sophisticated device used for evaluation of groundwater flow direction and velocity is shown in Figure 3.37 [3.20]. The approximately 1.5m (5-feet) long tracer probe can be installed in a 4- to 8-in. diameter well screen or uncased borehole. Its construction is described in detail in Drost [3.20]. The probe allows for accurate estimation of groundwater flow rates ranging from 0.05 m/day up to 50 m/day with an accuracy of better than ± 10%, and of flow direction with an accuracy of better than ± 20%.

The estimates of direction and magnitude of the Darcy velocity often show a considerable range of variation. Characteristic frequency distributions are given in Drost [3.21]. Permeability coefficients estimated by tracer dilution methods are approximately 15% higher than those determined from aquifer pumping tests because these dilution tests will average the horizontal conductivity over the aquifer material, and because of local conductivity extremes due to well bore damage.

3.3.2.2 Multiple Borehole Tests

This method involves injection of a tracer into the undisturbed groundwater flow system by means of completely penetrating injection wells. The tracer must be injected into the groundwater system with minimal water in order to inhibit disturbance of the natural flow regime (see Figure 2.2). Tracer injection is performed in one of the following ways:

- by pulselike injection of a tracer with the specified mass m in a very short time interval Δt
- by injecting a constant tracer mass flux $\dot{m}$ in kg/s (third-type boundary condition)
- by maintaining a constant concentration, c_o, in the injection well (first-type boundary condition) differing significantly from the initial concentration c_I in the aquifer for the period $t > 0$
- by introducing a rectangular shaped pulse with $\dot{m}$ constant or c_o constant as above, but limited to the time period of $0 < t < t_E$

Pulselike injections have frequently proven unsuccessful. Initially, the concentrations around the well (see Figure 1.36) are generally high, which can cause the injected migrants to behave differently than migrants in the concentration range which has to be investigated (see

Figure 3 in the Introduction). On the other hand, if appropriate injection concentrations are used, the dilution which occurs during transport between wells is so high that detection difficulties seem to be unavoidable.

The disadvantages of the pulselike tracer injection are only avoidable by injecting a constant tracer rate, which, however, is much more expensive. Therefore, the rectangular-shaped pulse injection over a longer time period seems to be an appropriate compromise. Computing the solution of the more complicated mathematical model is not a serious problem with today's computers. Figure 3.38 shows two examples of technical solutions for tracer preparation and injection into a test well: $c_o = c_{well} =$ constant (see also [3.63]) and $\dot{m} = dm/dt =$ constant. Membrane, injector, and airlift/air-operated pumps may be used to inject the tracer mixture (see §3.1.2). Other technical solutions to tracer preparation and injection are imaginable and are used in practice. Special attention must be given if a tracer is injected at a constant rate ($\dot{m} =$ constant) because the concentration in the well can then often rise to values which exceed reasonable levels.

The tracer concentration is monitored as a function of time, $c = c(x_p, y_p, t)$, at one or more points (observation wells) downstream. Mixed samples should be taken. It has proven best to generate the vertically averaged concentration in monitoring wells by cycling. For this, the setup shown on the right-hand side of Figure 3.39 may be used.

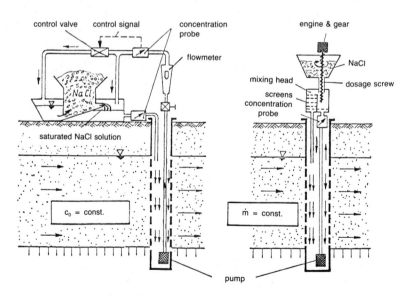

Figure 3.38. Examples of techniques for the tracer injection to implement $c_o =$ const. and $\dot{m} =$ const.

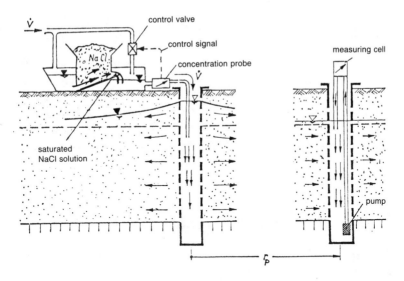

Figure 3.39. Example of techniques for field migration tests with an injection and a monitoring well.

For determination of the parameters θ_o or $\theta_f = \theta_o + \rho K_d$ (see Equation 1.78b), as well as the longitudinal and transverse dispersivities $\delta_\ell = \delta_x$ and $\delta_{tr} = \delta_y$, the analytical solutions 2.45b and 2.46b as deduced in §2.1.4.7 should be used. In Example D of §2.3.3.2, the determination of the aforementioned parameters by type-curve fitting has been demonstrated based on the mathematical process model 2.46b.

In reality, one must proceed as though y_p, the vertical distance of the observation well from the streamline running through the injection well, is not known. Therefore, y_p should be included in the parameter estimation procedure. In this case, more than one observation well needs to be operated. The evaluation can be done as before with the three unknown parameters θ_f, δ_ℓ, and δ_{tr} by varying $y_{p,i}$, that is, twisting the x,y-system, or by a least-square deviation analysis using all four unknowns (see §2.3.4).

If analytical solutions for specific boundary conditions are neither available nor deducible by convolution (see §2.1.5), numerical solutions of Equation 2.44 should be used (see Chapter 2.2). The computational effort, and thus costly computer time, however, increases significantly. Therefore, working with a constant tracer injection, $\dot{m}$ = constant (see Figure 3.38), is often advantageous. If there are noticeable advantages to limit the injection to the period $0 < t < t_E$, then for $t > t_E$, the solution becomes

$$c\,(x,y,t)\,=\,\frac{\dot{m}}{4\pi\sqrt{\overline{D}_x\overline{D}_y}}\,\exp\,\frac{qx}{2\overline{D}_x}\,[W(\sigma(t),b)\,-\,W(\sigma(t\,-\,t_E),b)]\quad(3.12)$$

As outlined in §1.3.3, the validity of vertically averaged concentration model concepts in aquifers is disputed (see also Figure 1.32). Thus, it may sometimes be appropriate to inject the tracer as a point source (see, e.g., Figure 2.14). In general, however, it is impossible to inject a tracer at a sufficiently high rate $\dot{m}$ at a point source without additional water (see §3.3.3). Therefore, line or areal injections appear to be more appropriate. The use of such line and areal pollution sources are discussed, for instance, in Kinzelbach [3.46].

3.3.3 PUMPING AND INJECTION TRACER TESTS

The natural groundwater flow field becomes perturbed when during injection of the tracer mass (m or $\dot{m}$), a quantity of water (V or $\dot{V}$ = Q) is also injected into the aquifer. We speak of pumping and injection tracer tests if this injection is accomplished through wells.

3.3.3.1 Injection Tracer Tests

In migration tests carried out by completely penetrating injection wells, the tracer is normally mixed externally with the injected water $\dot{V}$. In this case, the flow model is axisymmetric near the well, and the streamlines diverge. Commonly, quasi-steady flow conditions are created through injection at a rate $\dot{V}$ without the tracer. Steady state is established when the rate of rise of the water table in the injection well is equal to the rates of rise in the monitoring wells.

Example A of §2.3.2.1 demonstrated the evaluation of such a migration test. Figure 3.39 shows an example of a typical setup. Vertical mixing of the tracer is given in the injection well. In the monitoring well, mixing is accomplished by artificial cycling.

Unfortunately, it is not usually easy to obtain the required $\dot{V}$ from an independent water source. If $\dot{V}$ has to be withdrawn from the same aquifer, the distance (E) between the pumping well and the injection well must be more than five times the distance between the injection well and the furthest monitoring well (see Figure 2.5). Under this condition, the axisymmetric flow between the injection well and the monitoring wells remains relatively undisturbed.

Injection well tests provide reliable parameters for the longitudinal dispersivity δ_ℓ and the storage coefficient $\theta_f = \theta_o + \rho K_d$. The monitoring well should be drilled no further than 10 m away from the injection well. Injection tests should be designed by means of process models based on estimates of θ_o, ρ, K_d, and δ_ℓ. Examples and field experiences are described in Beims [3.7]. In these examples, sodium chloride, tritium, phenol, aniline, and heated water have been used as tracers. Significant conservation of tracers was attained when the tracers were only applied during the limited period t_E, that is, applied as a rectangular pulse. Then, instead of the mathematical process model 2.49a, the following analytical solution may be used as process model:

$$\bar{c}\,(r,t) = \frac{1}{2}\left(\text{erfc}\ \frac{r^2 - r^2_{F|t}}{\sqrt{5.33\delta_\ell^3 r_{F|t}}} - \delta^*\text{erfc}\ \frac{r^2 - r^2_{F|t-t_E}}{\sqrt{5.33\delta_\ell r^3_{F|t-t_E}}} \right) \quad (3.13)$$

where $\delta^* = 0$ for $t \leq t_E$
          $\delta^* = 1$ for $t > t_E$

It is advisable to evaluate the test data by straight-line (see §2.3.2) or type-curve fitting (see §2.3.3). Depending on the well volume V_w and the infiltration flux $\dot{V}$, it may also be valuable to forecast and to monitor the declining slope of the step or rectangular pulse concentration function in the injection well, for example, for the technical solution shown in Figure 3.39. If $V_w/\dot{V}$ exceeds about 2–3% of the test time, then the zero point $t = 0$ should be shifted.

If the influence of the natural groundwater flow is significant, the flow pattern around the well is no longer radially symmetric, but appears as shown in Figure 2.2a. In such a case, the tracer spreads, neglecting dispersion, according to Figure 2.2b.

Injection tracer tests with $\dot{V}$ = constant and $\dot{m}$ = constant during a definite time period $0 < t < t_E$—that is, rectangular pulse injection— are very effective. They have the decisive advantage, as compared with tests in the natural groundwater flow, of permitting sufficient tracer quantities to be injected without exceeding reasonable tracer concentrations in the vicinity of the injection well.

In evaluating the test data it would be appropriate to proceed as follows (see also Chapter 2.3):

1. Before starting the test, the expected tracer concentrations must be determined (forecasted) by computer simulations in the form of scenarios at the monitoring wells for the planned injection procedure and for various parameter combinations of the groundwater flow direction α, dispersivities δ_x, δ_y, and storage coefficient θ_f (see Figures 2.34a, 2.35a, and 2.38a).
2. Later the data recorded during the test should be plotted together with the forecast, and while the test is still proceeding, a plausibility analysis should be made. This analysis will permit decisionmaking during the experiment, allowing additional control measurements, technical alterations, or repetitions of test procedures.
3. After the test is finished, the identification of the parameters α, δ_x, δ_y,

and θ_f can be accomplished by the methods described in §2.3.4 using an efficient computer program.

The distance between the monitoring well and the injection well (x_p), and thus the test space, can be made comparatively large in this test if the required test time is available. Thus, the following test regime would seem to be appropriate in an aquifer with $T = 2 \cdot 10^{-2} \, m^2/s$, $M = 10 \, m$, $S = 0.2$, a slope of the groundwater table of $I = 0.1\%$, and a θ_f value of 1.0 or a retardation coefficient of $R = 3.0$:

- $\dot{V} = 3 \, L/s$ = constant during $0 < t < 1$ day, thereafter the injection is terminated
- $\dot{m} = c_o\dot{V} = 3 \, g/s$, with c_o < limiting concentration
- monitoring wells at (x_p, y_p) coordinates: (10 m, 0 m), (20 m, 0 m), (30 m, 0 m), (30 m, 5 m) and (30 m, -5 m)
- time required: 250 days

For each monitoring well, the tracer arrival should be forecasted with estimated parameters and then qualified by means of data gained at the nearest monitoring well, later by means of data gained at the second well, and so on to the furthest monitoring well. The values for T, M, S, θ_f, and I may be estimated, for example, with the aid of a single borehole test (k, α), water level measurements in the injection and monitoring wells (α, I), and soil sampling (M) combined with sample analysis in the laboratory (θ_f). Figure 3.40 shows the expected state of tracer distribution for the described test regime forecast by means of the MORW simulation program (see §2.2.5.2, Figure 2.30, and Prickett et al. [3.84]) and by means of the analytical solution of Equation 3.12.

3.3.3.2 Pumping Tracer Tests

In migration tests carried out in completely penetrating and screened wells, the tracer is fed into an injection well that is located at a distance r_p from a pumping well. Under such conditions, the flow regime is axisymmetric with converging streamlines, as long as the tracer injection is proceeded without additional water ($\dot{V} = 0$). This holds even if natural groundwater flow is present. As shown in §3.3.2, it is appropriate to apply continuous tracer injection $\dot{m}$ as long as c_o does not exceed limiting concentrations near the injection well.

The tracer concentration distribution during this test is, contrary to the distribution during an injection well test, not axisymmetric, $c = c(r,\varphi,t)$, (see Figure 2.28). However, radial symmetry will be obtained, if c is integrated over the angle coordinate φ for r = constant. The average concentration c* obtained by this procedure:

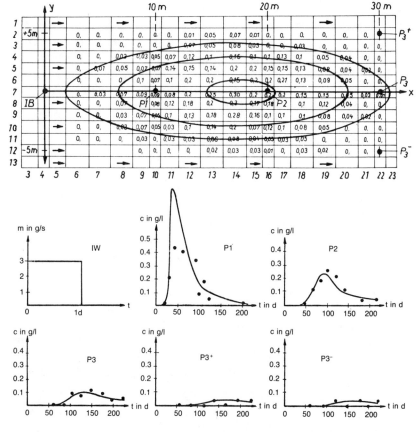

Figure 3.40.   Tracer distribution to be expected for the following conditions. Aquifer: M = 10 m, k = $2 \cdot 10^{-4}$m/s, S = 0.2, α = 0 ($\vec{v}$ = v_x), and I = 0.01. Parameters: θ_f = 1.0, see Equation 1.78b; δ_ℓ = δ_x = 0.1 $\bar{x}$, see Equation 1.30; δ_{tr} = δ_y = 0.1 δ_ℓ; v = restricting the flow model. Test regime: V = 3 L/s = const., $\dot{m}$ = c_oV = 3 g/s in the period of 0 < t < 1 day (m $\approx$ 300 kg). Injection and monitoring well diameters equal 0.5 m; all are completely penetrating and screened.

$$c^* = \frac{1}{2\pi r} \int_0^{2\pi} c\, r\, d\varphi$$

is then only a function of r and t, c^* = c^* (r,t), and Equation 2.48 is the appropriate migration process model.

If the well capacity of the observation well is neglected, the value c_o = c^* is given as follows:

$$c_o = c^*(r_p) = \dot{m}/(2\pi r_p M v_p) = \dot{m}/Q(r_p) \qquad (3.14)$$

where $Q(r_p)$ is the pumping rate Q under steady-state flow conditions. The integral outflow concentration from the pumping well $c^*(r_w)$ is immediately formed by the turbulent mixing processes in the well screen and riser, and only this average concentration c^* can be measured.

The approximate analytical solution given with Equation 2.49a has proven reliable for this migration problem if the several r-terms are defined as follows (compare Figure 3.41):

$$c^* \, (r_w,t) \; = \; c_I \; + \; \frac{c^*(r_p) - c_I}{2} \; \mathrm{erfc} \; \underbrace{\frac{r_F^2 - r_w^2}{\sqrt{5.33\delta_\ell r_F^{*3}}}}_{\alpha} \qquad (3.15a)$$

with

$$r_F \; = \; \sqrt{r_p^2 - Qt/(\pi\theta_f M)} \; \text{ and } \; r_F^* \; = \; Qt/(\pi\theta_f M) \qquad (3.15b)$$

Applying this mathematical process model, the simple procedure demonstrated in §2.3.2 (Example A) may be used to determine the migration parameters δ_ℓ and θ_f for $\dot m$ = constant. The use of type-curve fitting (see §2.3.3) would, of course, also be appropriate. The procedure described in §2.3.4, using an analytical or numerical process model, should be applied if $\dot m$ is not maintained constant during the test period or the injection is carried out in two or more injection wells. The process model can then be derived by superposition using Equation 3.15a. For example, the solution for two injection wells yields

$$2 \, \frac{c_w^* - c_I}{c^*(r_p) - c_I} \; = \; \mathrm{erfc} \, (\alpha)_{r1,t} - \delta^* \, \mathrm{erfc} \, (\alpha)_{r1,t-t_E} \qquad (3.15c)$$

$$+ \, \mathrm{erfc} \, (\alpha)_{r2,t} - \delta^* \, \mathrm{erfc} \, (\alpha)_{r2,t-t_E}$$

where $\delta^* = 1$ for $t > t_E$, $\delta^* = 0$ for $t \le t_E$, and α acc. to Equation 3.15a.

If numerical models are used, the boundary condition for the injection well should have the following form:

$$\dot m \; = \; Q(r_p)c^*(r_p) \; + \; \delta_\ell \, Q \, \partial c^*/\partial r|_{r_p} \qquad (3.16a)$$

Alternatively an inner source-sink term may be formulated. For the pumping well, the appropriate form is given by:

$$\dot{V c}^*|_{r_w} \; = \; (\delta_\ell \dot V \partial c^*/\partial r + \dot V c^*)|_{r_w} \qquad (3.16b)$$

Pumping tracer tests are quite effective in determination of θ_f and δ_ℓ for $r_p < 10$ m. It would be advisable to install, for instance, two injection wells at a distance of $r_{p1} = 5$ m and $r_{p2} = 10$ m, and then simultane-

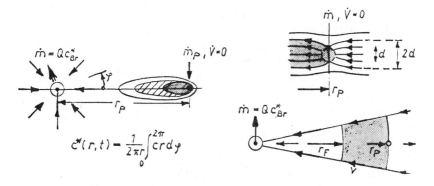

Figure 3.41. Pumping tracer test.

ously trace with $\dot{m}_5 = \dot{m}_{10}$ until the first migrants from the nearest injection well arrive in the pumping well. Contrary to the respective injection well test, no independent water source is required for this pumping tracer test. The aquifer domain influenced by the pumping tracer test is, however, comparatively small (see Figure 3.41), and $c^*(r_w)$ is frequently quite low because c must not exceed definite limits near the injection well. Therefore, planning is required, and the tracer rate $\dot{m}$ to be injected should be determined as follows (compare Figure 3.41):

$$\dot{m} < \frac{Qd}{\pi r_p} c_{limit} \qquad (3.17)$$

where d is the effective diameter of the injection well (see Figure 3.37) and c_{limit} the tracer concentration which must not be exceeded in the vicinity of the injection well.

3.3.3.3 Combined Tests

In combined tests, one tries to avoid the disadvantages of the injection tracer tests and pumping tracer tests. Three typical cases are considered.

Combined Test A

In the course of this test, an injection tracer test with an independent water source is carried out in the first time period, as explained previously. In the second time period, water is pumped from the same well, which allows removal of the tracer now distributed in the aquifer.

In both periods, the flow model as well as the quality model are axisymmetric such that Equation 2.48a is the valid migration model. In general, the flow model is transient (see Equation 2.47c). However, in the first period, a steady-state flow model may be used when flow is allowed to reach steady state through injection of nontraced water during a preliminary period ($\dot{V}$ = Q = constant; $\dot{m}$ = 0). At the beginning of the second period, the flow field is transient — which, however, is frequently neglected. Nevertheless, if a numerical migration model is used to reflect the dynamic process of this combined test, transient flow conditions should be considered according to Equation 2.47c. The required additional expenditure in the numerical solution is unimportant. For migration tests of this type, an injection pumping well and at least one monitoring well should be available.

Without any monitoring well, only δ_ℓ can be identified because the procedure is relatively insensitive to θ_f. Field tests of this type are described in, for instance, Beims [3.7], Beims and Mansel [3.8], and Mansel et al. [3.65].

Based on field experiences gathered till now, the following test setup is recommended: The injection/pumping well should be completely penetrating and screened ($\phi \approx$ 500 mm including gravel filter pack). The installation of four monitoring wells (P1-P4) seems to be appropriate, as with the hydrogeological pumping tests (see Figure 3.42). The injection of a "tracer packet" (see Figure 3.31) should be applied in the well with $\dot{m}_i$ = constant over a few hours (rectangular pulse injection). Measurements for data acquisition of the tracer breakthrough $c_i = c_i(t)$ then have to be carried out in all monitoring wells. After all tracers of the packet have passed at least the nearest monitoring wells, the pumping phase may be initiated. Data collection of the tracer breakthrough curve is now carried out in the pumping well (withdrawal flow rate $\dot{V}_w$).

The data analysis of such a migration test should be based on a numerical migration process model. For parameter identification, the method of minimizing an objective function may be recommended (see §2.3.4). When applying this model, a third-type boundary condition $\dot{m}_s$ = Qc_o = ($Qc - \delta_\ell Q\partial c/\partial r_{|r_w}$) should be used for the central well during the first time

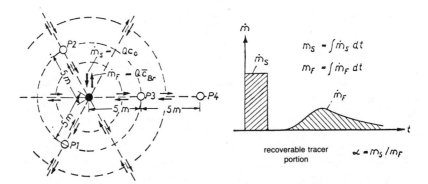

Figure 3.42. Example of a technical solution for a combined injection and pumping test (type A).

period, and a boundary condition in conformity with Equation 3.16b during the second period. With the migration parameters δ_ℓ and ϵ or θ_f (see Equation 2.48a) estimated from the measured data of the first period, the initial tracer distribution $c_i(r)$ for the second period may be calculated. An important characteristic is ultimately also the recoverable tracer portion α (see Figure 3.42). Only if α is greater than 0.8 to 0.9 should one regard the migration model Equation 2.48a as valid.

Combined Test B

If there is no independent water source available, the injection in the injection/pumping well is carried out with $\dot{m} > 0$ and $\dot{V} = Q = 0$ in the first time period (see §3.3.2 and Figure 3.38). A rectangular pulse injection is recommended; thus the migration process is reflected by Equation 3.12. The tracer will flow out of the well and migrate into the aquifer driven by the natural groundwater gradients.

During the second period, this tracer plume is pumped back. One must pay attention that this plume does not drift so far downstream that it is not recoverable by pumping at reasonable rates. Therefore, it would be appropriate to allow the tracer cloud to drift only so far that it will not get beyond the axisymmetric flow domain which will be formed around the well during the second period. These conditions may be attained with fair accuracy if the center of the cloud does not drift further away than $x_o = Q/(10\pi q_o)$ (see Figures 3.43 and 2.2). A monitoring well should be installed within a distance x_o of the combined well.

For this test analysis, an analytical solution of the migration model has been derived in Bachmat et al. [3.5, Appendix I] reflecting a pulselike injection. The equations for the parameter estimation may be found in Jahresbericht 1982 des Instituts für Radiohydrometrie [3.128]. The mathematical solution demands high standards, and its application is not simple.

Owing to the disadvantages of the pulselike injection mentioned above, the following new migration model is recommended:

Period 1 (ending with $t = t_{1E}$):

$$\int_{-\infty}^{+\infty} c\,dy \approx \int_0^{2\pi} cr\,d\varphi \approx \frac{\dot{m}}{2q_o}\left(\text{erfc}\ \frac{\theta_f Mr - q_o t_{1E}}{\sqrt{4\delta q_o \theta_f M t_{1E}}} - \text{erfc}\ \frac{\theta_f Mr - q_o t^*}{\sqrt{4\delta q_o \theta_f M t^*}} \right) \qquad (3.18a)$$

where $t^* = t_{1E} - t_E$ (see Figure 3.43)

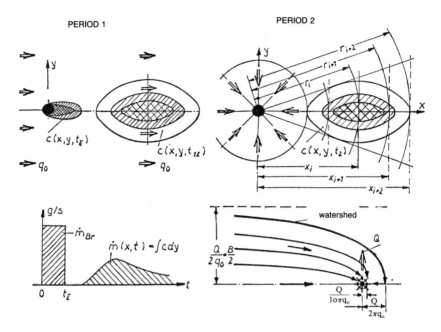

Figure 3.43. Schematic representation of the combined tracer test B.

Period 2 (Equation 2.48a)

$$\frac{\partial}{\partial r}\left(\delta Q\frac{\partial c^*}{\partial r}\right) - \frac{\partial}{\partial r}(Qc^*) = 2\pi r M\theta_f\frac{\partial c^*}{\partial t} \qquad (3.18b)$$

where $c^* = \frac{1}{2\pi r}\int_0^{2\pi} crd\varphi$

with the initial concentration:

$$c_I^*(r) = \frac{\dot{m}}{2Q}\left(\text{erfc}\ \frac{\theta_f Mr - q_o t_{1E}}{\sqrt{4\delta q_o\theta_f Mt_{1E}}} - \text{erfc}\ \frac{\theta_f Mr - q_o t^*}{\sqrt{4\delta q_o\theta_f Mt^*}}\right) \qquad (3.18c)$$

and the boundary conditions according to Equation 3.16b.

The parameters $\delta = \delta_\ell$ and θ_f should be determined by type-curve fitting. However, without a monitoring well, the sensitivity of Equation 3.18 to both parameters δ and θ_f is poor. With measured data from a monitoring well during the first period, both parameters can be estimated using Equation 3.12 and then applied to compute the initial concentration for the analysis of the second period. The analysis of the process proceeding subsequently in the second period with the concentration measured in the discharge pumped from the well $c^*(r_w)_{|t_i}$ results in the final identification of δ and θ_f. Iterating on this solution may increase accuracy [3.64]. The recoverable tracer portion is an important characteristic of this test. The recoverable portion provides a check on the assumptions made.

Combined Test C

During the first stage of this test, a steady-state flow field is generated between the pumping well and the injection well with $|Q_l| = |Q_p|$ (see Figure 2.5). A distance between the wells (E) of 10 m to 20 m is advisable. Q_p is directed from the pumping well into the injection well; thus no independent water source is required. This steady-state flow field can be continuously described by an analytical solution ($v_x = f(x,y)$; $v_y = f^*(x,y)$).

During the second stage, a constant tracer flux $\dot{m}_S = \dot{m}_z$ is injected into the injection well forming $\dot{m}_S = Q_l c_o$. A restricted injection period of $0 \leq t_2 \leq t_{2E}$ will be sufficient.

Finally, during the third stage, injection of the tracer is terminated, and the concentration of the discharge from the pumping well $c_p(t)$ is measured. Although the reinjected fluid from the pumping well will contain some tracer, $\dot{m}_p = Q_l c_p = \dot{m}_S$, this tracer does not have a measurable impact on $c_p(t)$ for a longer time. To avoid such an impact, the test should be terminated when approximately $m_S/3$ of the tracer has been recovered. It may be beneficial to install a few observation wells—for instance, the three shown in Figure 3.44.

For this migration test, a two-dimensional or coupled one-dimensional simulation model (coupled by transverse dispersion) should be used. With the help of the model, type curves may be generated prior to running the test (see §2.3.3), and the migration parameters δ_ℓ, δ_{tr}, and θ_f may be identified by means of these curves or by minimizing an objective function during test analysis (see §2.3.4).

3.3.4 SEEPAGE TESTS

Seepage tests are carried out either in undisturbed or disturbed soils or rocks. Seepage tests of interest include

- infiltration tests in natural structured soils (upper 1 to 2 m)
- seepage tests in the vadose zone extending from the soil surface to the groundwater surface, often 10 to 100 m deep
- seepage tests in loose rockfills, for example, spoils from open-pit mining activities or other tipped material

Seepage tests are carried out either by

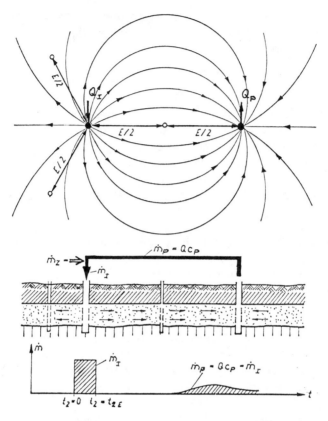

Figure 3.44. Example of a technical solution for a combined injection and pumping well test (type C).

- ponding the soil surface, causing a saturated/unsaturated water flow in the soil, predominantly directed vertically downward
- controlled sprinkling of the surface, leading to an unsaturated soilwater movement directed vertically downward
- exposing the soil surface to natural weather conditions

3.3.4.1 Ponding Recharge Tests

These tests are carried out to investigate artificial groundwater recharge below ponded areas. Such tests are, in general, imperative for prediction of quality changes in the infiltrating fluid when passing through the vadose zone. The zone below ponded areas becomes saturated with water if the bottom of the ponded area is not clogged, or if the permeability coefficient k_v does not increase with depth. Water pressure (p_w) under these conditions is higher than atmospheric pressure, thus allowing the infiltrate to freely drain into simple soilwater samplers, for instance into a simple minifilter.

Vacuum-operated minifilters with a membrane separating the soil air from the withdrawn soilwater sample, such as a ceramic porous cap, must be installed and monitored when the ponded area is small, when the pond is intermittently operated, when the bottom of the pond is clogged, or when k_v values increase with depth (see Figures 3.10, 3.22, and 3.23). Figure 3.45 illustrates an example of such a test site equipped for soilwater and groundwater sampling at specific depths.

In the top soil, tracer tests may be carried out with infiltrometers having circular infiltration areas of 1 dm² to 1 m² (see Figure 3.45).

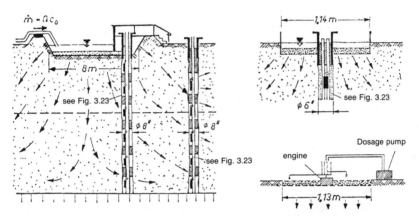

Figure 3.45. Examples of technical solutions for seepage tests with ponded or sprinkled infiltration areas.

3.3.4.2 Sprinkling Irrigation Tests

These tests allow a dosed infiltration. They can be carried out with agricultural and horticultural sprinklers or by small test sprinklers. Figure 3.45 shows such a small sprinkler described in Jahresbericht 1982 des Instituts für Radiohydrometrie [3.128]. Experience with these tests indicates that both vertical and significant lateral water flow and migration should be expected in the presence of natural stratification. The minifilters presented in §3.1.2 have proven efficient for soilwater sampling during these tests.

3.3.4.3 Lysimeter Tests

Lysimeter tests provide representative data sets for the study of migration processes. Although lysimeter installations exist in large numbers, there has been only limited application of lysimeters for the investigation of migration processes and migration parameters. Retrofitting lysimeters with devices suited for collection of soilwater samples (see §3.1.2) may often be of great value.

Because the flow model for a lysimeter experiment is relatively complicated, a quasi–steady-state flow field is often established before the tracer is injected. With few exceptions, digital simulation models are required for analyzing such tests.

3.3.5 VALUE OF PARAMETERS DETERMINED BY FIELD TEST METHODS

In §3.2.3 it was concluded that field and laboratory tests should be planned and operated jointly. Field tests have a number of significant advantages over laboratory tests:

1. They investigate a significantly larger scale in the subsurface.
2. The test space is generally minimally disturbed.
3. The natural conditions are much less perturbed.

Nevertheless, the transfer of migration parameters determined by such in situ tests to processes proceeding at larger scales may not be reliable. However, the extrapolation of dispersion parameters, through Equation 1.30 or Figure 1.33, is much more accurate when based on dispersivities (δ) which have been determined in laboratory tests and in field tests than when based solely on laboratory determinations.

Scale dependence is also critical for migration parameters of the storage-, exchange-, and reaction-process models. However, when only small deviations of these parameters are obtained by comparison of laboratory and field tests, extrapolation to larger scales seems quite reasonable.

It can certainly be stated that field migration tests are always important tools for studying migration processes and for determining model parameters. However, the applicability (transferability) of parameters determined at these scales to larger scales of space and time (see §1.2.1) involves significant uncertainties which are different and more profound than those of flow processes. These uncertainties should always be taken into account.

These problems of scale can only be overcome by real-scale investigations of migration processes. These are discussed in Chapter 3.4. However, the time required for the execution of real-time tests in the field scale is often not available. Therefore, it is advisable to regard the field tests methods discussed in this chapter as significant tools for studying and forecasting migration processes.

3.3.6 CONCLUSION ASSESSMENTS

The geophysical test methods presented are of great value. They should be applied if the conditions mentioned exist. The test equipment required is comparatively simple, thus permitting economical collection of important information which otherwise would have to be obtained via more costly alternative methods.

The value of tests carried out in undisturbed groundwater flow fields lies in the fact that the flow conditions remain unchanged. Thus, these tests cause no additional disturbances or uncertainties. Nevertheless, their field application is often subjected to comparatively strong restrictions. Thus, single borehole tests, which have become quite sophisticated, are not applicable without problems; notably, the uncertainties in the determination of the actual inflow width ($2\alpha r$) and the interferences caused by preferential vertical flows in and around the well screens lead to this assessment (see also Moser and Rauert [3.71]). Single borehole testing has not been widely applied to date; only a few institutions use this method.

Problems associated with multiple borehole testing arise from the fact that tracer plumes may remain small relative to the distance between observation wells. Hence it may happen that plumes pass between observation wells without being detected. To reduce this danger, at least three to five observation wells should be installed in a circular arch perpendicular to the estimated tracer transport direction. This practice, however, significantly increases the cost of multiple borehole testing.

The application of pumping and injection tracer tests has increased immensely in recent years. These tests allow reliable determination of dispersivities and storage capacities in characteristic space dimensions of about 10 m. The expense of these tests is dependent on the depth and thickness of the aquifer, the availability of independent water sources, the available investigation time, the number of monitoring wells, the drilling costs, and the available computer software.

If an independent water source or a second well located at a sufficient distance is available, the simple injection well test (see Figure 3.39) with rectangular pulse injection of migrants would be optimal. It is always advisable in these tests to retrieve the tracer by pumping and to analyze the data from the pump outflow above surface.

The value of the test regime shown in Figure 3.40 arises from the fact that there is no water demand, and a bigger aquifer domain can be subjected to the migration test than during the injection well test. Sometimes the combination test B is also a useful alternative. This test should gain more attention in the future.

In most cases, if no independent water source is available, pumping well tests appear to be optimal. In the case of tracers showing a sufficiently large range between the upper and lower limiting detection concentrations — as for example, in the case of $NH_4{}^{82}Br$ — pumping well tests are relatively simple and economically feasible.

Seepage tests for the determination of migration parameters have been performed only rarely. In the future, it will be necessary to use these tests more often.

This short assessment of migration field test methods shows that they are still in their developing stage and that there is too little knowledge available from previous efforts to allow production of reliable information. On the other hand, the theoretical and technical basis for running these tests has now matured, and their practical application will soon become more widespread.

3.4

Migration Process Observation and Interpretation

Monitoring migration processes in the application-scale can sometimes allow the scale problem to be overcome for estimation of the migration parameters (see §3.2.3 and §3.3.5). These processes usually have not been initiated for the purpose of parameter estimation, such as an unplanned spill or migration of natural chemical labels, but can be utilized to monitor migration at the scale application. However, the disadvantages as compared to the methods explained in §3.2.2 are the following:

1. The expense of the required data acquisition is large.
2. The complexity of the processes frequently camouflages the effects of single parameters.
3. The computing expenses for identification procedures may become high.

Parameter estimation requires the inversion of the mathematical process model (MPM), as shown in Chapter 2.3. The influence exerted by the model user and builder (MUB) (see §2.3.1 and Figure 2.32) upon the measurement procedures and data acquisition is of great importance. Experimental design requires

- an appropriate arrangement of the observation points, i.e., fixing their horizontal and vertical positions
- an appropriate schedule for the measurements and for sampling
- an appropriate selection of variables to be measured

These selections can only be reached with the aid of a hypothetically quantified MPM. With such model-aided scenario analyses, the space and time domains where the parameters significantly impact the process may be reliably determined (see §2.3.1). Only by using this information can the costs of observation and the complexity of the parameter identification model (PIM) be kept within reasonable limits.

The following explanations concentrate on aspects of monitoring environmental migrants (§3.4.1) and on technical monitoring facilities (§3.4.2).

3.4.1 MONITORING ENVIRONMENTAL MIGRANTS

Environmental migrants include radioactive environmental isotopes, atmospheric contaminants, and a variety of agrochemicals. Using environmental migrants for parameter estimation requires that

- the input functions of the environmental system are known
- the initial concentrations in the system are given
- the transport and fate of these substances within the system are reliably reflectable by definite migration models
- the concentrations in the system can be adequately measured at selected locations and points in time

This means that an MPM has to be set up in conformity with Figure 2.32, including the boundary and initial conditions, and that the migration process must be recorded at a sufficient number of space/time points so that an appropriate PIM may be derived.

3.4.1.1 Input Functions and Initial Concentrations

The radioactive environmental isotopes ^{3}H, ^{14}C, ^{39}Ar, and ^{85}Kr have been applied to the determination of groundwater age. These isotopes have the following half-lives t_H:

tritium	^{3}H	12.35 years
carbon	^{14}C	5730 years
argon	^{39}Ar	570 years
krypton	^{85}Kr	10.76 years

The mathematical model of the radioactive decay process is a first-order kinetic reaction $r = -\lambda c$. The decay or velocity constant or rate coefficient may be derived from the half-life t_H as

$$\lambda = \ln 2/t_H = 0.693/t_H \tag{3.19}$$

The concentrations of these isotopes are characterized by the following units:

- for ^{3}H in TU (tritium units); 1 TU = 0.118 Bq/L water and corresponds to a ratio of ^{3}H/^{1}H = 10^{-18})
- for ^{14}C in percent modern or percent mod; 1% mod = 2.26 m Bq/g carbon, where C exists in the dissolved phase primarily as HCO_3^-, CO_3^{2-}, and $H_2O \cdot CO_2$
- for ^{39}Ar and ^{85}Kr in dpm (decays per minute) per L or per mol Ar or Kr; 1 dpm = 16.67 mBq
- for all isotopes in a uniform way by the SI basic unit Becquerel; 1 Bq = 1 s^{-1}

Tritium, generated from nitrogen by cosmic rays in the atmosphere, is oxidized to form water and reaches the earth's surface with precipitation as $^1H^3HO$. This washout process proceeds so fast that only a limited mixing or equalization of the concentration occurs in the troposphere. Therefore, the ^{3}H concentration frequently varies greatly for different precipitation events. Seasonal variations of air exchange between the stratosphere and troposphere also lead to seasonal fluctuations. Winter input is of primary importance because winter precipitation is the major contributor to natural groundwater recharge in the northern latitudes.

Before 1953, the average annual input amounted to 6 TU, but was limited to 4 TU during the winter months. Following atmospheric testing of atomic bombs in the mid-1960s this concentration increased to more than 1,000 to 2,000 TU. In the 1980s the concentration fell to between 30 and 100 TU. This level is approximately stable at present due to emissions from nuclear power plants.

Figure 3.46 shows the typical ^{3}H concentration of precipitation for the Vienna/Munich gauging station [3.98]. This function may also be used as a first approximation for the input concentration of the natural groundwater recharge in the whole central part of Europe. Worldwide the IAEA (International Atomic Energy Agency) and the WMO (World Meteorological Organization) make available the exact monthly ^{3}H concentration data of precipitation through their joint monitoring network.

The current ^{85}Kr concentration in the atmosphere is greatly influenced by the increasing use of various nuclear engineering techniques. Unlike the ^{3}H concentration, ^{85}Kr shows only minimal seasonal variations. Furthermore, no continental effect has been detected. By an equalization function, $t^c \exp(a/t + b)$ where a, b, and c are empirical parameters, the atmospheric concentration measured between 40 and 60° N latitude can be predicted with sufficient accuracy.

Figure 3.46 shows the ^{85}Kr input function (given in Salvamoser [3.97; 3.98]) produced from 450 measurements between 40° and 60° N. Additional data for the Northern and Southern Hemispheres are given in

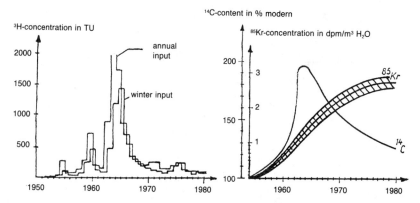

Figure 3.46. Input functions of radioactive environmental isotopes: ^{3}H concentration of precipitation for annual and winter input according to Salvamoser [3.98]; ^{85}Kr concentration of atmospheric air near the soil according to Salvamoser [3.98]; ^{14}C concentration of atmospheric air in the northern hemisphere according to Eichinger et al. [3.26].

Moser [3.71]. This function also shows that before 1956 an initial ^{85}Kr concentration of zero may be assumed ($\approx 10^{-7}$ of the present level).

The inert gas atoms of the atmosphere and those dissolved in precipitation and shallow surface water bodies maintain a thermodynamic equilibrium with one another. In the atmosphere, the mole fraction of krypton is $x_{Kr} = 1.1 \cdot 10^{-6} = 1.1$ ppm. At 10°C and 1 bar (atmospheric) pressure, $80\mu L$ Kr are dissolved in 1 m^3 water. Thus, the following approximate conversion may be applied:

$$1 \text{ Bq/mL Kr in the atmosphere} = 0.080 \text{ Bq/m}^3 \text{ water, or}$$

$$1 \text{ dpm/mL Kr in the atmosphere} = 0.080 \text{ dpm/m}^3 \text{ water} \quad (3.20)$$

The ^{39}Ar concentration of atmospheric air is only slightly anthropogenically affected. This impact is estimated to be less than 7% [3.28], thus allowing the assumption of a constant ^{39}Ar input function equal to

$$1.9 \text{ mBq/L Ar in the atmosphere, or}$$

$$0.5 \text{ mBq/m}^3 \text{ in water which is in contact with the atmosphere} \quad (3.21)$$

Therefore, the distribution of the initial concentration results from the steady-state solution of the migration process model.

As with ^{3}H, ^{14}C is formed from the interaction of cosmic rays and nitrogen in the upper atmosphere. It then oxidizes to form $^{14}CO_2$. Before nuclear tests were carried out, the ^{14}C concentration amounted to approximately 100% modern = 0.226 Bq/g C. Subsequently, it increased to about 200% mod in the mid-1960s and has since decreased to approximately 130% of the natural concentration. ^{14}C is used for age dating in the range of 1,000 to 10,000 years. The constant input function $c_{c,in}$ = 0.226 Bq/g C of atmospheric CO_2 is therefore assumed to be appropriate.

^{14}C enters the soilwater and groundwater primarily as dissolved CO_2 formed by respiration of plants and microorganisms. A ^{14}C concentration of 85 ± 5% modern is common in the groundwater recharge and about 80–120% modern in the soil air, decreasing with depth. Before 1954, 65% modern ^{14}C was a common input concentration for groundwater recharge [3.27].

Atmospheric contaminants such as acid waste gases or halogenated hydrocarbons, as well as characteristic agrochemicals including specific herbicides, insecticides, fungicides, or agents for the biological control of plant growth, may be utilized as natural tracers if the quantities and periods of application are known. Therefore, comparable input functions may be deduced in specific cases. Lysimeter test data are often an important basis for determining these input functions.

3.4.1.2 Migration Process Model (MPM)

The radioactive isotopes ^{3}H, ^{14}C, ^{39}Ar, and ^{85}Kr are again used as examples. From the explanations given in §1.8.1, efforts are commonly based on investigations of the single-phase model approach. This is reasonable as long as the exchange processes can be regarded as equilibrium processes. The migration process equations of the mixphases are then added to form the following model:

$$[TR_\Sigma] = [S_\Sigma] + [iR_\Sigma] + [eS/Si] \qquad (3.22)$$

Given these conditions, coupling the migrant species under consideration with other migrants is accomplished through the internal reaction term $[iR_\Sigma]$ and the external source-sink term $[eS/Si]$.

The environmental migrants ^{3}H, ^{39}Ar, and ^{85}Kr are conservative. They do not react with other substances and are not decomposed biochemically. This implies that a single-migrant model can describe the migration process. Omitting the source-sink terms, the following typical MPM is obtained for ^{3}H, ^{39}Ar, or ^{85}Kr in the subsurface:

$$[TR_\Sigma] = \phi \partial c / \partial t + \phi \lambda c \qquad (3.23)$$

The MPM is much more complicated for the migrant ^{14}C, since carbon is contained in many components of the subsurface. Considering carbonate–carbon acid equilibrium, one has to deal with six coupled migrants containing inorganic C (see Equation 1.89). Figure 3.47 shows an example of the total dissolved inorganic carbon (DIC) in water in relation to the partial CO_2 pressure in the adjacent gaseous phase.

In general, complex thermodynamic equilibrium or even nonequilibrium models must be used to describe inorganic carbon migration (see, e.g., Equation 1.55 and the associated text on computer programs). The model calculations published in Jahresbericht 1982 des Instituts für Radiohydrometrie [3.128] for the components CO_3^{2-}, HCO_3^-, H_2CO_3, $CaCO_3$, $CaHCO_3^+$, $MgCO_3$, $MgHCO_3^+$, $NaCO_3^-$, $NaHCO_3$, and $MnHCO_3^+$ emphasize these requirements. These calculations were performed with the aid of the computer programs PHREEQE, WATEQF, and WATEQ.

The chemical reactions which influence ^{14}C during its migration have been summed up in a schematic representation in Eichinger [3.27] (see also Figure 3.48). Therefore, approximations have to be made for each particular case to obtain reliable migration process models which remain tractable in application.

Transport processes are reflected quite differently in environmental migrant models. Frequently, they are completely neglected. Under these conditions the following model results:

$$[TR_\Sigma] = 0 \quad (mixing\ reactor\ model) \qquad (3.24a)$$

Ignorance or neglect of transport processes means that the input should be mixed in the subsurface system under consideration after a relatively

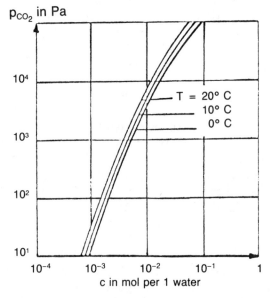

Figure 3.47. Dependence of C concentration upon partial CO_2 pressure and the temperature in the $CO_2/CaCO_3/H_2O$ system.

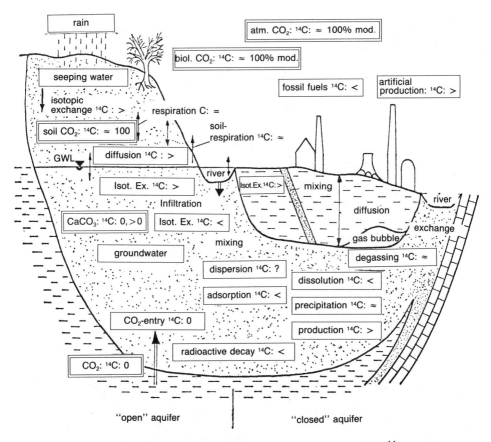

Figure 3.48. Schematic representation of the processes to which ^{14}C may be subjected during its migration in the soil and groundwater zone according to Eichinger [3.27]. Double frames mean "measurable." > or < means ^{14}C concentration increasing or decreasing. $\approx$ or = means slight or no change of ^{14}C concentration.

short time. Such a mixing reactor model approach, which is often used for unstratified surface water bodies, can be applied to soil and groundwater investigations only in a few exceptional cases, for example, in shallow aquifers with diffuse solute sources.

Consideration of the pure convective transport, $[TR_\Sigma] = \text{div} \, \Sigma \, (\vec{v}_j c_j)$, results in a single-phase model which takes the form:

$$[TR_\Sigma] = - \text{div} \, (\vec{v}c) \quad \textit{(convection or piston flow model)} \quad (3.24b)$$

This equation is a clear improvement over the mixing reactor model Equation 3.24a. However, it still does not reflect the true character of groundwater transport processes because it does not consider mixing by dispersion. Therefore, hydrodynamic dispersion should also be included in the model:

$$[TR_\Sigma] = - \text{div} (\vec{v}c - \phi D \text{ grad} c) \quad \textit{(convection/dispersion model)} \quad (3.24c)$$

Frequently, the natural flow field is also strongly approximated. A steady-state flow tube approach is common. In migration models based on such an approach, the transverse hydrodynamic dispersion, which is responsible for the migrant exchange with adjacent flow tubes, is usually neglected:

$$[TR_\Sigma] = -\partial(\vec{v}c - v\delta_\ell \partial c/\partial s)/\partial s \quad (3.24d)$$

Through the example shown in Figure 3.49, the relative impacts of transport models on prediction of migrant transport are illustrated. Based upon Equation 3.23, the following three models are compared:

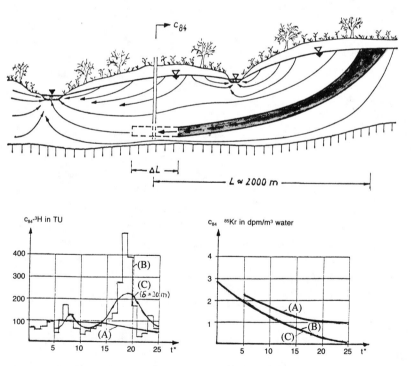

Figure 3.49. Example of the significance of migration modeling in determining the groundwater age.

(A): $\quad dc/dt = -\lambda c$ $\hfill$ (3.25a)

(B): $\quad \dfrac{\partial}{\partial \ell}(-vc) = \phi \dfrac{\partial c}{\partial t} + \phi\lambda c$ $\hfill$ (3.25b)

(C): $\quad \dfrac{\partial}{\partial \ell}\left(v\delta_\ell \dfrac{\partial c}{\partial \ell} - vc\right) = \phi \dfrac{\partial c}{\partial t} + \phi\lambda c$ $\hfill$ (3.25c)

For the case of constant parameters of the flow rate $v = v_o$, porosity ϕ, rate coefficient λ, and longitudinal dispersivity δ_ℓ, these mathematical process models provide the following solutions:

Model A—Mixing Reactor Model

$$\dot{V} = \dot{V}_{in} = \dot{V}_{out} = \text{const.}; \quad V = \text{const.}; \quad \tau = V/\dot{V}; \quad t_M - t = \tau$$

If the concentration c at time $t = t_M - \tau$ is given as $c(t_M - \tau)$ and the inflow concentration from $t_M - \tau$ through t_M is zero ($c_{in} = 0$), then the concentration at time t_M may be derived from the following equation:

$$\dot{V}dc/dt = -\dot{V}\lambda c - \dot{V}c \rightarrow -\tau \Rightarrow \frac{1}{(\lambda + 1/\tau)} \ln \frac{c(t_M)}{c(t_M - \tau)}$$

$$c(t_M)/c(t_M - \tau) = \exp(-\lambda\tau)\exp(-\tau/\tau) \hfill (3.26a)$$

If $c(t_M)/c(t_M - \tau)$ is now convoluted with a time-varying input $c_{in} = c_{in}|_{t_M - \tau}$ and related to the average residence time $\bar{\tau}$, then $c(t_M)$ according to Equation 3.26b is given as follows:

$$c(t_M) = \frac{1}{\bar{\tau}} \int_{\tau=0}^{t_M - t_I} c_{in}|_{t_M - \tau} \cdot \exp(-\tau(\lambda + 1/\bar{\tau}))d\tau \hfill (3.26b)$$

Note that c_{in} should be greater than zero at some point during the period $t_I < t < t_M$ if $c(t_M)$ is to be greater than zero.

This model, also designated as the *hydrological exponential model* (see, e.g., Moser and Rauert [3.71]), is only applicable to groundwater problems under exceptional conditions because it is based on the following assumptions:

- that complete and fast mixing in the whole system takes place
- that the real values V and $\dot{V}$ can be replaced by $\bar{\tau}$

Figure 3.49 shows the contribution to the 1984 concentration (c_{84}) of ^{3}H and ^{85}Kr which entered the groundwater at the time $1984 - \tau$. These calculations are based on Equation 3.26b with inputs as shown in Figure 3.46. The values of τ are often interpreted as groundwater age.

Model B—Convection Model

The solution of Equation 3.25b is obtained in conformity with Equation 2.17c (see also Figure 2.1) as follows:

$$c(t_M) = c_{in}|_{t_M - \tau} \cdot \exp(-\lambda\tau) \qquad (3.27)$$

The groundwater age is again assumed to be $t^* = \tau$. Figure 3.49 also shows this function for the same time point of measurement $t_M = 1984$ and the input functions ^{3}H and ^{85}Kr, according to Figure 3.46.

Model C—Convection/Dispersion Model

The solution of Equation 3.25c is obtained for a pulselike input as follows (pulse-response function):

$$h(t) = \frac{0.5(L + vt/\phi)}{2t\sqrt{\pi t \delta_\ell v/\phi}} \exp\left[\frac{-(L-vt/\phi)^2}{4t\partial_\ell v/\phi} - \lambda t\right] \qquad (3.28a)$$

This solution is obtained by derivation of Equation 2.39f. Therefore, convoluted with the time-varying input c_{in}, it yields

$$c(t_M) = \int_{\tau=0}^{t_M - t_I} c_{in}|_{t_M - \tau} \cdot h(\tau)\, d\tau \qquad (3.28b)$$

Such pulse-response functions may, of course, also be deduced from numerical solutions $h(\tau)$ if more complicated migration processes are considered.

For $t \to \infty$ and thus $\partial c/\partial t \to 0$ (steady-state conditions), Equation 3.28a and Equation 3.28b yield the following solution, which is useful for instance in ^{14}C age dating:

$$c(L) = \exp\left[L\left(\frac{1}{2\partial_\ell} - \frac{\sqrt{1 + 4\delta_\ell \lambda\phi/v)}}{4\delta_\ell^2}\right)\right] \qquad (3.28c)$$

Figure 3.49 also shows the c_{84} concentration computed by Equation 3.28b as a function of $t^* = L\phi/v$. The input functions of ^{3}H and ^{85}Kr are again taken from Figure 3.46.

The several functions $c_{84} = f(t^*)$ shown in Figure 3.49 underline the importance of model selection. A specific measured c_{84} value of ^{3}H or

^{85}Kr will result in three age estimates from these three models. This gives modeling a key role in estimating reliable results. This conclusion also applies to utilization of other environmental migrants.

3.4.1.3 Measurements

Soil and groundwater sampling has been explained in §3.1.2. A problem in measuring environmental migrants may arise from the low concentrations at which these chemicals exist and the large sample volumes required to enrich the chemicals before analyzing. The concentration of radioactive environmental isotopes in water samples, for instance, must exceed

$$c_{min} = (F/A)\sqrt{N_o/\tau} \tag{3.29}$$

to detect them with a probability of 95% [3.28], where A is an enrichment factor characterizing the increase of the radionuclide concentration in the sample before measurement, N_o is the counting rate of the zero sample, τ is the time of measurement, and F is a characteristic equipment factor (calibration factor). As may also be seen from Equation 3.29, the application of low-level measurement techniques for radionuclides is always connected with long times of measurement, appropriate methods of enriching migrants, and minimization of the counting rate of the zero sample, which may be attained by screening, establishment of deep basement laboratories, and avoiding laboratory contamination. This refers notably to the ^{14}C and ^{85}Kr analysis.

^{3}H is a β emitter. The sample volume required for the determination of ^{3}H concentration in water amounts to only 0.01 to 1.5 L; in other words, special requirements are not needed.

The simplest measurement is performed with the aid of standard fluid scintillation spectrometers. Using direct measurement, 10 to 15 mL samples of water are mixed with 13 to 85 mL scintillator liquid. This mixture is poured into small plastic bottles of an analyzing automaton. In general, the duration of one measurement (pulse counting) amounts to about 1000 minutes. Concentrations greater than 70 TU guarantee the highest accuracy. Concentrations in the range of 2 to 70 TU require an electrolytic enrichment [3.28], which may take several additional days.

If only small sample volumes are available, which is typical for soilwater samples, comparatively high accuracies are attained with gas tube counters. In these devices, hydrogen is formed by reduction and then catalytically converted into propane gas. The duration of one measurement ranges from 1000 to 4000 minutes. It is advisable to adopt this detection method in the low-level concentration range of $0.2 < c < 2$ TU. For these low concentrations, sample volumes of 0.5 or 1.5 L of water are required to enrich the tritium concentration to 20 or 50 times. The ^{3}H concentration of soil air can also be determined with a high accuracy after catalytic oxidation to form water or propane gas.

This brief overview provides a basic orientation for preparing field tests. It cannot replace exact arrangements with the responsible analytical chemist. Only the chemist can ultimately give precise information on the sample volume required for the available technique and the measuring errors to be expected (for further information, see, e.g., Eichinger et al. [3.28]).

^{14}C is a β emitter (^{14}C $\rightarrow$ ^{14}N $+ \beta$). The quantity of about 1 to 5 g C is required for radiometric ^{14}C measurements. It may be withdrawn from 200 to 500 L water, depending on the HCO_3^- concentration c_{HCO_3}. Thus, sampling from soilwater is practically impossible. Further, efficient equipment is required for ^{14}C analysis of saturated groundwater (see §3.1.2).

In situ or in the laboratory, carbon is extracted from the groundwater sample by precipitation, for example in the form of $BaCO_3$. It can be also adsorbed on ion exchange resins. CO_2 of soil air samples may be washed out by means of NaOH alkaline solutions. The method to be used should be selected by a responsible analytical chemist in the ^{14}C laboratory.

The CO_2 gas required for the analysis will be obtained from the carbonate sample, NaOH lye, or an ion exchange resin. Organic material is oxidized in the oxygen flow to form CO_2 by combustion. The ^{14}C concentration may be determined after cleaning the CO_2 by counting in proportional gas counters. Other laboratories convert the CO_2 into other gases (e.g., CH_4) or into liquids (e.g., benzol synthesis into C_6H_6, a commonly used method), or absorb it by means of special scintillator solutions, which is the simplest method. Standard liquid scintillation spectrometers have recently become of equal value to gas counter setups in terms of measuring accuracy and detection limits.

A good overview of analytical laboratory devices is given in Moser and Rauert [3.71] and Eichinger et al. [3.28]. For instance, in Eichinger et al. [3.28], it is shown that in this laboratory 3.2 g C are required for the benzol synthesis. In the subsequent 1000 minutes of liquid scintillation measurement, a ^{14}C minimal detection level of 0.6% modern (corresponding to an age of 40,000 years) and an accuracy of $\pm 1.5\%$ modern for a sample of approximately 100% modern are obtainable.

^{85}Kr is also a β emitter. The groundwater quantity required for a radiometric measurement ranges from 200 to 1,000 L. For analysis, the gases dissolved in the groundwater sample must first be extracted. It is preferable to perform the extraction in situ by continuously spraying the sample jet into a vacuum vessel. From there, the mixture of gases should be withdrawn and injected into small steel bottles. Such equipment is described in Eichinger et al. [3.28].

In the laboratory, O_2, N_2, and CO_2 are chemically separated from the gas mixture; thereupon krypton is separated by gas chromatography from the remaining sample. H_2, which is used in the gas chromatograph as a carrier gas, must also be separated before the Kr is transferred into the gas counter. This procedure takes 1 to 2 days.

Finally, the Kr sample is mixed with other inert gases until reaching a total gas pressure of 1 bar. It is then measured in the gas tube counter over a period of 1000 to 10,000 minutes. In Eichinger et al. [3.28], the minimal detection level of ^{85}Kr concentration is given as 3% of the present atmospheric concentration.

High technical and personnel expenditure and long analysis times are typical requirements for monitoring environmental migrants. Therefore, careful planning is imperative.

3.4.2 MONITORING OF MIGRATION PROCESSES

The term *monitoring* implies much more rigorous effort than does the term *observation*. In general, we understand monitoring to be:

1. a scientifically founded program for continuously observing the significant dynamic process features
2. the analysis and scientifically based explanation of the changes of the state within the system known to have taken place in the past
3. the scientifically based prediction of future changes of state

The value of information gained from a monitoring program has to justify the costs; that is, the value of information must exceed the expenses of the program. Therefore, monitoring design and the end use of the information gained (e.g., for decisionmaking or control) are inseparable. Basic monitoring of migration processes include

- monitoring the present state of the groundwater resource to ensure an adequate supply for a given task (in general drinking water supply) over the period of prediction
- monitoring soil and soilwater to ensure appropriate soil-soilwater interactions during the period of prediction
- monitoring potential contamination sources threatening the soilwater and groundwater resources (see Figure 2 in the Introduction)
- monitoring the development of the soil and groundwater resources (see Figure 2 in the Introduction)

Apart from resource-oriented monitoring (i.e., monitoring of a potentially endangered resource—see the Introduction), monitoring of engineered migration processes is also important. These monitoring tasks include

- monitoring of artificial groundwater recharge facilities
- monitoring of surface water treatment stations
- monitoring of aquifer rehabilitation measures
- monitoring of subsurface heat storage facilities

The focal point of any monitoring program will always remain observation. The development of scientifically based observation programs (e.g., measurement programs), based on modeling and simulations, is a task of the highest priority. The information base required to develop an appropriate monitoring system with regard to staff, technology, and economy can only be developed through consideration of the six "w" questions:

1. Why measure?
2. Who will measure?
3. What is to be measured?
4. Where to measure?
5. When to measure and how often?
6. Which technologies to measure with?

The problems discussed in Chapters 3.1 and 3.2 lay a solid foundation for answering these questions.

Model-aided, scientifically based simulations of historic processes is primarily aimed at identifying the appropriate model and identifying the model parameters. Verifying a model, and thus describing the governing processes phenomenologically, including quantification of model parameters, is an indispensable prerequisite for designing a scientifically based

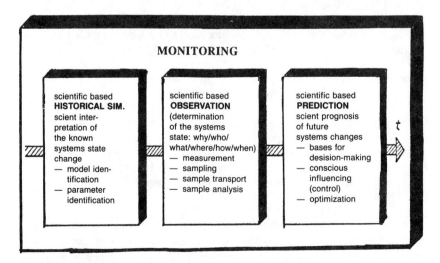

Figure 3.50. Monitoring block diagram.

measurement program and for prediction of the future changes of state of the system. This prediction is a definitive basis for decisionmaking, planning and design, and optimal process control and optimization. Simulation of historic processes and prediction of future systems states thus go hand-in-hand with observation, as shown in Figure 3.50.

References for Part 3

1. Andersen, L. J. Techniques for groundwater sampling. Prague Int Symp. IAH, Impact of Agricultural Activities on Groundwater 1982, Memoires Vol. XVI, Part 1, pp. 115–124.
2. Arche, S. A review of sensors for remote continous automatic monitoring of water quality. Chemistry and Industry 15 (1980), pp. 613–617.
3. Assmann, H.-J., et al. Tiefe Grundwassermeβstellen im Lockergestein. Erfahrungen und Weiterentwicklungen. Brunnenbau, Bau von Wasserwerken, Rohrleitungsbau 34 (1983) 2, pp. 45–50.
4. Assmann, W., et al. Gewinnung von Lockergesteinsproben ohne Störung des natürlichen Milieus – Methodik, Erfahrungen, Ergebnisse. Brunnenbau, Bau von Wasserwerken, Rohrleitungsbau 34 (1983) 10, pp. 363–368.
5. Bachmat, Y., et al. Entwicklung von Einbohrlochtechniken zur quantitativen Grundwassererkundung. Berlin: Umweltbundesamt, Forschungsbericht 82-10205 111, 1982.
6. Behrens, H., and G. Teichmann. Untersuchungen zum Lichteinfluβ auf Fluoreszenztracer. GSF-Bericht; R 290 des Institutes für Radiohydrometrie, München 1982, pp. 13–16.
7. Beims, U. Planung, Durchführung und Auswertung von Gütepumpversuchen. Geohydrodynamische Erkundung 25, Zeitschr. für angewandte Geologie 29 (1983) 10.
8. Beims, U., and H. Mansel. Assessment of groundwater resources by computer-aided designed and operated pumping tests. Proc. Intern. Symposium on Groundwater Monitoring and Management, Dresden, 1987.
9. Bergmann, H. Hydromechanische Fragen zur Interpretation von Tracermessungen. GSF-Bericht; R 36, des Institutes für Radiohydrometrie, München, 1971.
10. Bernheiden, W., K. Förster, and S. Fricke. Instruktion für die Durchführung von geophysikalischen Bohrlochmessungen mit der Apparatur KAT-150 und zur Interpretation der erhaltenen Meβergebnisse 1. Teil: KAT-150, Methodische Instruktion für Messung und Interpretation; 2. Teil:

KAT-150, Bedienungs-und Wartungsanleitung, Leipzig: Geophysik Gmbh, 1981.

11. Bianchi, W. C., et al. A positive action pump for sampling small bore wells. Soil Sci. Soc. of America Proc. 26 (1962) 1.

12. Busch, K.-F., and L. Luckner. Geohydraulik. Leipzig: Deutscher Verl. für Grundstoffindustrie, 1972.

13. Cammann, K. Das Arbeiten mit ionensensitiven Elektroden. Berlin: Springer Verl., 1973.

14. Cammann, K. Fehlerquellen bei Messungen mit ionenselektiven Elektroden. Analytiker-Taschenbuch, Band 1. Berlin: Akademie-Verl., 1980, pp. 245-267.

15. Chow, T. Fritted glass bead materials of tensiometers and tension plates. Soil Sci. Soc. of America Jour. 41 (1977) 1, pp. 19-22.

16. Cosby, R., and J. W. Keely. Sampling groundwater for organic contaminants. Groundwater 19 (1981) 2.

17. Dässler, H.-G. Einfluß von Luftverunreinigungen auf die Vegetation — Ursachen, Wirkungen, Gegenmaßnahmen. 2. Aufl. Jena: Gustav Fischer Verl., 1981, p. 211.

18. Dohr, G. Applied geophysics. Stuttgart: Enke Verl., 1974.

19. Drost, W. Das Tracerlog, ein neues Bohrlochverfahren zur Grundwassermessung mit radioaktiven Isotopen. Brüssel: Strahlungs-und Isotopenanwendung im Bauwesen, Informationsh. 55 des Büro Euroisotop der Kom. der EG, pp. 335-378.

20. Drost, W. Die Tracersonde zur Bestimmung der Filtergeschwindigkeit und der Fließrichtung des Grundwassers. GSF-Bericht; R. 290, des Institutes für Radiohydrometrie. München, 1982, pp. 37-50.

21. Drost, W. Zur geostatischen Auswertung der Ergebnisse von Einbohrlochmessungen. Beiträge zur Geologie der Schweiz. Hydrologie 28 II (1982), pp. 275-281.

22. Drost, W., and F. Neumaier. Application of single borehole methods in groundwater research. Isotope Techniques in Groundwater Hydrology (1974) 2.

23. Dunlap, W.-J., et al. Sampling for organic chemicals and microorganisms in the subsurface. U.S. Environmental Protection Agency 600/2-77-176, 1977.

24. Dunger, W. Tiere im Boden. Die Neue Brehm-Bücherei. Wittenberg-Lutherstadt: A. Ziemser Verl., 1983.

25. Eckardt, A. Nutzung von Pocket-Computern zur Lösung von Migrationsproblemen. Techn. Univ. Dresden, Sektion Wasserwesen, Abschlußarbeit im postgr. Studium, 1985.

26. Eichinger, L., et al. ^{14}C-Messungen an Weinen und Baumringen. In GSF-Bericht; R 250, des Instituts für Radiohydrometrie, München, 1980, pp. 105-118.

27. Eichinger, L. Zur Problematik des ^{14}C-Anfangsgehaltes neu gebildeter Grundwässer im Hinblick auf die Grundwasserdatierung. In GSF-Bericht; R 290, des Instituts für Radiohydrometrie, München, 1982, pp. 102-112.

28. Eichinger, L., et al. Meßmethoden für radioaktive Umweltisotope. In GSF-

Bericht; R 290, des Instituts für Radiohydrometrie, München, 1982, pp. 74–94.

29. Folkens, K., and G. Miersch. Erste Ergebnisse eines neu entwickelten Flügelradflowmeters in Filterbrunnen bzw. Filterbohrungen. Freiberger Forschungshefte C 378, Leipzig: Dt. Verl. f. Grundstoffindustrie, 1982.

30. Funk, W. Neue Techniken in der Dünnschichtchromatographie. Vom Wasser 54 (1980), p. 207.

31. Garvis, D. G. A well-head instrument package for multi-parameter measurement during well-water sampling. Water Res. Research 14 (1980) 10, pp. 1525–1527.

32. Genuchten, M. T. van. Non-equilibrium transport parameters from miscible displacement experiments. U.S. Salinity Laboratory, Research Report 119 (1981).

33. Gerschler, L. J. Probleme der Vorbehandlung von Wasserproben der Grundwasserüberwachung für die Untersuchung auf organische Spurenstoffe. Veröffentl. des Instituts für Stadtbauwesen der Techn. Univ. Braunschweig; Antropogene Einflüsse auf die Grundwasserbeschaffenheit in Niedersachsen. Braunschweig: Fallstudien 34 (1982), pp. 305–319.

34. Grünke, U., and P. Hartmann. Ionensensitive Festkörpermembran-Elektroden; Eigenschaften und Anwendungsmöglichkeiten. Hermsdorfer Technische Mitteilungen (1978).

35. Gudernatsch, H. Probenahme und Probenaufbereitung von Wässern. Analytiker-Taschenbuch, Bd. 3. Berlin: Akademie-Verl., 1982, pp. 23–35.

36. Haines, B. L., et al. Soil solution nutrient concentrations sampled with tension and zero-tension lysimeters. Soil Sci. Soc. of America Journ. 46 (1982), pp. 658–660.

37. Hitchmann, M. L. Measurement of dissolved oxygen. New York: Wiley & Sons and Orbisphese Corp. Geneva, 1978.

38. Hoffmann, W. et al. Chemisch-sensitive Halbleitersensoren – Chemiesensoren. Berlin: AdW der DDR, ZfK Rossendorf, 1980.

39. Hofreiter, G. Ein Gerät zur Entnahme von Grundwasserproben. GWF, Wasser/Abwasser 118 (1977), pp. 384–385.

40. Janata, J. Ion-selective field effect transistors: Principles and application in clinical chemistry and biology. Cardiff: Int. Symp. on Electroanalysis in Clinical, Environmental and Pharmaceutical Chemistry, 1981.

41. Jordan, H.-P., et al. Kleindimensionale Probenahmetechnik in der Hydrogeologie. Berlin: Wasserwirtsch. Wassertechn. 32 (1982) 11, p. 394.

42. Käss, W. Eine vielseitig verwendbare Kleinstpumpe für hydrologische Zwecke. GWF, Wasser/Abwasser 119 (1978), pp. 81–82.

43. Kaden, S., et al. Vorrichtung zur Entnahme repräsentativer Wasserproben aus GWBR. WP GO 1 N/2522075, Ausgabetag 02.01.85.

44. Karmbrock, F., and K. Haberer. Einsatzmöglichkeiten der kombinierten GC-MS im Wasserlabor. Vom Wasser 60 (1983), p. 237.

45. Khalil, H., and P. D. Mancy. New water quality sensor technology employing automated wet chemicals. Automated in situ water quality sensor workshop, U.S. Env. Protection Agency, Las Vegas, 1978.

46. Kinzelbach, W. Analytische Lösungen der Schadstofftransportgleichung und ihre Anwendung auf Schadensfälle mit flüchtigen Chlorkohlenwasser-

stoffen. Mitteilungen der Universität Stuttgart, Institut für Wasserbau (1983) 54, pp. 115–200.

47. Klotz, D. Verhalten hydrologischer Tracer in ausgewählten Sanden und Kiesen. GSF-Bericht; R 290, des Instituts für Radiohydrometrie. München, 1982, pp. 17–29.

48. Klotz, D., and F. Oliv. Eine einfache Methode zur Bestimmung der Verteilungskoeffizienten von Radionukliden im Grundwasser. GWF, Wasser/Abwasser 124 (1983) 3, pp. 139–141.

49. Krätzschmar, H., and L. Luckner. Analoge elektrische Strömungsuntersuchungen zur Klärung des Zusammenhanges zwischen der meßbaren Konzentrationsabnahme eines Tracers in einem Pegel und der natürlichen Grundwasserfließgeschwindigkeit. Berlin: Wasserwirtsch. Wassertechn. 16 (1966) 4 and 5, pp. 122–125 and 151–154.

50. Kurzmann, G. E. Instrumentelle Analytik von Oxydationsmitteln. In Brunnenbau, Bau von Wasserwerken, Rohrleitungsbau 34 (1983) 3, pp. 91–98.

51. Lang, H. Zur Bestimmung von Verteilungsgleichgewichten zwischen Lockergestein und wäßrigen Radionuklid-Lösungen mittels verschiedener Batch-Verfahren. Deutsche Gewässerkundl. Mitteilungen 26 (1982), pp. 69–73.

52. Lange, B., and Z. J. Vejdelek. Photometrische Analyse. Weinheim: Verl. Chemie, 1982.

53. Leichnitz, K. Prüfröhrchen. Analytiker Taschenbuch, Bd. 1. Berlin: Akademie-Verl., 1980, pp. 205–216.

54. Leichnitz, K. Prüfröhrchentaschenbuch — Luftuntersuchungen und technische Gasanalyse mit Dräger-Röhrchen, 5. Ausg. Lübeck: Drägerwerk AG, 1982.

55. Lienig, D. Wasserinhaltsstoffe, 2. Aufl. Berlin: Akademie-Verl., 1983.

56. Lienig, D., and B. Wiemer. Zur Bestimmung des im Wasser gelösten organisch gebundenen Kohlenstoffs (TOC). Berlin: Wasserwirt. Wassertechnik 26 (1976) 9, p. 304.

57. Löffler, H. Ein neuartiger Wasserschöpfer. Techn. Univ. Dresden, Sektion Wasserwesen, Techn. Bericht, 1965.

58. Luckner, L., A. Eckhardt, et al. Handbuch zur Lösung von Migrationsproblemen im Boden- und Grundwasserbereich mit Taschenrechnern (SHARP PC 1401). Techn. Univ. Dresden, Sektion Wasserwesen, 1987.

59. Luckner, L., et al. Vorrichtung und Verfahren zur teufenbezogenen Ermittlung von Wasserkennwerten. WP G 01 N/2538820; Anm.: 11.08.83, Ausgabetag: 02.01.85.

60. Luckner, L., C. Nitsche, et al. Teufengerechte Wasserdruckmessung und repräsentative Wasserprobenahme mit neuer Technik. Brunnenbau/Bau von Wasserwerken/Rohrleitungsbau 40 Jg.(1989) 5, pp. 261–264.

61. Luckner, L., and C. Nitsche. Bildung systembeschreibender Modelle der Migrationsprozesse in der Aerationszone und ihre digitale Simulation. Geodätische und geophysikalische Veröffentl. Berlin: Reihe IV (1980) 32, pp. 99–109.

62. Luckner, L., and H. Reißig. Estimation of the longitudinal dispersion and sorption coefficients in saturated soils by straightline methods. Hydrological Sciences-Bull. Sciences Hydrol. 24 (1979) 6, pp. 229–238.

63. Luckner, L., and K. Tiemer. Mathematische Modellbildung der Geofiltra-

tion und Migration. Technische Universität Dresden. Lehrheft im postgrad. Studium Grundwasser, 1981, H 1. S. 78.

64. Mansel, H. Beitrag zur mathematischen Modellbildung und digitalen Simulation rotationssymmetrischer Migrationsprozesse unter besonderer Berücksichtigung der Ermittlung von Migrationsparametern durch Gütepumpversuche. Dissertation A, TU Dresden, Section Wasserwesen, 1987.

65. Mansel, H., W. Kritzner, U. Beims, and Schwan, M. Migrationsparameterermittlung durch Feldversuche im Oberen Elbtal. Berlin, Wasserwirtschaft Wassertechnik, 1990, H.1.

66. Militzer, H., J. Schön, et al. Angewandte Geophysik im Ingenieur-Bergbau. Leipzig: Deutscher Verl. für Grundstoffindustrie, 1982.

67. Minear, A., and L. H. Keith. Water analysis. Vol. I, Inorganic species; Part 1. New York; Academic Press, 1982.

68. Miotke, F. D. Die Messung des CO_2-Gehaltes der Bodenluft mit dem Dräger-Gerät und die beschleunigte Kalklösung durch höhere Fließgeschwindigkeiten. Zeitschrift Geomorph. N. F. 16 (1972) 1, pp. 93–102.

69. Moser, H., et al. Vergleich verschiedener Verfahren zur Bestimmung der Sorption wäßriger Radio-Strontiumlösungen an einem Sand. GSF-Bericht des Instituts für Radiohydrometrie, München, 1982.

70. Moser, H., et al. Die Anwendung radioaktiver Isotope in der Hydrologie: Ein Verfahren zur Ermittlung der Ergiebigkeit von Grundwasserströmungen. Atomenergie 2 (1957), pp. 225–231.

71. Moser, H., and W. Rauert. Isotopenmethoden in der Hydrologie. Lehrbuch der Hydrologie, Bd. 8. Berlin: Verl. Gebrüder Borntraeger, 1980.

72. Müller, G. Bodenkunde, 1. Aufl. Berlin: Deutscher Landwirtsch. Verl., 1980.

73. Nestler, W., et al. Anwendungsdokumentation für den Pegeltiefenschöpfer. WAB Gmbh Frankfurt/Oder, 1983.

74. Nilson, D., and R. Phillips. Small fritted glass bead plates for determination of moisture retention. Soil Sci. Soc. of Am. Journ. 22 (1958), pp. 574–579.

75. Nitsche, C. Beitrag zur mathematischen Modellbildung und digitalen Simulation von Stofftransport-, Stoffaustausch-, Stoffspeicher- und Stoffumwandlungsprozessen in der Aerationszone. Dresden, Technische Univers., Fakultät für Bau-, Wasser- und Forstwesen, Diss. A, 1979.

76. Nitsche, C. Einfluß von Tensiometermaterialien auf Beschaffenheitsänderungen von Bodenproben. Techn. Univ. Dresden, Sektion Wasserwesen, Laborbericht, 1985.

77. Nitsche, C., and W. Kritzner. Entwicklung einer Injektorprobenahmepumpe für 2″ Grundwasserbeobachtungsrohre. Techn. Univ. Dresden, Sektion Wasserwesen, Technischer Bericht, 1984.

78. Obermann, P. Möglichkeiten der Anwendung des Doppelpackers in Beobachtungsbrunnen bei der Grundwassererkundung. In Brunnenbau, Bau von Wasserwerken, Rohrleitungsbau 27 (1976), pp. 93–96.

79. Obermann, P. Hydrochemische/hydromechanische Untersuchungen zum Stoffgehalt von Grundwasser bei landwirtschaftlicher Nutzung. Bes. Mitteilungen zum Deutschen Gewässerkundl. Jahrbuch, Nr. 42, Bonn, 2. Aufl., 1982.

80. Perkampus, H.-H. Analytische Anwendungen der UV-VIS-Spektroskopie. Analytiker Handbuch, Bd. 3. Berlin: Akademie-Verl., 1982, pp. 279–316.
81. Pettyjohn, W. A., W. J. Dunlop, et al. Sampling ground water for organic contaminants. Ground Water 19 (1981) 2, pp. 180–187.
82. Pinder, G. F., and W. G. Gray. Finite element simulation in surface and subsurface hydrology. New York: Academic Press, 1977.
83. Prager, R., et al. Technisches Handbuch Pumpen. Berlin: Verl. Technik, 1984.
84. Prickett, T. A., et al. A random-walk solute transport model for selected groundwater quality evaluations. Illinois State Water Survey, Urbana, Illinois 65 (1981), p. 149.
85. Raquet, V. Methodische Untersuchungen zur labor-und feldmäßigen Ermittlung von Wasserkennwerten, insbesondere zum Einfluß von Filtermaterialien auf die Wasserbeschaffenheit. Diplomarbeit der Techn. Univ. Dresden, Sektion Wasserwesen, 1984.
86. Reasoner, D. J. Microbiology of potable water and groundwater. In Journal Water Pollution Control Federation 55 (1983) June, pp. 891–895.
87. Repsold, H., and H. Friedrich. über eine elektromechanische Sonde zur Wasserprobenahme in Bohrlöchern. Hannover: Geolog. Jahrbuch E9, 1976, pp. 35–39.
88. Reuter, G. Gelände- und Laborpraktikum der Bodenkunde, 3. Aufl. Berlin: Deutscher Landwirtsch.-Verl., 1976.
89. Richards, L. A. Diagnosis and improvement of saline and alkali soils. U.S. Deparment of Agriculture, Agriculture Handbook No. 60, 1969.
90. Riha, M. Multi-stage sampling and testing of groundwater—a prerequisite for maximum utilization of aquifer systems. Noordwijkerhoul, Netherlands: Proc. of Int. Symp. Quality of Groundwater, 1981, pp. 737–741.
91. Riha, M. Groundwater sampling and testing. National Water Well Assoc. of Australia, Conf. Cowes, 1980, pp. 1–9.
92. Rinnert, B. Hydrodynamische Dispersion in porosen Medien: Einfluß von Dichteunterschieden auf die Vertikalvermischung in horizontaler Strömung. Mitteilungen der Univers. Stuttgart, Institut für Wasserbau (1983) 52, p. 161.
93. Rommel, K., and E. Seelos. Leitfähigkeitsmessung—Automatische Temperaturkompensation unter Berücksichtigung "natürlicher Wässer". Wasser, Luft und Betrieb (1980) 9, pp. 14–17.
94. Runge, H. Gasspurenanalyse, Messen von Emissionen und Immissionen. Analytiker Taschenbuch, Bd. 3, Berlin: Akademie-Verl., 1982, pp. 317–360.
95. Ryan, J. P. Batch and column strontium coefficients with water-saturated soil strata from the Savannah River Plant Burial Ground. IAEA-SM-257/69, pp. 133–145.
96. Salacz, K. Gesichtspunkte zur hygienischen Beurteilung in der Technikwasserversorgung angewandten Kunststoffe. Acta Hydrochimica et Hydrobiologica 10 (1982) 6, pp. 575–581.
97. Salvamoser, J. Modellüberlegungen zur hydrogeologischen Verwendung von ^{85}Kr-Messungen. GSF-Bericht; R 250, des Instituts für Radiohydrometrie, München, 1980, pp. 98–104.

98. Salvamoser, J. über die Genauigkeit der Bestimmung der mittleren Verweilzeit nach dem Exponentialmodell anhand von ^{3}H-und ^{85}Kr-Gehalten im Grundwasser. GSF-Bericht; R 290, des Instituts für Radiohydrometrie, München, 1982, pp. 95–101.

99. Scale, M. R., et al. Manual of ground-water sampling procedures. Worthington: National Water Well Association, 1981, p. 93.

100. Schenk, V. Erfahrungen beim Bau tiefer Grundwassermeßstellen und bei der Bestimmung des Probenahmezeitpunktes. Brunnenbau, Bau von Wasserwerken, Rohrleitungsbau 2 (1983), pp. 5–56.

101. Schestakow, W. M. Dinamika podzemnyk vod. Moskva: Ind. Moskovskogo Universiteta, 1979.

102. Schmidt, T. Repräsentative Wasserprobenahme zur Gewinnung von Steuerparametern für UNEIS-Anlagen. Techn. Univ. Dresden, Diplomarbeit, 1984.

103. Schneider, H. Die Wassererschließung. 2. Aufl. Essen: Vulkan-Verl., 1973.

104. Schön, J. Petrophysik, physikalische Eigenschaften von Gesteinen und Mineralen. Berlin: Akademie-Verl., 1983.

105. Schönborn, C. Probleme der Schwermetallanalytik, Beispiele aus dem Bereich der Grundwasserüberwachung. Veröffentl. des Instituts für Stadtbauwesen der TU Braunschweig: Anthropogene Einflüsse auf die Grundwasserbeschaffenheit in Niedersachsen, Fallstudien 34 (1982), pp. 305–319.

106. Schuler, P. Ein neues mikroprozessor-gesteuertes Ionenmeter. Laborpraxis 4 (1980).

107. Schuller, R., et al. Recommended sampling, procedures for monitoring wells. Ground Water Monitoring Review 1 (1981), pp. 42–46.

108. Schurig, H. Einsatz von Mammutpumpen in Grundwasserbeobachtungsrohren (GWBR). Techn. Univ. Dresden, Sektion Wasserwesen, Diplomarbeit, 1984.

109. Schwabe, K. pH-Meßtechnik. Dresden/Leipzig: Steinkopff Verl., 1976.

110. Silliman, S. E., L. F. Konikow, and C. I. Voss. Laboratory investigation of longitudinal dispersion in anisotropic porous media. Water Res. Research. 23 (1987) 11, pp. 2145–2151.

111. Silliman, S. E., and R. Robinson. Identifying fracture interconnections between boreholes using natural temperature profiling − I. Conceptual basis. Ground Water 27 (1989) 3, pp. 393–402.

112. Silliman, S. E., and E. S. Simpson. Laboratory evidence of the scale effect in dispersion of solutes in porous media. Water Res. Research 23 (1987) 8, pp. 1667–1673.

113. Sneiting, H. Mini-screens sampling system. National Institute for Water Supply, Netherlands. Quarterly Report, March 1979, No. 16.

114. Stegmann, P. Combibox CB 570 − Tragbares Meßgerät für Feld und Ambulanz zur Wasser-und Abwasserüberwachung: wetterfest, wasserdicht, robust. Weilheim: Wiss.-techn. Werkstätten, N 3/81.

115. Tinlin, R. M. Monitoring groundwater quality: illustrative examples. U.S. EPA-600/4-76-036, 1976, 11.

116. Turnow, E. Beitrag zur Gewinnung von Wasserproben aus der gesättigten und ungesättigten Bodenzone zur Ermittlung von physikalischen und

chemischen Kennwerten. Techn. Univ. Dresden, Sektion Wasserwesen, Diss. A, 1984.

117. Urban, D., and G. Schettler. Untersuchungsergebnisse zur Gewinnung repräsentativer Grundwasserproben für die chemische Analyse aus Pegelbrunnen. Berlin: Wasserwirtsch. Wassertechn. 30 (1980) 12, pp. 425–430.

118. Warden, A. Durchführung der chemischen Wasseruntersuchung nach den Erfordernissen aktueller Verordnungen und Richtlinien. Brunnenbau, Bau von Wasserwerken, Rohrleitungsbau 34 (1983) 9, pp. 324–331.

119. Wiemer, B. Vorschlag für eine Routinemethode zur Bestimmung polycyclischer aromatischer Kohlenwasserstoffe. Acta Hydrochimica et Hydrobiologica 9 (1981) 3, p. 337.

120. Wilson, L. G. Monitoring in the vadose zone: A review of technical elements and methods. GE79TMP-55, General Electric Company – TEMPO, Centre for Advanced Studies, 1979.

121. Ausgewählte Methoden der Wasseruntersuchung. Jena: Gustav Fischer Verl., Band 1, Loseblattsammlung, 1971–76; Band II, 2. Aufl., 1982.

122. Deutsche Einheitsverfahren zur Wasser-, Abwasser- und Schlammuntersuchung. 11. Lief. Weinheim/Bergstr.: Chemie-Verl., 1982.

123. DVWK Merkblätter zur Wasserwirtschaft. Entnahme von Proben für hydrogeologische Grundwasseruntersuchungen, Blatt 203. Hamburg: Verl. Paul Parey, 1982.

124. Environmental testing instrumentation for air, water and soil. Illinois: Soiltest, Catalog 484.

125. Halogenkohlenwasserstoffe in Grundwässern. DVGW-Schriftenreihe, Wasser 29 (1981), p. 139.

126. Horiba Ion Meters N-8 Series. Tokyo: Nichimen Elektronics Dept., 11–1 Nihonbashi 3-Chome Chuo-ku, Tokyo, 103 Japan.

127. ISCO's new well sampler. Journal Water Pollution Control Federation, Vol. 55 (1953) Sept., p. 409.

128. Jahresbericht 1982 des Instituts für Radiohydrometrie der Gesellschaft für Strahlen-und Umweltforschung mbH. GSF-Bericht; R 328, des Instituts für Radiohydrometrie, München, 1983.

129. Küvetten-Test-Photometer für CSB und Wasseranalysen. Wasser und Boden 36 (1984) 5, p. 181.

130. Mark XV water quality microprocessor. In Journal Water Pollution Control Federation 55 (1983) June.

131. Pneumatik-Unterwasserpumpe für Grundwasserprobenahme. Brunnenbau, Bau von Wasserwerken, Rohrleistungsbau 34 (1983) 1, Industrieberichte.

132. SEBA-Hydrometrie. Probenahmegeräte für Grundwasser. Kaufbeuren: Prospekt der SEBA Hydrometrie GmbH.

133. Standard methods for the examination of water and wastewater. 14th Ed. American Public Health Association, 1980.

Appendix 1

Physicochemical Values and Their SI-Units

		Basic Units		
Amount of substance	n	Mole	mol	basic unit: 1 mol contains $6.023 \cdot 10^{23}$ particles
Equivalents or	$n_{eq} = n_c$	Mole	mol	$n_c = n \cdot z$ (1 val = charges 1 mol/z); z = effective valency
Molar mass	M		kg/mol	$M = m/n$
Molar volume	V_{m},v		m³/mol	$V_m = V/n$
Mass	m	Kilogram	kg	basic unit
Density	ρ		kg/m³	
Pressure	p	Pascal	Pa	$1Pa = 1N/m^2 = 1\ kg/(ms^2)$; $1\ bar = 10^5\ Pa = 100\ kPa$; $1\ atm = 1.013\ bar$
Work	A	Joule	J	$1\ J = 1\ Nm = 1\ Ws$
Energy	E	Watt-second	Ws	$1\ Ws = 1\ J = 1\ kgm^2/s$
Electrical current	I	Ampere	A	basic unit
Electrical voltage	U	Volt	V	$1\ V = 1\ W/A$
Electrical resistance	R	Ohm	Ω	$1\Omega = 1V/A = 1W/A^2$
Electrical charge	Q	Coulomb	C	$1\ C = 1\ As$
Electrical capacity	C	Farad	F	$1\ F = 1\ C/V = 1\ As/V$
Temperature	T	Kelvin	K	basic unit
	t	Celsius	°C	$T = 273.15 + t$
Amount of heat	W	Joule	J	also internal energy, enthalpy and reaction heat

Flux Units

Fluid flux	$\dot{V}, Q$	$\dot{V} = dV/dt$	m^3/s, L/s, m^3/d
Fluid flux rate (Darcy's velocity or flux density)	v	$v = \dot{V}/A$	$m^3/(m^2 s) = m/s$
Mass flux	$\dot{m}$	$\dot{m} = dm/dt$	kg/s
Mass flux rate	i	$i = \dot{m}/A$	$kg/(m^2 s)$
Substance flux	$\dot{n}$	$\dot{n} = dn/dt$	mol/s
Substance flux rate	f	$f = \dot{n}/A$	$mol/(m^2 s)$
Heat flux	$\dot{W}$	$\dot{W} = dW/dt$	$W = J/s$
Heat flux rate	q	$a = \dot{W}/A$	$J/(m^2 s) = W/m^2$

Constants

Gas constant	$R = 8.314$ J/(mol K)
Avogadro's Number	$N_L = 6.023 \cdot 10^{23}$ mol^{-1}
Faraday's constant	$F = e \cdot N_L$
Atomic mass unit	$m_u = (1/12)\ m(^{12}C)$
	$1.66 \cdot 10^{-27}$ kg
Electronic charge	$e = 1.6 \cdot 10^{-19}$ As

ENGLISH-SI CONVERSION TABLE

Length	1 inch	= 2.54 cm	Mass	1 oz	= 28.35 g
	1 ft	= 0.3048 m		1 lb_m	= 0.4536 kg
	1 mi	= 1.609 km		1 s. ton	= 907 kg
				1 l. ton	= 1016 kg
Area	1 inch^2	= 6.4516 cm^2		1 lb_m/ft^3	= 16.02 kg/m^3
	1 ft^2	= 0.0929 m^2			
	1 acre	= 0.4047 ha	Force	1 lb_f	= 4.448 N
		= 0.4047 $\cdot$ 10^4 m^2	Temperature	x°F	= (9/5)x°C + 32
	1 mi^2	= 2.590 km^2			
			Stress and pressure	1 lb_f/foot^2	= 47.88 Pa
Volume	1 US ft oz	= 29.54 cm^3		1 psi	= 6.895 $\cdot$ 10^3 Pa
	1 ft^3	= 2.832 $\cdot$ 10^{-2} m^3		1 atm	= 1.013 $\cdot$ 10^5 Pa
		= 28.32 liter		1 bar	= 10^5 Pa
	1 US gal	= 3.785 $\cdot$ 10^{-3} m^3			= 0.1 MPa
		= 3.785 liter			
	1 UK gal	= 4.546 $\cdot$ 10^{-3} m^3	Work or energy	1 ft lb_f	= 1.356 J
		= 4.546 liter		1 calorie	= 4.185 J
	1 US bushel	= 3.524 $\cdot$ 10^{-2} m^3		1 BTU	= 1.055 $\cdot$ 10^3 J
		= 35.24 liter			
	1 oil barrel	= 0.156 m^3	Hydraulic conductivity	1 ft/s	= 0.3048 m/s
		= 156 liter		1 US gal/day ft^2	= 4.720 $\cdot$ 10^{-7} m/s
Fluid flux	1 cubic ft/s	= 2.832 $\cdot$ 10^{-2} m^3/s	Transmissivity	1 ft^2/s	= 9.290 $\cdot$ 10^{-2} m^2/s
		= 28.32 liter/s		1 US gal/day ft	= 1.438 $\cdot$ 10^{-7} m^2/s
	1 US gal/min	= 6.309 $\cdot$ 10^{-5} m^3/s			
		= 6.309 $\cdot$ 10^{-2} liter/s	Intrinsic permeability	1 ft^2	= 9.290 $\cdot$ 10^{-2} m^2
	1 UK gal/min	= 7.576 $\cdot$ 10^{-5} m^3/s			= 9.412 $\cdot$ 10^{10} darcy
		= 7.576 $\cdot$ 10^{-2} liter/s		1 darcy	= 0.987 $\cdot$ 10^{-12} m^2

PREFIXES FOR MULTIPLYING AND DIVIDING OF SI UNITS

10^{-1}	tenth	deci	d		10^{1}	ten	deca	da
10^{-2}	hundredth	centi	c (%)		10^{2}	hundred	hecto	h
10^{-3}	thousandth	milli	m $^{o}/_{oo}$		10^{3}	thousand	kilo	k
10^{-6}	millionth	micro	μ (ppm)		10^{6}	million	mega	M
10^{-9}	billionth	nano	n		10^{9}	billion	giga	G
10^{-12}	trillionth	pico	p (ppb)		10^{12}	trillion	tera	T

ppm (parts per million) – particles per million particles
ppb (parts per billion) – particles per billion particles

CHEMICAL SYMBOLS

Mass number   Ion charge
SYMBOL
Number of protons   Number of atoms

Examples
$^{12}_{6}C$, Ca^{2+}, O_2

Appendix 2

PERIODIC TABLE OF THE ELEMENTS

	Group I A	Group I B	Group II A	Group II B	Group III A	Group III B	Group IV A	Group IV B
1. Period	1 H Hydrogen 1.0073							
2. Period	3 Li Lithium 6.941		4 Be Beryllium 9.01218		5 B Boron 10.81		6 C Carbon 12.011	
3. Period	11 Na Sodium 22.98977		12 Mg Magnesium 24.305		13 Al Aluminum 26.98154		14 Si Silicon 28.0855	
4. Period	19 K Potassium 39.0983		20 Ca Calcium 40.08			21 Sc Scandium 44.9559		22 Ti Titanium 47.88
		29 CU Copper 63.546		30 Zn ZInc 65.38	31 Ga Gallium 69.72		32 Ge Germanium 72.59	
5. Period	37 Rb Rubidium 85.4678		38 Sr Strontium 87.62			39 Y Yttrium 88.9059		40 Zr Zirconium 91.22
		47 Ag Silver 107.868		48 Cd Cadmium 112.41	49 In Indium 114.82		50 Sn Tin 118.69	
6. Period	55 Cs Cesium 132.9054		56 Ba Barium 137.33			57 La[x] Läthanum 138.90055		72 Hf Hafnium 178.49
		79 Au Gold 196.9665		80 Hg Mercury 200.59	81 Tl Thallium 204.383		82 Pb Lead 207.2	
7. Period	87 Fr Francium (223)		88 Ra Radium 226.0254			89 Ac[xx] Actinium 227.0278		104 Ku Kurtschatowium (261)

PERIODIC TABLE OF THE ELEMENTS (continued)

Group V		Group VI		Group VII		Group VIII			
A	B	A	B	A	B	A	B		
						2 He Helium 4.00260			
7 N Nitrogen 14.0067		8 O Oxygen 15.9994		9 F Fluorine 18.998403		10 Ne Neon 20.179			
15 P Phosphorus 30.97376		16 S Sulfur 32.06		17 Cl Chlorine 35.453		18 Ar Argon 39.948			
	23 V Vanadium 50.9415		24 Cr Chromium 51.996		25 Mn Manganese 54.9380		26 Fe Iron 55.847	27 Co Cobalt 58.9332	28 Ni Nickel 58.69
33 As Arsenic 74.9216		34 Se Selenium 78.96		35 Br Bromine 79.904		36 Kr Krypton 83.80			
	41 Nb Niobium 92.9064		42 Mo Molybdenum 95.94		43 Tc Technetium (98)		44 Ru Ruthenium 101.07	45 Rh Rhodium 102.9055	46 Pd Palladium 106.42
51 Sb Antimony 127.75		52 Te Tellurium 127.60		53 I Iodine 126.9045		54 Xe Xenon 131.29			
	73 Ta Tantalum 180.9479		74 W Yungsten 183.85		75 Re Rhenium 186.207		76 Os Osmium 190.2	77 Ir Iridium 192.22	78 Pt Platinum 195.08
83 Bi Bismuth 208.9804		84 Po Polonium (209)		85 At Astatine (210)		86 Rn Radon (222)			

Appendix 3

2-dimensional field $i = 1,2; j = 1,2$ (i.e., $\xi_1 = x, \xi_2 = y$)(see Equation 1.27)

$$\nabla \overset{=}{D} \nabla c = \frac{\partial}{\partial \xi_i}\left(D_{ij}\, \frac{\partial c}{\partial \xi_j} \right) = \frac{\partial}{\partial x}\left(D_{xx}\, \frac{\partial c}{\partial x} + D_{xy}\, \frac{\partial c}{\partial y} \right) + \frac{\partial}{\partial y}\left(D_{yx}\, \frac{\partial c}{\partial x} + D_{yy}\, \frac{\partial c}{\partial y} \right)$$

D_0 molecular diffusion

$$D_{xx} = D_1\cos^2 \alpha + D_{tr}\sin^2 \alpha + D_0 = \frac{D_1 v_x^2 + D_{tr} v_y^2}{v_x^2 + v_y^2} + D_0$$

$$= \delta_{tr}|v| + (\delta_1 - \delta_{tr})\, v_x^2/|v|$$

$$D_{yy} = D_{tr}\cos^2 \alpha + D_1\sin^2 \alpha + D_0 = \frac{D_{tr} v_x^2 + D_1 v_y^2}{v_x^2 + v_y^2} + D_0$$

$v_x^2 + v_y^2 = |v|^2$

$$= \delta_{tr}|v| + (\delta_1 - \delta_{tr})v_y^2/|v|$$

$$D_{xy} = D_{yx} = (D_1 - D_{tr})\cos \alpha \sin \alpha = \frac{(D_1 - D_{tr})v_x v_y}{v_x^2 + v_y^2}$$

$\vec{v}$ velocity vector to which $\overset{=}{D}$ is referred to $\overset{=}{D} = \vec{v}\vec{\delta}$

$$= (\delta_1 - \delta_{tr})\, v_x v_y/|v|$$

The treatment of $D_{xx}, D_{xy} = D_{yx}, D_{yy}$ proceeds as above with the scalar quantity a 2-dimensional axisymmetric field ($\xi_1 = r, \xi_2 = z$)

$$\nabla \overset{=}{D} \nabla c = \frac{1}{r}\, \frac{\partial}{\partial r}\left(rD_{rr}\, \frac{\partial c}{\partial r} + rD_{rz}\, \frac{\partial c}{\partial z} \right) - \frac{\partial}{\partial z}\left(D_{rz}\, \frac{\partial c}{\partial r} + D_{zz}\, \frac{\partial c}{\partial z} \right)$$

The determination of $D_{rr}, D_{rz} = D_{zr}$ and D_{zz} proceeds as for $D_{xx}, D_{xy} = D_{yx}$ and D_{yy}

457

Appendix 4

APPENDIX 4a
FD-TERMS OF THE MODEL

$$\frac{\partial c}{\partial t} + u\frac{\partial c}{\partial x} + w\frac{\partial c}{\partial z} - D\frac{\partial^2 c}{\partial x^2} + D\frac{\partial^2 c}{\partial z^2} + kc = 0$$

for a uniform rectangular grid according to [3.78]

Term	$\dfrac{\partial c}{\partial t}$
Finite Element Representation	$\dfrac{\Delta x \Delta z}{36}\left(\left[\dfrac{dc_{-1,1}}{dt} + 4\dfrac{dc_{0,1}}{dt} + \dfrac{dc_{1,1}}{dt}\right] + 4\left[\dfrac{dc_{-1,0}}{dt} + 4\dfrac{dc_{0,0}}{dt} + \dfrac{dc_{1,0}}{dt}\right] \right.$ $\left. + \left[\dfrac{dc_{-1,-1}}{dt} + 4\dfrac{dc_{0,-1}}{dt} + \dfrac{dc_{1,-1}}{dt}\right]\right)$
Numerical Integration $0(\Delta h^4)$ (1–dimensional Simpson rule)	$\displaystyle\int \frac{\partial c}{\partial t}\, dx\, dz = \frac{\Delta x \Delta z}{9}\left(\overbrace{\left[\dfrac{\partial c_{-1,1}}{\partial t} + 4\dfrac{\partial c_{0,1}}{\partial t} + \dfrac{\partial c_{1,1}}{\partial t}\right]}\right.$ $+ 4\left[\dfrac{\partial c_{-1,0}}{\partial t} + 4\dfrac{\partial c_{0,0}}{\partial t} + \dfrac{\partial c_{1,0}}{\partial t}\right]$ $\left. + \left[\dfrac{\partial c_{-1,-1}}{\partial t} + 4\dfrac{\partial c_{0,-1}}{\partial t} + \dfrac{\partial c_{1,-1}}{\partial t}\right]\right)$ $- \dfrac{\Delta x \Delta z^5}{45}\dfrac{\partial^4}{\partial z^4}\left(\dfrac{\partial c}{\partial t}\right) - \dfrac{\Delta z \Delta x^5}{45}\dfrac{\partial^4}{\partial x^4}\left(\dfrac{\partial c}{\partial t}\right)$ (2-dimensional Simpson rule)

461

Term	$-D \dfrac{\partial^2 c}{\partial x^2}$
Finite Element Representation	$-D \dfrac{\Delta x \Delta z}{6} \left(\left[\dfrac{c_{1,1} - 2c_{0,1} + c_{-1,1}}{\Delta x^2} \right] + 4 \left[\dfrac{c_{1,0} - 2c_{0,0} + c_{-1,0}}{\Delta x^2} \right] \right.$ $\left. + \left[\dfrac{c_{1,-1} - 2c_{0,-1} + c_{1,-1}}{\Delta x^2} \right] \right)$
Numerical Differentiation $0(\Delta h^2)$	$\dfrac{\partial^2 c_{0,k}}{\partial x^2} = \dfrac{c_{1,k} - 2c_{0,k} + c_{-1,k}}{\Delta x^2} - \dfrac{1}{12}\, \Delta x^2 \dfrac{\partial^4 c}{\partial x^4}$ $k = -1, 0, 1$
Numerical Integration $0(\Delta h^4)$ (1-dimensional Simpson rule)	$\displaystyle\int \dfrac{\partial^2 c_{0,k}}{\partial x^2}\, dz = \dfrac{\Delta z}{3} \left\{ \dfrac{\partial^2 c_{0,-1}}{\partial x^2} + 4 \dfrac{\partial^2 c_{0,0}}{\partial x^2} + \dfrac{\partial^2 c_{0,1}}{\partial x^2} \right\}$ $- \dfrac{\Delta x^5}{90} \dfrac{\partial^6 c}{\partial x^2 \partial z^4}$

Term	$u\dfrac{\partial c}{\partial x}$	$w\dfrac{\partial c}{\partial z}$
Finite Element Representation	$u\dfrac{\Delta x\Delta z}{6}\left\{\left[\dfrac{c_{1,1}-c_{-1,1}}{2\,\Delta x}\right]+4\left[\dfrac{c_{1,0}-c_{-1,0}}{2\,\Delta x}\right]\right.$ $\left.+\left[\dfrac{c_{1,-1}-c_{-1,-1}}{2\,\Delta x}\right]\right\}$	$w\dfrac{\Delta x\Delta z}{6}\left\{\left[\dfrac{c_{1,1}-c_{1,-1}}{2\,\Delta z}\right]+4\left[\dfrac{c_{0,1}-c_{0,-1}}{2\,\Delta z}\right]\right.$ $\left.+\left[\dfrac{c_{-1,1}-c_{-1,-1}}{2\,\Delta x}\right]\right\}$
Numerical Differentiation $0(\Delta h^2)$	$\dfrac{\partial c_{0,k}}{\partial x}=\dfrac{c_{1,k}-c_{-1,k}}{2\,\Delta x}-\dfrac{1}{6}\Delta x^2\dfrac{\partial^3 c_{0,k}}{\partial x^3}$ $k=-1,0,1$	$\dfrac{\partial c_{j,0}}{\partial z}=\dfrac{c_{j,1}-c_{j,-1}}{2\,\Delta x}-\dfrac{1}{6}\Delta z^2\dfrac{\partial^3 c_{j,0}}{\partial z^3}$ $j=-1,0,1$
Numerical Integration $0(\Delta h^2)$ (1-dimensional Simpson rule)	$\displaystyle\int\dfrac{\partial c_{0,k}}{\partial x}\,dz=\dfrac{\Delta z}{3}\left\{\dfrac{\partial c_{0,-1}}{\partial x}+4\dfrac{\partial c_{0,0}}{\partial x}+\dfrac{\partial c_{0,1}}{\partial x}\right\}$ $-\dfrac{\Delta z^5}{90}\dfrac{\partial^5 c}{\partial x\partial z^4}$	$\displaystyle\int\dfrac{\partial^2 c}{\partial z^2}\,dx=\dfrac{\Delta x}{3}\left\{\dfrac{\partial^2 c_{-1,0}}{\partial z^2}+4\dfrac{\partial^2 c_{0,0}}{\partial z^2}+\dfrac{\partial^2 c_{1,0}}{\partial z^2}\right\}$ $-\dfrac{\Delta x^5}{90}\dfrac{\partial^5 c}{\partial x^4\partial z^2}$

Term	$-D\dfrac{\partial^2 c}{\partial z^2}$	kc
Finite Element Representation	$-D\dfrac{\Delta x\Delta z}{6}\left\{\left[\dfrac{c_{1,1}-2c_{1,0}+c_{1,-1}}{\Delta z^2}\right]+4\left[\dfrac{c_{0,1}-2c_{0,0}+c_{0,-1}}{\Delta z^2}\right]\right.$ $\left.+\left[\dfrac{c_{-1,1}-2c_{-1,0}+c_{-1,-1}}{\Delta z^2}\right]\right\}$	$k\dfrac{\Delta x\Delta z}{36}\{[c_{-1,1}+4c_{0,1}+c_{1,1}]$ $+4[c_{-1,0}+4c_{0,0}+c_{1,0}]$ $+[c_{-1,-1}+4c_{0,-1}+c_{1,-1}]\}$
Numerical Differentiation $O(\Delta h^2)$	$\dfrac{\partial^2 c_{j,0}}{\partial z^2}=\dfrac{c_{j,1}-2c_{j,0}+c_{j,-1}}{\Delta z^2}-\dfrac{1}{12}\Delta z^2\dfrac{\partial^4 c}{\partial z^4}$ $j=-1,0,1$	
Numerical Integration $O(\Delta h^4)$ (1-dimensional Simpson rule)	$\displaystyle\int\dfrac{\partial c_{1,0}}{\partial z}\,dx=\dfrac{\Delta x}{3}\left\{\dfrac{\partial c_{-1,0}}{\partial z}+4\dfrac{\partial c_{0,0}}{\partial z}+\dfrac{\partial c_{1,0}}{\partial z}\right\}$ $-\dfrac{\Delta x^5}{90}\dfrac{\partial^5 c}{\partial z\partial x^2}$	$\displaystyle\int c\,dx\,dz=\dfrac{\partial x\partial z}{9}\{[c_{-1,1}+4c_{0,1}+c_{1,1}]$ $+4[c_{-1,0}+4c_{0,0}+c_{1,0}]$ $+[c_{-1,-1}+4c_{0,-1}+c_{1,-1}]\}$ $-\dfrac{\Delta x\Delta z^5}{45}\dfrac{\partial^4 c}{\partial z^4}-\dfrac{\Delta z\Delta x^5}{45}\dfrac{\partial^4 c}{\partial x^4}$

for a now uniform rectangular grid according to [3.78]

$$\frac{\partial c}{\partial t} + u\,\frac{\partial c}{\partial x} + w\,\frac{\partial c}{\partial z} - D\,\frac{\partial^2 c}{\partial x^2} + D\,\frac{\partial^2 c}{\partial z^2} + kc = 0$$

Term	
$\dfrac{\partial c}{\partial t}$	

Finite Element Representation

$$\frac{\Delta x \Delta z}{36}\left\{\left[\frac{dc_{-1,1}}{dt} + 2(1+\beta)\,\frac{dc_{0,1}}{dt} + \beta\,\frac{dc_{1,1}}{dt}\right]\right.$$

$$+ 2(1+\alpha)\left[\frac{dc_{-1,0}}{dt} + 2(1+\beta)\,\frac{dc_{0,0}}{dt} + \beta\,\frac{dc_{1,0}}{dt}\right]$$

$$\left. + \alpha\left[\frac{dc_{-1,-1}}{dt} + 2(1+\beta)\,\frac{dc_{0,-1}}{dt} + \beta\,\frac{dc_{1,-1}}{dt}\right]\right\}$$

Numerical Integration $O(\Delta h)$

$$\iint \frac{\partial c}{\partial t}\,dx\,dz = \frac{\Delta x \Delta z}{9}\left\{\left[\frac{\partial c_{-1,1}}{\partial t} + 2(1+\beta)\,\frac{\partial c_{0,1}}{\partial t} + \beta\,\frac{\partial c_{1,1}}{\partial t}\right]\right.$$

$$+ 2(1+\alpha)\left[\frac{\partial c_{-1,0}}{\partial t} + 2(1+\beta)\,\frac{\partial c_{0,0}}{\partial t} + \beta\,\frac{\partial c_{1,0}}{\partial t}\right]$$

$$\left. + \alpha\left[\frac{\partial c_{-1,-1}}{\partial t} + 2(1+\beta)\,\frac{\partial c_{0,-1}}{\partial t} + \beta\,\frac{\partial c_{1,-1}}{\partial t}\right]\right.$$

$$+ \frac{(\beta^2 - 1)(1+\alpha)\,\Delta x^2\,\Delta z}{6}\,\frac{\partial}{\partial x}\left(\frac{\partial c}{\partial t}\right)$$

$$+ \frac{(1-\alpha^2)(1+\beta)\,\Delta x\,\Delta z^2}{6}\,\frac{\partial}{\partial z}\left(\frac{\partial c}{\partial t}\right)$$

Term	$u \dfrac{\partial c}{\partial x}$	$w \dfrac{\partial c}{\partial z}$
Finite Element Representation	$u \dfrac{(\beta+1)\,\Delta x\,\Delta z}{12}\left\{\left[\dfrac{c_{1,1}-c_{-1,1}}{(\beta+1)\,\Delta x}\right] + 2(\alpha+1)\left[\dfrac{c_{1,0}-c_{-1,0}}{(\beta+1)\,\Delta x}\right] + \alpha\left[\dfrac{c_{1,-1}-c_{-1,-1}}{(\beta+1)\,\Delta x}\right]\right\}$	$w \dfrac{(\alpha+1)\,\Delta x\,\Delta z}{12}\left\{\beta\left[\dfrac{c_{1,1}-c_{1,-1}}{(\alpha+1)\,\Delta z}\right] + 2(\beta+1)\left[\dfrac{c_{0,1}-c_{0,-1}}{(\alpha+1)\,\Delta z}\right] + \left[\dfrac{c_{-1,1}-c_{-1,-1}}{(\alpha+1)\,\Delta z}\right]\right\}$
Numerical Differentiation $O(\Delta h)$	$\dfrac{\partial c_{0,k}}{\partial x} = \dfrac{c_{1,k}-c_{-1,k}}{(\beta+1)\,\Delta x} - \dfrac{\partial^2 c_{0,k}}{\partial x^2}\dfrac{(\beta-1)\,\Delta x}{2}$ $k = -1, 0, 1$	$\dfrac{\partial c_{j,0}}{\partial z} = \dfrac{c_{j,1}-c_{j,-1}}{(\alpha+1)\,\Delta z} - \dfrac{\partial^2 c_{j,0}}{\partial z^2}\dfrac{(1-\alpha)\,\Delta x}{2}$ $j = -1, 0, 1$
Numerical Integration $O(\Delta h)$	$\displaystyle\int \dfrac{\partial c_{0,k}}{\partial x}\,dz = \dfrac{\Delta z}{3}\left\{\alpha\dfrac{\partial c_{0,-1}}{\partial x} + 2(\alpha+1)\dfrac{\partial c_{0,0}}{\partial x} + \dfrac{\partial c_{0,1}}{\partial x}\right\} + \dfrac{(1-\alpha^2)\,\Delta z^2}{6}\dfrac{\partial^2 c}{\partial z\,\partial x}$	$\displaystyle\int \dfrac{\partial c_{j,0}}{\partial z}\,dx = \dfrac{\Delta x}{3}\left\{\dfrac{\partial c_{-1,0}}{\partial z} + 2(\beta+1)\dfrac{\partial c_{0,0}}{\partial z} + \beta\dfrac{\partial c_{1,0}}{\partial z}\right\} + \dfrac{(\beta-1)\,\Delta x^2}{6}\dfrac{\partial^2 c}{\partial x\,\partial z}$

Term	
	$-D\dfrac{\partial^2 c}{\partial x_2}$

Finite Element Representation

$$-D\frac{(\beta+1)\Delta x\,\Delta z}{12}\left\{\left[\frac{\dfrac{c_{1,1}-c_{0,1}}{\Delta x}-\dfrac{c_{0,1}-c_{-1,1}}{\beta\,\Delta x}}{(\beta+1)\,\Delta x/2}\right]\right.$$

$$+2(\alpha+1)\left[\frac{\dfrac{c_{1,0}-c_{0,0}}{\Delta x}-\dfrac{c_{0,0}-c_{-1,0}}{\beta\,\Delta x}}{(\beta+1)\,\Delta x/2}\right]$$

$$+\left.\left[\frac{\dfrac{c_{1,-1}-c_{0,-1}}{\Delta x}-\dfrac{c_{0,-1}-c_{-1,-1}}{\beta\Delta x}}{(\beta+1)\,x/2}\right]\right\}$$

Numerical Differentiation O(Δh)

$$\frac{\partial^2 c_{0,k}}{\partial x^2}=\frac{\dfrac{c_{1,k}-c_{0,k}}{\Delta x}-\dfrac{c_{0,k}-c_{-1,k}}{\beta\Delta x}}{(\beta+1)\Delta x/2}-\frac{\partial^3 c_{0,k}}{\partial x^3}\frac{(\beta-1)\,\Delta x}{3}$$

$$k=-1,\,0,\,1$$

Numerical Integration O(Δh)

$$\int\frac{\partial^2 c_{0,k}}{\partial x^2}\,dz=\frac{\Delta z}{3}\left\{\alpha\frac{\partial^2 c_{0,-1}}{\partial x^2}+2(\alpha+1)\frac{\partial^2 c_{0,0}}{\partial x^2}+\frac{\partial^2 c_{0,1}}{\partial x^2}\right\}$$

$$+\frac{(1-\alpha^2)\,\Delta x^2}{6}\frac{\partial^3 c}{\partial z\,\partial x^2}$$

Index